VOLUME SIX HUNDRED AND FORTY SIX

METHODS IN ENZYMOLOGY

Liquid-liquid Phase Coexistence and Membraneless Organelles

METHODS IN ENZYMOLOGY

VOLUME SIX HUNDRED AND FORTY SIX

METHODS IN ENZYMOLOGY

Liquid-liquid Phase Coexistence and Membraneless Organelles

Edited by

CHRISTINE D. KEATING

Department of Chemistry, Pennsylvania State University, University Park, PA, United States

ELSEVIER

ACADEMIC PRESS

An imprint of Elsevier

Academic Press is an imprint of Elsevier
50 Hampshire Street, 5th Floor, Cambridge, MA 02139, United States
525 B Street, Suite 1650, San Diego, CA 92101, United States
The Boulevard, Langford Lane, Kidlington, Oxford OX5 1GB, United Kingdom
125 London Wall, London, EC2Y 5AS, United Kingdom

First edition 2021

Notices
Knowledge and best practice in this field are constantly changing. As new research and experience broaden our understanding, changes in research methods, professional practices, or medical treatment may become necessary.

Practitioners and researchers must always rely on their own experience and knowledge in evaluating and using any information, methods, compounds, or experiments described herein. In using such information or methods they should be mindful of their own safety and the safety of others, including parties for whom they have a professional responsibility.

To the fullest extent of the law, neither the Publisher nor the authors, contributors, or editors, assume any liability for any injury and/or damage to persons or property as a matter of products liability, negligence or otherwise, or from any use or operation of any methods, products, instructions, or ideas contained in the material herein.

ISBN: 978-0-12-821159-5
ISSN: 0076-6879

For information on all Academic Press publications
visit our website at https://www.elsevier.com/books-and-journals

Publisher: Zoe Kruze
Editorial Project Manager: Shellie Bryant
Production Project Manager: James Selvam
Cover Designer: Christian J. Bilbow

Typeset by SPi Global, India

Contents

Contributors

Samim Ali
Materials Science and Engineering Division, Material Measurement Laboratory, National Institute of Standards and Technology, Gaithersburg, MD, United States

Ibraheem Alshareedah
Department of Physics, University at Buffalo, Buffalo, NY, United States

Wiggert J. Altenburg
Department of Biomedical Engineering; Department of Chemical Engineering and Chemistry, Institute for Complex Molecular Systems, Eindhoven University of Technology, Eindhoven, The Netherlands

Alain A.M. André
Institute for Molecules and Materials, Radboud University, Nijmegen, The Netherlands

Priya R. Banerjee
Department of Physics, University at Buffalo, Buffalo, NY, United States

Philip C. Bevilacqua
Department of Chemistry; Department of Biochemistry, Microbiology, and Molecular Biology; Center for RNA Molecular Biology, Pennsylvania State University, University Park, PA, United States

Markus Bleuel
Department of Materials Science and Engineering, University of Maryland, College Park; Center for Neutron Research, National Institute of Standards and Technology, Gaithersburg, MD, United States

Whitney C. Blocher McTigue
Department of Chemical Engineering, University of Massachusetts Amherst, Amherst, MA, United States

Jesse B. Hopkins
The Biophysics Collaborative Access Team (BioCAT), Department of Biological Sciences, Illinois Institute of Technology, Chicago, IL, United States

Taranpreet Kaur
Department of Physics, University at Buffalo, Buffalo, NY, United States

Christine D. Keating
Department of Chemistry, Pennsylvania State University, University Park, PA, United States

Suzanne Lafon
Université de Bordeaux, CNRS, Centre de Recherche Paul Pascal, UMR5031, Pessac; Ecole Normale Supérieure d'Ulm, Paris, France

Yuanchi Ma
Materials Science and Engineering Division, Material Measurement Laboratory, National Institute of Standards and Technology, Gaithersburg, MD, United States

Roshan Mammen Regy
Department of Chemical and Biomolecular Engineering, Lehigh University, Bethlehem, PA, United States

Yimin Mao
Department of Materials Science and Engineering, University of Maryland, College Park; Center for Neutron Research, National Institute of Standards and Technology, Gaithersburg, MD, United States

Amanda B. Marciel
Department of Chemical and Biomolecular Engineering, Rice University, Houston, TX, United States

Erik W. Martin
Department of Structural Biology, St. Jude Children's Research Hospital, Memphis, TN, United States

Nicolas Martin
Université de Bordeaux, CNRS, Centre de Recherche Paul Pascal, UMR5031, Pessac, France

Alexander F. Mason
Department of Biomedical Engineering; Department of Chemical Engineering and Chemistry, Institute for Complex Molecular Systems, Eindhoven University of Technology, Eindhoven, The Netherlands

McCauley O. Meyer
Department of Biochemistry, Microbiology, and Molecular Biology; Center for RNA Molecular Biology, Pennsylvania State University, University Park, PA, United States

Tanja Mittag
Department of Structural Biology, St. Jude Children's Research Hospital, Memphis, TN, United States

Jeetain Mittal
Department of Chemical and Biomolecular Engineering, Lehigh University, Bethlehem, PA, United States

Begoña Monterroso
Centro de Investigaciones Biológicas Margarita Salas, Consejo Superior de Investigaciones Científicas (CSIC), Madrid, Spain

Gregory A. Mountain
Department of Chemistry, Pennsylvania State University, University Park, PA, United States

Karina K. Nakashima
Institute for Molecules and Materials, Radboud University, Nijmegen, The Netherlands

Sarah L. Perry
Department of Chemical Engineering; Institute for Applied Life Sciences, University of Massachusetts Amherst, Amherst, MA, United States

Raghav R. Poudyal
Department of Chemistry; Center for RNA Molecular Biology, Pennsylvania State University, University Park, PA, United States

Vivek M. Prabhu
Materials Science and Engineering Division, Material Measurement Laboratory, National Institute of Standards and Technology, Gaithersburg, MD, United States

Andrea Putnam
Department of Molecular Biology and Genetics, Howard Hughes Medical Institute, Johns Hopkins University, School of Medicine, Baltimore, MD, United States

Germán Rivas
Centro de Investigaciones Biológicas Margarita Salas, Consejo Superior de Investigaciones Científicas (CSIC), Madrid, Spain

Miguel Ángel Robles-Ramos
Centro de Investigaciones Biológicas Margarita Salas, Consejo Superior de Investigaciones Científicas (CSIC), Madrid, Spain

Geraldine Seydoux
Department of Molecular Biology and Genetics, Howard Hughes Medical Institute, Johns Hopkins University, School of Medicine, Baltimore, MD, United States

Shidong Song
Department of Biomedical Engineering; Department of Chemical Engineering and Chemistry, Institute for Complex Molecular Systems, Eindhoven University of Technology, Eindhoven, The Netherlands

Evan Spruijt
Institute for Molecules and Materials, Radboud University, Nijmegen, The Netherlands

Samanvaya Srivastava
Department of Chemical and Biomolecular Engineering, University of California, Los Angeles, Los Angeles, CA, United States

Jeffrey M. Ting
Pritzker School of Molecular Engineering, University of Chicago, Chicago; Center for Molecular Engineering and Materials Science Division, Argonne National Laboratory, Lemont, IL, United States

Matthew V. Tirrell
Pritzker School of Molecular Engineering, University of Chicago, Chicago; Center for Molecular Engineering and Materials Science Division, Argonne National Laboratory, Lemont, IL, United States

Jan C.M. van Hest
Department of Biomedical Engineering; Department of Chemical Engineering and Chemistry, Institute for Complex Molecular Systems, Eindhoven University of Technology, Eindhoven, The Netherlands

Marleen van Stevendaal
Department of Biomedical Engineering; Department of Chemical Engineering and Chemistry, Institute for Complex Molecular Systems, Eindhoven University of Technology, Eindhoven, The Netherlands

Wenwei Zheng
College of Integrative Sciences and Arts, Arizona State University, Mesa, AZ, United States

Silvia Zorrilla
Centro de Investigaciones Biológicas Margarita Salas, Consejo Superior de Investigaciones Científicas (CSIC), Madrid, Spain

Preface

Intracellular compartmentalization by liquid–liquid phase separation (LLPS) is both an old and a new idea. In 1929, colloid chemists Bungenberg de Jong and Kruyt coined the term "coacervation" from the Latin *acervus*, for aggregation, and the prefix co, to denote combining together, to describe associative LLPS they observed in polymer solutions. They noted at the time that, "On closer inspection of the ground mass of the protoplasm, it strikes one that this has some properties in common with the coacervates, so that there is a possibility that this ground mass may be considered as a coacervate or as a system of coacervates" (Bungenberg de Jong, H. G., & Kruyt, H. R. (1929). *Proceedings of the Koninklijke Nederlandse Akademie van Wetenschappen, 32*, 849–856). Since then many physical studies of biomacromolecules such as proteins and nucleic acids supported the possibility of their phase separation under a variety of conditions. Over roughly the past decade, it has become clear that many intracellular membraneless organelles are in fact formed by coacervation inside living cells. These organelles maintain compositions distinct from the surrounding fluids despite their lack of a surrounding membrane and the often dynamic nature of their constituent molecules, and serve important roles in cell physiology. Understanding phase separation can provide mechanistic insight into how membraneless organelles form, and how living systems use these structures to organize themselves and their biochemistries.

Compartmentalization by phase coexistence is a general paradigm, relevant far beyond the membraneless organelles of living cells. It is also a powerful approach to concentrating low-abundance molecules in prebiotic scenarios, generating functional subcompartments in artificial cells, encapsulating biomacromolecules for drug delivery, food science, or other biotechnological applications. The wide-ranging applicability of aqueous macromolecule demixing phenomena attracts considerable interest from the physics, chemistry, biology, and engineering communities. Expertise and inspiration in phase separation can thus be found across seemingly disparate fields that may not normally interact at conferences or publish in the same journals. We use different terminology to describe our phase-separated droplet systems, and different techniques to study them. My goal for this volume was to bring together experts across these fields to share their methods as broadly as possible. Some chapters describe methods performed using phase-separating proteins, while others used synthetic polymers; often the general method could be adapted to nearly any coacervating system. Methods described include coarse-grain modeling of phase separation, experimental preparation of coexisting phases, and several approaches for characterization of phase-separating systems, followed by methods for biomolecule encapsulation and enzymatic activity, and control over phase separation via light and enzymatic reactions. Readers, I encourage you to peruse all the chapters with an eye toward new ways to learn about your own phase-separating systems of interest. The authors have included not only protocols but also their advice for making your experiments a success.

My heartfelt thanks to all of the authors for sharing their knowledge and their insights in these chapters. I am also grateful to graduate student and postdoctoral colleagues who provided valuable feedback on the chapters, helping to make the Methods as clear as possible for junior scientists. Finally, thanks to Shellie Bryant and the entire *Methods in Enzymology* team for their support and efficiency in bringing this volume to completion.

Christine D. Keating
Department of Chemistry
Penn State University
University Park, PA, United States

CHAPTER ONE

Using a sequence-specific coarse-grained model for studying protein liquid–liquid phase separation

Roshan Mammen Regy[a], Wenwei Zheng[b], and Jeetain Mittal[a,*]
[a]Department of Chemical and Biomolecular Engineering, Lehigh University, Bethlehem, PA, United States
[b]College of Integrative Sciences and Arts, Arizona State University, Mesa, AZ, United States
*Corresponding author: e-mail address: jeetain@lehigh.edu

Contents

Abstract

The formation of membraneless organelles (MLOs) via liquid–liquid phase separation (LLPS) of biomolecules is a topic that has garnered significant attention in the scientific community recently. Experimental studies have revealed that intrinsically disordered proteins (IDPs) may play a major role in driving the formation of these droplets via LLPS by forming multivalent interactions between amino acids. To quantify these interactions is an arduous task as it is difficult to investigate these interactions at the amino acid level using currently available experimental tools. It becomes necessary to complement experimental studies using appropriate computational methods such as coarse-grained models of IDPs that can allow one to simulate biomolecular LLPS using general-purpose hardware. Here, we summarize our coarse-grained modeling framework that uses a single bead per amino acid resolution and the co-existence sampling technique to study sequence-specific protein phase separation using molecular dynamics simulations. We further discuss the caveats and technicalities, which one must consider while using this method to obtain thermodynamic phase diagrams. To ease the learning curve, we provide our implementations of the coarse-grained potentials in the HOOMD-Blue simulation package and associated python scripts to run such simulations.

Methods in Enzymology, Volume 646
ISSN 0076-6879
https://doi.org/10.1016/bs.mie.2020.07.009

1. Introduction

Membraneless organelles (MLOs) are concentrated compartments consisting of biomolecules that lack a membrane separating their contents from the surrounding cytosol (Brangwynne et al., 2009; Brangwynne, Mitchison, & Hyman, 2011; Burke, Janke, Rhine, & Fawzi, 2015; Elbaum-Garfinkle et al., 2015). These MLOs are formed via liquid–liquid phase separation (LLPS), where the loss of entropy upon demixing of a homogeneous mixture can be accounted for by the favorable enthalpy due to intermolecular interactions between different components (Brangwynne, Tompa, & Pappu, 2015). Previous studies have shown that for most MLOs, intrinsically disordered proteins (IDPs) play a significant role in driving or promoting phase separation. IDPs are proteins that lack a well-defined three-dimensional structure but are known to adopt a range of conformations (Tompa, 2012; Wright & Dyson, 1999). An ensemble description of IDPs makes it difficult for any single experimental technique to provide a complete view of the structural properties (Fisher & Stultz, 2011). The use of computational methods can, therefore, offer complementary information to describe the phase separation properties (Dignon, Zheng, Kim, et al., 2018; Lin et al., 2016; Vitalis & Pappu, 2008).

The investigation of the intermolecular interactions driving protein phase separation has been an uphill task for experimental studies as these IDPs are known to have multiple different modes of interactions among themselves, which is challenging to quantify solely based on experimental studies. One would have to look for solutions in the computational domain where methods like all-atom simulations (Best et al., 2014; Huang et al., 2017; Piana et al., 2015) would provide a detailed view of all possible interactions driving or disrupting phase separation. Currently, it is computationally prohibitive to directly simulate protein phase separation using an all-atom representation, and these methods are instead used to provide molecular information on the atomic modes of interactions (Conicella et al., 2020, 2016; Murthy et al., 2019; Schuster et al., 2020). Hence, one must turn to simpler models, which can capture the sequence-specific details of proteins deemed necessary to study their phase behavior while still providing molecular or sequence-level information (Dignon, Zheng, & Mittal, 2019). These methods range from analytical and empirical theories, bioinformatics, to coarse-grained (CG) molecular simulations. Every model has its merit based on the questions one would like to ask. For instance, analytical or numerical mean-field polymer models have the advantage of quickly providing valuable information on the chain-length dependent multivalent interactions, the contribution of charges, and other sequence-descriptors based on amino acid properties (Brangwynne et al., 2015; de Gennes, 1978; Dignon, Zheng, Best, et al., 2018; Flory, 1953; Sawle & Ghosh, 2015; Zheng et al., 2018; Zheng et al., 2020). However, such models often require experimental input in a system-specific manner and are limited in their ability to provide details on molecular structures and dynamics. Molecular simulations based on highly simplified particle-based CG models can further bridge this gap and allow one to study phase behaviors, molecular interactions responsible for stabilizing the liquid-like state, and the dynamics within the condensed phase. The simplest CG model that can capture the IDP sequences is based on representing each amino acid with a single bead, and such a model was proposed recently by us to study sequence-dependent LLPS of IDPs (Dignon, Zheng, Kim, et al., 2018).

In this chapter, we discuss the technical details of this off-lattice CG model, henceforth referred to as the HPS (*H*ydro*P*athy *S*cale) model, which represents IDPs as flexible polymers with 20 different types of monomer beads and solvent interactions captured implicitly in terms of nonbonded inter-residue interactions. Even with such a highly CG representation, it is almost impossible to compute the phase behavior (co-existence envelope between the high-density and low-density phases) using commonly employed sampling techniques such as replica exchange molecular dynamics, as expected based on the nature of the phase transition involved (Kim et al., 2010). We have successfully overcome this challenge by using the co-existence method (Blas et al., 2008), in which the two phases are simulated next to each other in a single simulation cell. Here, we provide more details on how to construct the system using the HPS model and practically conduct MD simulations on a publicly available software package HOOMD-Blue (Anderson et al., 2020). We also provide scripts using which the readers would be able to set up such simulations for a system of their interest. Our primary goal in writing this chapter is to help the reader understand the underlying details of the model, the methods employed in our previous work and what challenges to expect when conducting such simulations.

2. HPS model

Our CG modeling approach for IDPs uses one bead per amino acid level of resolution (Fig. 1A), and IDP sequence details are preserved using

A

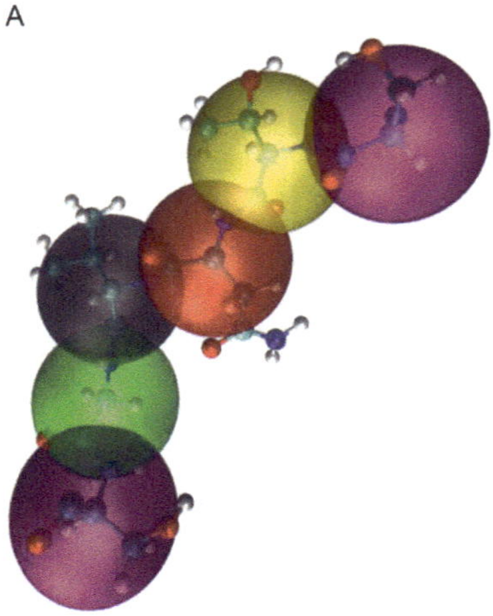

B

Type	Mass (amu)	Charge	σ (Å)	λ
ALA	71.08	0	5.04	0.730
ARG	156.20	1	6.56	0.000
ASN	114.10	0	5.68	0.432
ASP	115.10	-1	5.58	0.378
CYS	103.10	0	5.48	0.595
GLN	128.10	0	6.02	0.514
GLU	129.10	-1	5.92	0.459
GLY	57.05	0	4.50	0.649
HIS	137.10	0.5	6.08	0.514
ILE	113.20	0	6.18	0.973
LEU	113.20	0	6.18	0.973
LYS	128.20	1	6.36	0.514
MET	131.20	0	6.18	0.838
PHE	147.20	0	6.36	1.000
PRO	97.12	0	5.56	1.000
SER	87.08	0	5.18	0.595
THR	101.10	0	5.62	0.676
TRP	186.20	0	6.78	0.946
TYR	163.20	0	6.46	0.865
VAL	99.07	0	5.86	0.892

C

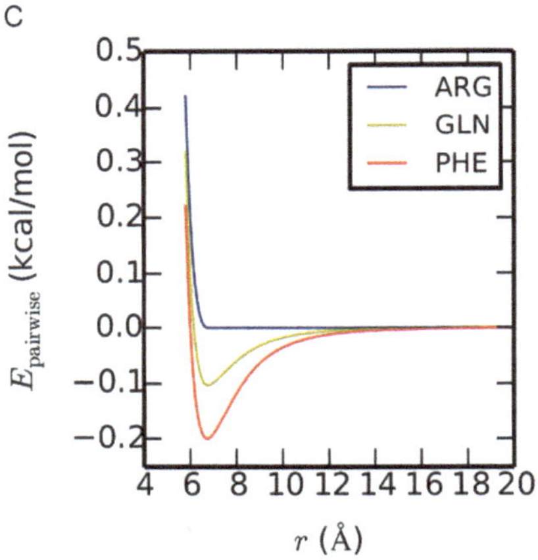

Fig. 1 HPS model. (A) One residue per bead coarse-grained (CG) representation of a peptide. (B) Parameters used in the HPS model. (C) The energy function of short-range pairwise interaction for representative amino acids.

different parameter values for all the 20 naturally occurring amino acids (Fig. 1B). The following parameters completely define the identity of each amino acid in our model: the mass of the amino acid, the charge of the amino acid at near neutral pH, the diameter of the amino acid (σ), and the hydropathy (λ) that we believe can capture the short-range interactions between amino acids. We compute σ by mapping the van der Waals volume of an amino acid to a sphere (Creighton, 1993). The λ value for each amino acid is estimated using an atomistic hydropathy scale proposed by Kapcha and Rossky (2014), which is based on partial charges of atoms belonging to an amino acid in an all-atom force field. For convenience, the λ values for all the amino acids are scaled to range from 0 (Arginine) to 1 (Phenylalanine).

In the CG model energy function, we have three different types of interactions, namely, bonded, electrostatic, and short-range pairwise interactions. The bonds between consecutive amino acids in the protein sequence are modeled using a harmonic potential with a spring constant of 10 kJ/Å^2 and a bond length of 3.8 Å. Hence the protein representation in our CG model resembles a freely jointed polymer chain where amino acids are represented by spheres that are connected by harmonic springs. The electrostatic interactions are modeled using a Debye-Hückel electrostatic screening term (Debye & Hückel, 1923),

$$E_{i,j} = \frac{q_i q_j}{4\pi D r} \exp(-r/\kappa) \tag{1}$$

in which κ is the Debye screening length and ionic strength dependent, and $D=80$ is the dielectric constant of the solvent. The Debye length is proportional to the square root of the ionic strength, and a Debye screening length of 10 Å corresponds to ~100 mM ionic strength at room temperature. This functional form is available in HOOMD-Blue using the function *hoomd.md.pair.yukawa* (Anderson et al., 2020).

The following functional form by Ashbaugh-Hatch is used for the short-range pairwise interactions (Ashbaugh & Hatch, 2008):

$$\Phi(r) = \begin{cases} \Phi_{LJ} + (1-\lambda_{ij})\epsilon, \text{if } r \le 2^{\frac{1}{6}}\sigma_{ij} \\ \lambda_{ij}\Phi_{LJ}, \text{otherwise} \end{cases} \tag{2}$$

where Φ_{LJ} is the standard 12–6 Lennard-Jones potential given by,

$$\Phi_{LJ} = 4\epsilon\left[\left(\frac{\sigma_{ij}}{r}\right)^{12} - \left(\frac{\sigma_{ij}}{r}\right)^{6}\right] \tag{3}$$

Both the hydropathy (λ_{ij}) and the diameter (σ_{ij}) for a pair of amino acids (i,j) are set to be the arithmetic average of the corresponding values of the two amino acids. Small values of λ_{ij} represent weak interactions between the two amino acids. In contrast, high values represent strong inter-residue interactions though the exact atomic modes of interactions such as hydrophobic, cation-π, hydrogen bonding, etc., are not explicitly included in our CG approach. The interaction parameter (ε) in Eq. 3 is a free parameter that sets the absolute scale of interactions among different amino acid pairs, which was adjusted to reproduce the experimentally measured radius of gyration of IDPs as 0.2 kcal/mol (Dignon, Zheng, Kim, et al., 2018). The resulting potential energy as a function of the distance for three representative amino acids is shown in Fig. 1C. In practice, the functional form for these nonbonded interactions can be realized using the function *azplugins.pair.ashbaugh()* provided within the "azplugins" component (plugin) for HOOMD-Blue found at https://github.com/mphowardlab/azplugins.

Next, we would like to point out some salient features of our model to the reader, which make it well suited for studying phase separation and the formation of MLOs compared to other existing coarse-grained models for IDPs and folded proteins (Ayton et al., 2007; Clementi, 2008). First of all, the sequence specificity of the HPS model allows us to investigate interactions at the amino acid level and therefore the effects of mutations and sequence patterning on phase behavior (Dignon, Zheng, Kim, et al., 2018; Monahan et al., 2017). Second, the HPS model is a transferable modeling framework, which does not require experimental input on a system of interest after parameterization, with potential extensibility to non-natural amino acids and other biomolecules such as nucleic acids. Given the limited number of free parameters in the HPS model, it should be relatively easier to optimize model parameters further as more experimental data is available (Latham & Zhang, 2019). Third, additional features such as temperature dependence of short-range interactions can be introduced in the model to capture the thermoresponsive LLPS behavior of IDPs accurately (Dignon, Zheng, Kim, & Mittal, 2019). These features of the CG modeling framework make it a nifty tool to utilize when studying LLPS in protein systems of interest in biology or material science.

3. Simulation strategy for obtaining phase diagrams

To construct a thermodynamic phase diagram, estimation of the protein saturation concentrations in the protein-rich and protein-deficient phases is required as a function of control variables such as temperature. Obtaining these estimates using commonly used simulation techniques such as the Gibbs ensemble method or grand canonical simulations for phase behavior calculations of simple soft matter systems is a herculean task that cannot be accomplished in a reasonable time even with a CG off-lattice model (Panagiotopoulos, 2000; Potoff & Panagiotopoulos, 1998). An alternate sampling method is, therefore, necessary to overcome the enthalpy gap between the two phases (Blas et al., 2008; Kim et al., 2010). One efficient way of achieving this in brute-force molecular dynamics simulations is to construct the system in a way by which the two phases coexist (commonly referred to as the co-existence method) with a common interface to allow for exchanges of molecules between the two phases for sufficient equilibration of densities in these phases. The use of planar interfaces in a slab configuration is effective in reducing the system-size effects as opposed to a spherical droplet configuration to sample the interfacial properties and saturation densities of co-existing phases (Blas et al., 2008; Silmore et al., 2017).

The following section provides detailed information on how to set up and analyze co-existence simulations to construct phase diagrams and is organized into three separate subsections, each explaining the steps the reader would have to take to implement our methodology, i.e., (1) how to create the initial configuration of protein chains for a co-existence simulation, (2) how to calculate density profiles over the simulation box and check for convergence, and (3) how to construct the phase diagram given the density profiles as a function of the temperature.

3.1 Preparing an initial slab configuration

In the first step, we select an equilibrated IDP conformation based on the HPS model. This can be done by using the HPS model and HOOMD-Blue, or by using any other software packages such as LAMMPS (Plimpton et al., 2007) or ABSINTH (Vitalis & Pappu, 2008). Then we place required copies of the protein chain in a selected conformation or multiple conformations on a periodic lattice inside a cubic simulation box large enough to avoid clashes between monomers of different protein chains (Fig. 2A). The selection of the number of chains for such simulations is a balance between the accuracy (system-size dependent effects) and the computational power needed to conduct simulations to sufficient timescales. We provide some practical guidance in this direction later in this chapter.

Once the proteins are arranged in a low-density configuration, we use it as the initial configuration for an NPT simulation to condense these protein chains into a smaller cubic periodic box of the desired side length to generate a high-density configuration at protein concentrations higher than observed experimentally for many IDPs, e.g., 200–500 mg/mL (Fig. 2B). This step can be efficiently done in HOOMD-Blue using the MTK barostat (Martyna et al., 1994) provided in the function *hoomd.md.integrate.npt*. One can iteratively select a target pressure for a given temperature to condense protein chains into the appropriate box size. From our experience, this process is usually rapid and it happens within tens of nanosecond simulation

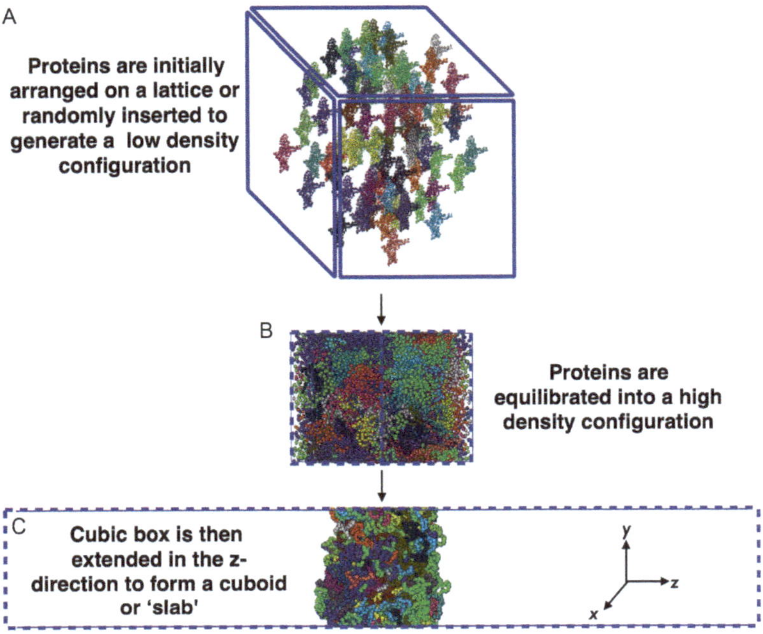

Fig. 2 Illustration of preparing a slab configuration. (A) IDPs in arbitrary lattice positions. (B) Configurations equilibrated with NPT simulations in a periodic cubic box. (C) The z-axis of the cubic box is extended to make a slab configuration.

time. An alternative method to generate a high-density protein condensed phase is to simply reduce the box size over a certain runtime using the function *hoomd.update.box_resize()*. The resulting configuration can be well-equilibrated in a high-density condensed phase before the next step by either allowing the box resizing to happen over a sufficiently long time or conducting an NVT simulation, over tens of nanoseconds, to help equilibrate protein chains and start the next step with a low potential energy configuration.

Selecting a condensed configuration within the cubic box from the NPT simulation or after resizing the box, we now extend the box length in the z-direction to create empty space to allow for redistribution of protein chains in two co-existing high- and low-density phases (Fig. 2C). To avoid boundary artifacts, as the periodic configuration from the previous step is placed in a new periodic box with a different z-dimension, care should be taken to move the monomers of a protein chain that are split across the periodic boundary in the z-direction to make each protein whole. This extended slab with protein chains in the center (Fig. 3A) would then serve as the initial configuration for our co-existence simulations, as described in the next section.

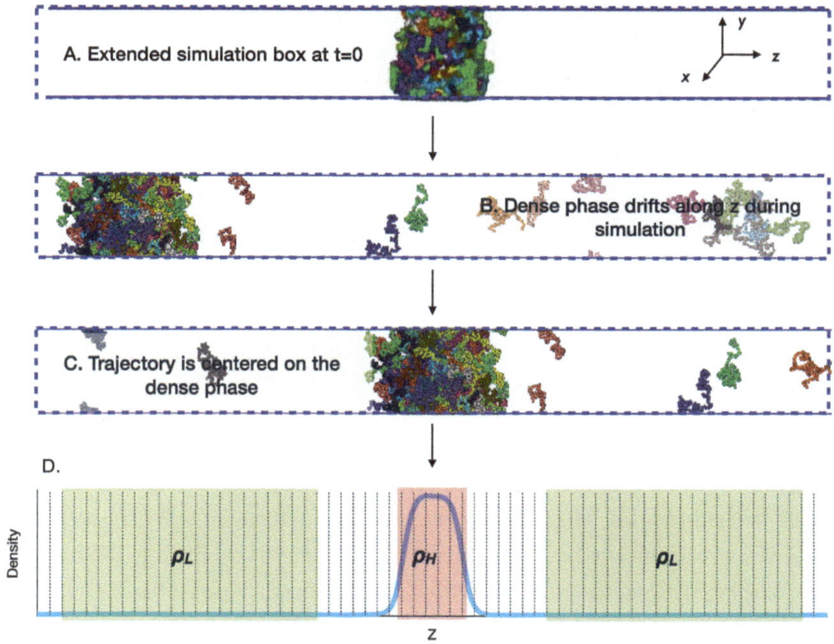

Fig. 3 Obtaining the density profile using the co-existence simulation method.

3.2 Obtaining density profiles

With the initial slab configuration (Fig. 3A), we conduct 5 μs NVT simulations at a range of temperatures so that we can obtain co-existence densities to make a phase diagram as a function of protein concentration and temperature. As we go from low to high temperatures, we see that protein chains start breaking off the dense central region and enter into the space created in the extended z-direction of the slab simulation box more frequently until a dynamic equilibrium is established after a sufficient simulation time (Fig. 3B). As we go to very high temperatures, we see all the proteins that were initially condensed at the center, breaking off and getting distributed throughout the simulation box in a single phase, which is indicative of the simulation temperature being higher than the critical temperature. The presence of this critical temperature above which system exists in a single phase is a characteristic of systems with an upper critical solution temperature (UCST). Also, many systems can display lower solution critical temperature (LCST) behavior, for which phase separation occurs above the critical temperature (Quiroz & Chilkoti, 2015). Some systems can also show both UCST and LCST behavior, which would present an hourglass phase diagram as a single phase is observed between the two critical temperatures.

We have recently proposed an extension of our coarse-grained modeling framework to capture the thermoresponsive behavior of proteins by introducing temperature-dependent nonbonded interactions (Dignon, Zheng, Kim, & Mittal, 2019). For the sake of simplicity, we will proceed to explain our methodology with the original model here (Dignon, Zheng, Kim, et al., 2018)).

During the co-existence NVT simulation, the interface between the low- and high-density phases tends to drift along the z-direction (Fig. 3B), which makes it difficult to determine the average concentration profile as a function of the z-direction without further processing to account for this slab mobility. To overcome this, we center all the saved trajectory frames to the initial configuration so that the dense phase is placed at $z = 0$ for all the frames. In practice, we first identify the largest protein cluster inside the simulation box by counting the number of protein chains in the clusters found in each frame. We then center the simulation box to the center of mass of the largest cluster for each frame, as shown in Fig. 3C. This gives us a processed trajectory file in which the dense slab phase is at the center of the simulation box and can be used for further analysis. Another method, which can be used to fix the location of the interface is by using a spatial autocorrelation function to align concentration profiles for each frame to have the highest peak of intensity at the center of the simulation box (Jung & Yethiraj, 2018). This "centered" trajectory is binned along the z-direction to calculate the average protein density as a function of the z-coordinate averaged over all the frames, as shown in Fig. 3D. To estimate the sampling errors and to provide guidance on the simulation runtime to achieve sufficient convergence of co-existing densities, we use block averages by dividing the trajectory into blocks of 1 μs each. Before we move on to the calculation of the co-existence densities based on the computed density profiles, we would like to make the reader aware of some issues that must be taken into account to use this simulation method properly.

The first issue is how to decide on the number of protein chains that can provide a trustworthy estimate of the co-existence densities. The most important indicators of a suitable system size are the ability to observe a flat density profile in the middle region corresponding to the high-density liquid phase and that the protein concentration in this phase is statistically independent of any further increase in the number of protein chains in the same size simulation box. In a previous study of the FUS low complexity (LC) disordered region (163 residues) with liquid phase densities between 300 and 600 mg/mL, we found that 100 FUS LC chains are sufficient to achieve

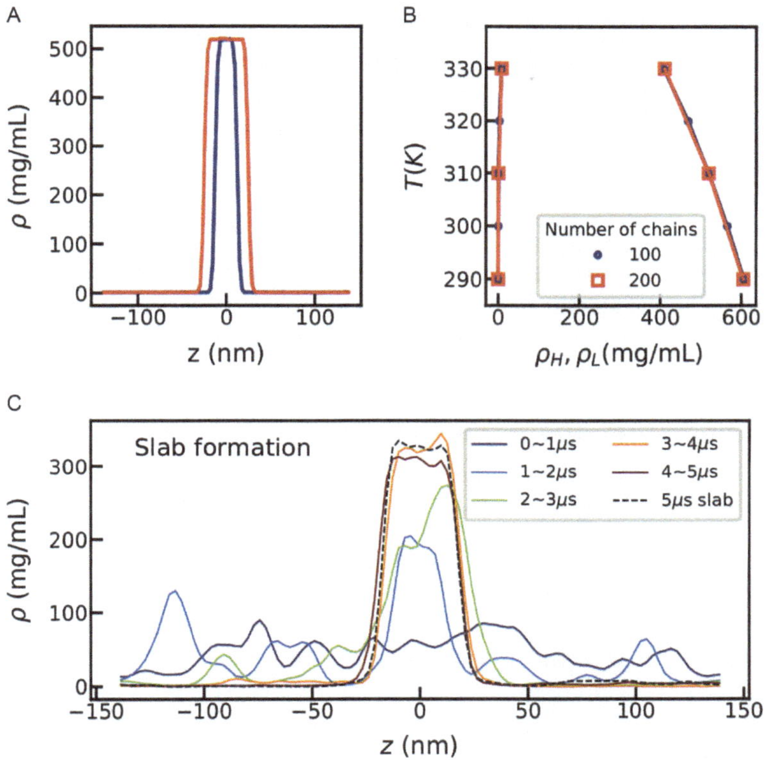

Fig. 4 Issues with obtaining density profiles from simulations. (A) Density profiles by using 100 (blue) and 200 (red) chains of proteins. (B) Phase diagram by using 100 and 200 chains of proteins. (C) Density profiles highlighting that condensed slab configuration can be obtained from a wholly dispersed configuration after several microseconds of simulation time.

reasonable accuracy and the results with 200 chains are almost identical (Fig. 4A and B) (Dignon, Zheng, Best, et al., 2018). One should perform a similar analysis on a system of interest to decide on a suitable system-size.

The second issue is deciding on the simulation run length to obtain a converged density profile. In our recent work, we used a simulation time of 5 μs because we stop seeing any significant deviations in the density profile when comparing 1 μs block averages over the full trajectory. The choice can be system-specific but careful attention should be paid to the evolution of the density profiles over time and if the information obtained from multiple independent trajectories or block averages over a long trajectory is within desirable statistical uncertainty values. We note that there is an inherent advantage in using the pre-formed slab configuration as opposed to waiting for fully dispersed chains in a long simulation box to form a desired planar condensed phase spontaneously. As shown in Fig. 4C starting from a dispersed configuration, the evolution of the density profile starts to look similar to the profile obtained from an initial slab configuration after about 3 μs of simulation time. This time can be saved by condensing a low-density configuration within tens of nanoseconds, as we discussed earlier in the chapter.

For benchmarking purposes, we were able to simulate a 100-chain system with a 163 amino acid long protein chain, for 5 μs with a wallclock time of approximately 36 hhours on a standard desktop computer using HOOMD-Blue (v.2.1.5) with a single Nvidia GTX1080 GPU.

3.3 Generating phase diagrams

Once we have the converged density profiles from the simulations at different temperatures, we can estimate the densities ρ_{H} (high-density liquid phase) and ρ_{L} (low-density dilute phase). In a well-sampled density profile from a converged simulation, we would see a flat peak at the center of the slab, i.e., at $z=0$ with either half of the density profile resembling a sigmoidal curve. And as shown in Fig. 4A, the density at the center $z=0$ would start decreasing, and densities away from the center would start increasing as we go to higher temperatures closer to the critical temperature. We can define the regions corresponding to the low and high-density phases along the z-axis and take averages over bins in the respective regions to get ρ_{H} and ρ_{L}. However, in practice, there can be sampling issues at specific temperatures, especially on the low-density side. Hence, the following sigmoidal function can also be used to fit the density profile (Blas et al., 2008)

$$\rho(z) = \frac{1}{2}(\rho_H + \rho_L) - \frac{1}{2}(\rho_H - \rho_L)\tanh\left(\frac{z - z_0}{d}\right) \tag{4}$$

Once we have ρ_{H} and ρ_{L} at different temperatures, we can estimate the critical temperature using the following equation,

$$\rho_H - \rho_L = A(T_c - T)^{\beta} \tag{5}$$

where β is the critical exponent ($\beta=0.325$ for universality class of 3D Ising model (Rowlinson & Widom, 2013)), A is a protein-specific fitting parameter, and T_{c} is the critical temperature (Fig. 5A). The critical concentration ρ_{c} can be further obtained by fitting to,

$$\frac{\rho_H + \rho_L}{2} = \rho_c + B(T_c - T) \tag{6}$$

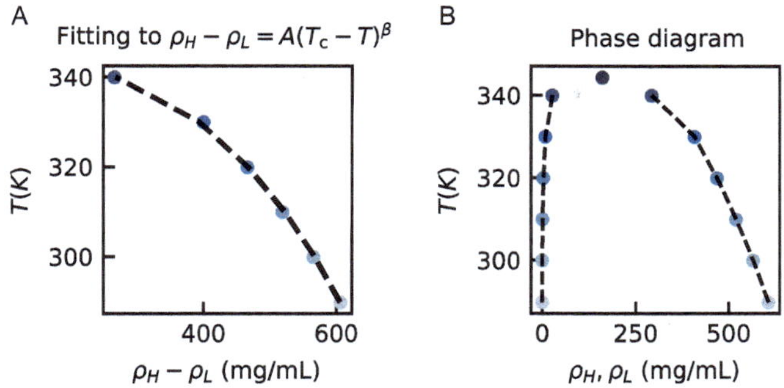

Fig. 5 Obtaining the phase diagrams from density profiles. (A) Fitting the density profile to obtain the critical temperature. (B) Drawing phase diagrams.

where B is a fitting parameter. We note that Eq. 5 is only asymptotically valid near the critical temperature, so this should be accounted for when selecting appropriate data for fitting (Blas et al., 2008). The fit results should be relatively independent of the maximum temperature near the T_c to a reasonable degree, but this should be checked for a specific system of interest carefully (Dignon, Zheng, Best, et al., 2018). Of course, additional simulations should be conducted if there is a large gap between the estimated T_c and the highest temperature simulated.

As it may be difficult to ascertain the suitable temperature range for a new system a priori, one can utilize empirical correlation between the theta-solvent temperature (T_θ), i.e., the temperature at which a protein behaves like a random coil, and T_c (Dignon, Zheng, Best, et al., 2018). As T_θ can be efficiently determined using HOOMD-Blue or other CPU-based codes such as LAMMPS (Plimpton et al., 2007) for a single-chain system that can be further facilitated using replica exchange molecular dynamics, this should result in a significant saving of computational effort in searching for a suitable temperature range to conduct co-existence simulations.

With ρ_H and ρ_L from density profiles at a range of temperatures, the critical temperature T_c and the critical concentration ρ_c that we obtain from fitting to Eqs. 5 and 6, we can plot a typical thermodynamic phase diagram with the densities on the x-axis and the temperature on the y-axis (Fig. 5B). The left arm of the phase diagram represents the protein concentration in the low-density phase, also referred to as the saturation concentration c_{sat} in the protein LLPS literature, whereas the right arm represents the protein concentration in the high-density liquid phase. The region inside these two curves and below T_c is the co-existence region; a protein solution prepared at concentrations within this region will undergo phase separation into two separate co-existing phases. The concentration of these two phases can be estimated by drawing a horizontal tie line which connects the two sides of the phase envelope and passes through the initial concentration point. In fact, these fundamental features of a thermodynamic phase diagram were used for setting up the co-existence simulations and the successful generation of phase diagrams. These phase diagrams provide a rigorous thermodynamic basis for comparison between different variants of a protein sequence and how underlying changes due to mutations or additions/deletions of domains may affect their phase behavior. In experimental studies, instead of temperature, salt concentration can also be used as a parameter of interest for constructing the phase diagram. In our coarse-grained modeling framework, changing the salt concentration will alter the electrostatic interactions by modifying the Debye length based on Eq. 1. One can also use our model to study the sequence-specific characteristics contributing to the phase separation propensity of a protein, including changes arising from point mutations (Conicella et al., 2020; Murthy et al., 2019).

4. Summary

In this chapter, we introduce the reader to a practical and efficient way of computing the phase behavior of an intrinsically disordered protein (IDP) using our coarse-grained (CG) but sequence-specific protein model together with a co-existence method in a molecular dynamics simulation. We also try to bring into light some of the nuances associated with our methodology, which the reader needs to be cautious with, while implementing this proposed method into practice. The computer codes/scripts needed for implementation are provided which work with HOOMD-Blue, which is a publicly available software package which uses graphics processing units (GPUs) to achieve significant computational speed in such calculations. Our CG modeling strategy is a flexible and extensible framework that is most importantly, also transferable without requiring any additional experimental input on a system of interest. This has already been applied to numerous important biological systems to provide critical molecular insights into the sequence-dependent liquid–liquid phase separation (LLPS) behavior. We hope that the information contained in this chapter will lower the barrier for others to integrate our modeling framework into their workflow as an efficient way of estimating phase behavior of IDPs on systems of their choice and help move the field forward. The python scripts and the instructions for using these scripts can be found at the following website: https://bitbucket.org/jeetain/hoomd_slab_builder.

Acknowledgments

Our work summarized in this book chapter was supported by the US Department of Energy, Office of Science, Basic Energy Sciences Award DE-SC00013979, the National Institutes of Health grants R01GM118530, R01NS116176, and R01GM120537, and the National Science Foundation (NSF) grants 2004796 and 1720530. W.Z. acknowledges the support from NSF MCB-2015030. This research used resources of the National Energy Research Scientific Computing Center, a DOE Office of Science User Facility supported under Contract No. DE-AC02-05CH11231. Use of the high-performance computing capabilities of the Extreme Science and Engineering Discovery Environment (XSEDE), which is supported by the National Science Foundation, project no. TG-MCB120014 is also gratefully acknowledged.

References

Anderson, J. A., Glaser, J., & Glotzer, S. C. (2020). HOOMD-blue: A Python package for high-performance molecular dynamics and hard particle Monte Carlo simulations. *Computational Materials Science*, *173*, 109363. https://doi.org/10.1016/j.commatsci.2019.109363.

Ashbaugh, H. S., & Hatch, H. W. (2008). Natively unfolded protein stability as a coil-to-globule transition in charge/hydropathy space. *Journal of the American Chemical Society*, *130*, 9536–9542.

Ayton, G. S., Noid, W. G., & Voth, G. A. (2007). Multiscale modeling of biomolecular systems: In serial and in parallel. *Current Opinion in Structural Biology*, *17*, 192–198.

Best, R. B., Zheng, W., & Mittal, J. (2014). Balanced protein-water interactions improve properties of disordered proteins and non-specific protein association. *Journal of Chemical Theory and Computation*, *10*, 5113–5124.

Blas, F. J., MacDowell, L. G., de Miguel, E., & Jackson, G. (2008). Vapor-liquid interfacial properties of fully flexible Lennard-Jones chains. *The Journal of Chemical Physics*, *129*, 144703. https://doi.org/10.1063/1.2989115.

Brangwynne, C. P., Eckmann, C. R., Courson, D. S., Rybarska, A., Hoege, C., Gharakhani, J., et al. (2009). Germline P granules are liquid droplets that localize by controlled dissolution/condensation. *Science*, *324*, 1729–1732. https://doi.org/10.1126/science.1172046.

Brangwynne, C. P., Mitchison, T. J., & Hyman, A. A. (2011). Active liquid-like behavior of nucleoli determines their size and shape in Xenopus laevis oocytes. *Proceedings of the National Academy of Sciences of the United States of America*, *108*, 4334–4339. https://doi.org/10.1073/pnas.1017150108.

Brangwynne, C. P., Tompa, P., & Pappu, R. V. (2015). Polymer physics of intracellular phase transitions. *Nature Physics*, *11*, 899.

Burke, K. A., Janke, A. M., Rhine, C. L., & Fawzi, N. L. (2015). Residue-by-residue view of in vitro FUS granules that bind the C-terminal domain of RNA polymerase II. *Molecular Cell*, *60*, 231–241.

Clementi, C. (2008). Coarse-grained models of protein folding: Toy models or predictive tools? *Current Opinion in Structural Biology*, *18*, 10–15. https://doi.org/10.1016/j.sbi.2007.10.005.

Conicella, A. E., Dignon, G. L., Zerze, G. H., Schmidt, H. B., D'Ordine, A. M., Kim, Y. C., et al. (2020). TDP-43 α-helical structure tunes liquid–liquid phase separation and

function. *Proceedings of the National Academy of Sciences of the United States of America, 117*, 5883–5894. https://doi.org/10.1073/pnas.1912055117.
Conicella, A. E., Zerze, G. H., Mittal, J., & Fawzi, N. L. (2016). ALS mutations disrupt phase separation mediated by α-helical structure in the TDP-43 low-complexity C-terminal domain. *Structure, 24*, 1537–1549. https://doi.org/10.1016/j.str.2016.07.007.
Creighton, T. E. (1993). *Proteins: Structures and molecular properties*. Macmillan.
de Gennes, P.-G. (1978). *Scaling concepts in polymer physics*. Cornell University Press.
Debye, P., & Hückel, E. (1923). De la theorie des electrolytes. I. abaissement du point de congelation et phenomenes associes. *Physikalishce Zeitschrift, 24*, 185–206.
Dignon, G. L., Zheng, W., Best, R. B., Kim, Y. C., & Mittal, J. (2018). Relation between single-molecule properties and phase behavior of intrinsically disordered proteins. *Proceedings of the National Academy of Sciences of the United States of America, 115*, 9929–9934. https://doi.org/10.1073/pnas.1804177115.
Dignon, G. L., Zheng, W., Kim, Y. C., Best, R. B., & Mittal, J. (2018). Sequence determinants of protein phase behavior from a coarse-grained model. *PLoS Computational Biology, 14*, e1005941. https://doi.org/10.1101/238170.
Dignon, G. L., Zheng, W., Kim, Y. C., & Mittal, J. (2019). Temperature-controlled liquid-liquid phase separation of disordered proteins. *ACS Central Science, 5*, 821–830. https://doi.org/10.1021/acscentsci.9b00102.
Dignon, G. L., Zheng, W., & Mittal, J. (2019). Simulation methods for liquid–liquid phase separation of disordered proteins. *Current Opinion in Chemical Engineering, 23*, 92–98. https://doi.org/10.1016/j.coche.2019.03.004.
Elbaum-Garfinkle, S., Kim, Y., Szczepaniak, K., Chen, C. C.-H., Eckmann, C. R., Myong, S., et al. (2015). The disordered P granule protein LAF-1 drives phase separation into droplets with tunable viscosity and dynamics. *Proceedings of the National Academy of Sciences of the United States of America, 112*, 7189–7194. https://doi.org/10.1073/pnas.1504822112.
Fisher, C. K., & Stultz, C. M. (2011). Constructing ensembles for intrinsically disordered proteins. *Current Opinion in Structural Biology, 21*, 426–431. https://doi.org/10.1016/j.sbi.2011.04.001.
Flory, P. J. (1953). *Principles of polymer chemistry*. Cornell University Press.
Huang, J., Rauscher, S., Nawrocki, G., Ran, T., Feig, M., de Groot, B. L., et al. (2017). CHARMM36m: An improved force field for folded and intrinsically disordered proteins. *Nature Methods, 14*, 71.
Jung, H., & Yethiraj, A. (2018). A simulation method for the phase diagram of complex fluid mixtures. *The Journal of Chemical Physics, 148*, 244903. https://doi.org/10.1063/1.5033958.
Kapcha, L. H., & Rossky, P. J. (2014). A simple atomic-level hydrophobicity scale reveals protein interfacial structure. *Journal of Molecular Biology, 426*, 484–498. https://doi.org/10.1016/j.jmb.2013.09.039.
Kim, J., Keyes, T., & Straub, J. E. (2010). Generalized replica exchange method. *The Journal of Chemical Physics, 132*, 224107.
Latham, A. P., & Zhang, B. (2019). Improving coarse-grained protein force fields with small-angle X-ray scattering data. *The Journal of Physical Chemistry B, 123*, 1026–1034. https://doi.org/10.1021/acs.jpcb.8b10336.
Lin, Y.-H., Forman-Kay, J. D., & Chan, H. S. (2016). Sequence-specific polyampholyte phase separation in membraneless organelles. *Physical Review Letters, 117*, 178101.
Martyna, G. J., Tobias, D. J., & Klein, M. L. (1994). Constant pressure molecular dynamics algorithms. *The Journal of Chemical Physics, 101*, 4177–4189. https://doi.org/10.1063/1.467468.
Monahan, Z., Ryan, V. H., Janke, A. M., Burke, K. A., Zerze, G. H., O'Meally, R., et al. (2017). Phosphorylation of FUS low-complexity domain disrupts phase separation, aggregation, and toxicity. *The EMBO Journal, 36*, 2951–2967.

Murthy, A. C., Dignon, G. L., Kan, Y., Zerze, G. H., Parekh, S. H., Mittal, J., et al. (2019). Molecular interactions underlying liquid–liquid phase separation of the FUS low-complexity domain. *Nature Structural & Molecular Biology*, *26*, 637–648. https://doi.org/10.1038/s41594-019-0250-x.
Panagiotopoulos, A. Z. (2000). Monte Carlo methods for phase equilibria of fluids. *Journal of Physics Condensed Matter*, *12*, R25. https://doi.org/10.1088/0953-8984/12/3/201.
Piana, S., Donchev, A. G., Robustelli, P., & Shaw, D. E. (2015). Water dispersion interactions strongly influence simulated structural properties of disordered protein states. *The Journal of Physical Chemistry B*, *119*, 5113–5123.
Plimpton, S., Crozier, P., & Thompson, A. (2007). LAMMPS-large-scale atomic/molecular massively parallel simulator. In (Vol. 18), Sandia Natl Lab, 43.
Potoff, J. J., & Panagiotopoulos, A. Z. (1998). Critical point and phase behavior of the pure fluid and a Lennard-Jones mixture. *The Journal of Chemical Physics*, *109*, 10914–10920. https://doi.org/10.1063/1.477787.
Quiroz, F. G., & Chilkoti, A. (2015). Sequence heuristics to encode phase behaviour in intrinsically disordered protein polymers. *Nature Materials*, *14*, 1164–1171. https://doi.org/10.1038/nmat4418.
Rowlinson, J. S., & Widom, B. (2013). *Molecular theory of capillarity*. Courier Corporation.
Sawle, L., & Ghosh, K. (2015). A theoretical method to compute sequence dependent configurational properties in charged polymers and proteins. *The Journal of Chemical Physics*, *143*, 085101. https://doi.org/10.1063/1.4929391.
Schuster, B. S., Dignon, G. L., Tang, W. S., Kelley, F. M., Ranganath, A. K., Jahnke, C. N., et al. (2020). Identifying sequence perturbations to an intrinsically disordered protein that determine its phase separation behavior. *Proceedings of the National Academy of Sciences*, *117*, 11421–11431. https://doi.org/10.1101/2020.01.06.894576.
Silmore, K. S., Howard, M. P., & Panagiotopoulos, A. Z. (2017). Vapour—liquid phase equilibrium and surface tension of fully flexible Lennard—Jones chains. *Molecular Physics*, *115*, 320–327.
Tompa, P. (2012). Intrinsically disordered proteins: A 10-year recap. *Trends in Biochemical Sciences*, *37*, 509–516. https://doi.org/10.1016/j.tibs.2012.08.004.
Vitalis, A., & Pappu, R. V. (2008). ABSINTH: A new continuum solvent model for simulations of polypeptides in aqueous solutions. *Journal of Computational Chemistry*, *30*, 673–699.
Wright, P. E., & Dyson, H. J. (1999). Intrinsically unstructured proteins: Re-assessing the protein structure-function paradigm. *Journal of Molecular Biology*, *293*, 321–331. https://doi.org/10.1006/jmbi.1999.3110.
Zheng, W., Dignon, G., Brown, M., Kim, Y. C., & Mittal, J. (2020). Hydropathy patterning complements charge patterning to describe conformational preferences of disordered proteins. *The Journal of Physical Chemistry Letters*, *11*, 3408–3415. https://doi.org/10.1101/2020.01.25.919498.
Zheng, W., Zerze, G. H., Borgia, A., Mittal, J., Schuler, B., & Best, R. B. (2018). Inferring properties of disordered chains from FRET transfer efficiencies. *The Journal of Chemical Physics*, *148*, 123329. https://doi.org/10.1063/1.5006954.

CHAPTER TWO

Reconstituting bacterial cell division assemblies in crowded, phase-separated media

Begoña Monterroso*,†, Miguel Ángel Robles-Ramos, Silvia Zorrilla*,†, and Germán Rivas*

Centro de Investigaciones Biológicas Margarita Salas, Consejo Superior de Investigaciones Científicas (CSIC), Madrid, Spain

*Corresponding authors: e-mail address: monterroso@cib.csic.es; silvia@cib.csic.es; grivas@cib.csic.es

Contents

Abstract

Here we have summarized several strategies to reconstruct complexes containing the FtsZ protein, a central element of the cell division machinery in most bacteria, and to test their functional organization in minimal membrane systems and cell-like containers, as vesicles and droplets produced by microfluidics. These synthetic systems have been devised to mimic elements of the intracellular complexity, as excluded volume effects due to natural crowding, and macromolecular condensation resulting from biologically regulated liquid-liquid phase separation, in media of known and controllable

† These authors contributed equally.

Methods in Enzymology, Volume 646
ISSN 0076-6879
https://doi.org/10.1016/bs.mie.2020.06.012

composition. This integrative approach has allowed to demonstrate that macromolecular phase separation and crowding may also help to dynamically organize FtsZ in the intracellular space thus modulating its functional reactivity in cell division.

1. Introduction

Bottom-up synthetic biology: The integration of individual molecular systems into functional modules toward the building of an artificial cell from the bottom up is one of the major challenges of synthetic biology (Ausländer, Ausländer, & Fussenegger, 2017). These efforts to attain or optimize functionality, require a profound mechanistic understanding of how the elements of essential molecular machines are organized, in time and space, and coordinated one to another, as a network of multiple interactions to function. Besides, the behavior of these machines is governed not only by specific molecular interactions but also by the physicochemical properties of the intracellular environment, as molecular crowding, surface interactions, and—more in particular—compartmentalization driven by liquid-liquid phase separation, an emergent mechanism of the intracellular organization (see below). It follows that these reconstitution studies should be performed under controlled conditions that mimic the local microenvironments in which these machines are located and have evolved to function *in vivo*.

The bacterial division machinery (the divisome) and the Z-ring: Cell division is one of the most fundamental hallmarks of life and, consequently, its reconstitution in artificial systems from its building blocks is a significant challenge. A full understanding of protein-based division machineries, particularly those existing in simple prokaryotic systems, will help to determine the minimum elements needed for dividing a synthetic cell (Tarnopol et al., 2019). The most extensively studied protein involved in prokaryotic division is FtsZ. This protein is a soluble GTPase, ancestor of eukaryotic tubulin, that serves as a central element of the divisome in most bacteria (Haeusser & Margolin, 2016). FtsZ interacts with additional proteins that contribute to its function forming a dynamic ring (Z-ring) at midcell that is essential to constrict the membrane (Fig. 1). The initial pathway of ring formation is the GTP-dependent polymerization of FtsZ. In *Escherichia coli*, two additional proteins, ZipA and FtsA, tether FtsZ to the membrane forming the proto-ring, the first molecular assembly of the divisome. These proteins serve to anchor the ring and, together with a set of Zap proteins (which promote FtsZ bundling), they stabilize it (Ortiz, Natale, Cueto, & Vicente, 2016).

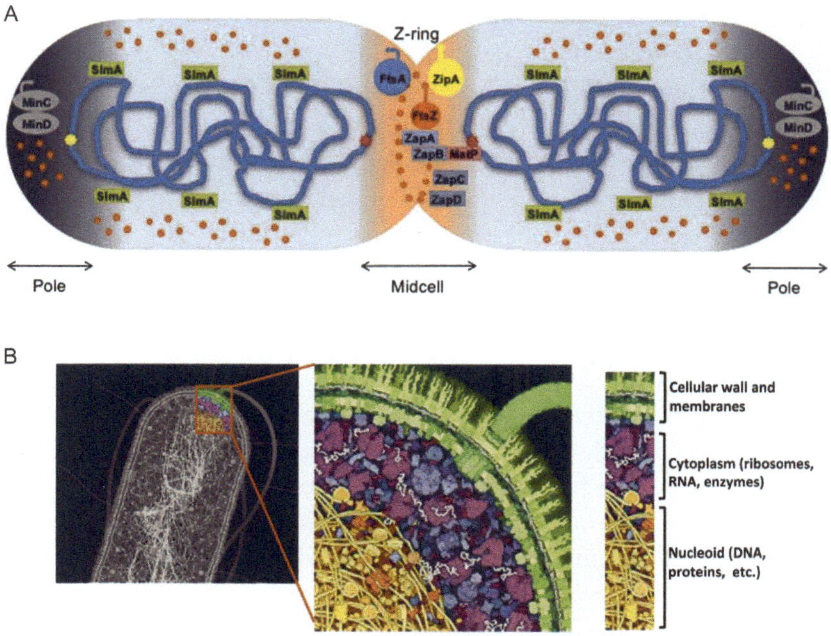

Fig. 1 Overview of *E. coli* cell division and crowding inside the cell. (A) Scheme showing the distribution of FtsZ and the positive and negative regulators of the *Z*-ring stability in a dividing cell. (B) Illustration of the interior of an *Escherichia coli* cell. *Left, reproduced with permission from Goodsell, D. S. (2009).* The machinery of life*, New York: Springer-Verlag. Right, adapted from Goodsell, D. S. gallery http://pdb101.rcsb.org/sci-art/goodsell-gallery.*

The spatiotemporal control of cell division is finely coordinated with chromosome segregation to maintain the correct order of events and safeguard genomic integrity and bacterial survival (Dewachter, Verstraeten, Fauvart, & Michiels, 2018). In this regard, the location of the *Z*-ring is precisely controlled by the concerted action of several site-selection mechanisms, the canonical ones being the nucleoid occlusion mediated by the SlmA protein, and the Min system, which prevent FtsZ from polymerizing at locations were septation is undesired, namely the vicinity of the nucleoid and the poles, respectively. SlmA is a DNA-binding protein that when bound to a specific DNA sequence (SBS—SlmA-binding sequence) blocks the *Z*-ring formation in the cell regions occupied by the nucleoid (Tonthat et al., 2013). The Min system comprises the MinC, MinD, and MinE proteins; MinC has a low FtsZ inhibitory activity that needs to be activated upon interaction with membrane-bound MinD, while MinE prevents the formation of MinCD complexes at midcell (Ramm, Heermann, & Schwille, 2019).

Reconstructing minimal divisomes—achievements and challenges: The functions, interactions, and assembly properties of proto-ring elements have been reconstituted in nanodiscs, coated-microbeads, bilayers, and vesicles as membrane scaffolds (reviewed in Martos, Jiménez, Rivas, and Schwille, 2012 and Rivas, Vogel, and Schwille, 2014). On the other hand, the oscillatory behavior of the Min system and the coupling between FtsZ wave formation and the oscillating Min have also been reproduced in bilayers, droplets, and in microchambers (Ramm et al., 2019). Despite these developments, a fully functional reconstruction of the *Z*-ring in artificial containers, beyond the already published assembly of FtsZ static structures with FtsA (Osawa & Erickson, 2013; Szwedziak, Wang, Bharat, Tsim, & Löwe, 2014) or relatively dynamic ones with ZipA (Cabré et al., 2013) has not been achieved. These observations suggest that some additional elements are still missing from the current reconstitution attempts, and further experimental and theoretical efforts are required to integrate biochemical and mechanical information on the systems within the complex architectures that have evolved to function in the cell.

The bacterial interior is a crowded and heterogeneous environment: The reconstruction of active minimal divisomes in the absence of cells needs to reproduce faithfully the intracellular media, which consists of a variety of distinct microenvironments characterized by high total macromolecular concentration, the immediate presence of quasi-static boundaries (membranes, fibers), and heterogeneity of composition (Vendeville, Larivière, & Fourmentin, 2011). From this overall picture (Fig. 1), we can state that FtsZ, in addition to crowding effects (related to the volume occupancy of 200–300 g/L of soluble macromolecules), will be subjected to additional non-specific interactions with elements of its immediate surroundings during its life cycle, as the ribosomes, the nucleoid (a large polyanionic structure within which FtsZ will encounter a high total concentration of DNA and will be subjected to phase separation effects), the inner plasma membrane, and large amounts of positively charged ions and osmolytes that can also alter the properties of FtsZ and its interactions. Several physicochemical phenomena are known to considerably impact the reactivity and spatiotemporal organization of proteins and macromolecular complexes in such media, as excluded volume effects due to crowding, interactions and adsorption at surfaces, and partitioning via macromolecular phase separation (Rivas & Minton, 2016, 2018). The latter can be driven by crowding and despite its relevance for the organization of intracellular biochemistry in eukaryotic systems (see below) it has been observed in substantially fewer cases in bacterial cells

(Abbondanzieri & Meyer, 2019), one of them reported from our laboratory (Monterroso et al., 2019). We hypothesize that the physical properties of the bacterial interior should be favorable to liquid-liquid phase separation.

Macromolecular phase separation as an organizer of intracellular biochemistry: Although membrane-less organelles that cluster specific biomolecules away from the surrounding milieu have long been theorized, recent breakthrough work has led to a major wave of discoveries of "biomolecular condensates" (Banani, Lee, Hyman, & Rosen, 2017). These dynamic structures contain a wide range of proteins and nucleic acids and assemble through the process of liquid-liquid phase separation—LLPS (Crowe & Keating, 2018; Shin & Brangwynne, 2017). Up to now these condensates have been primarily found in eukaryotic cells, where they play crucial roles in cell physiology (e.g., metabolic regulation, signaling, gene expression, stress adaptation, etc.) and pathology (e.g., diseases related to protein aggregation, amyloid formation, etc.), thus emerging as a fundamental mechanism to organize the biochemistry of the intracellular space. Proteins that contain multivalent domains, mostly involved in protein-protein and protein-nucleic acid interactions, and those containing intrinsically disordered regions are prone to form these condensates. As previously mentioned, volume exclusion due to natural crowding can promote these phase separation processes.

2. Experimental approaches and analysis

In this section we describe the main procedures followed to study the influence of phase separation on FtsZ, in bulk and encapsulated in cytomimetic containers, and for the characterization of FtsZ·SlmA·SBS biomolecular condensates. While the procedures can be applied to other systems as described, specific experimental conditions such as ligands concentrations, buffers and other additives correspond to the particular studies performed.

2.1 Material preparation

2.1.1 Proteins purification and labeling

- Follow the established protocols for purification of *E. coli* proteins FtsZ (Rivas et al., 2000) and SlmA (Cabré et al., 2015; Cho, McManus, Dove, & Bernhardt, 2011), and keep them stored at −80 °C until use.
- Follow the procedures described in Cabré et al. (2015), González et al. (2003) and Reija et al. (2011) for labeling of proteins at the amino groups with Alexa Fluor 488 or Alexa Fluor 647 carboxylic acid succinimidyl

ester dyes (Molecular Probes/Invitrogen/Thermo Fisher). Estimate the degree of labeling from the molar absorption coefficients of the fluorophore and the protein.

Tip: detailed protocols for labeling are available from the manufacturer.

2.1.2 Hybridization of SBS oligonucleotide specifically targeted by SlmA

- Hybridize single-stranded SBS oligonucleotides (5′-AAGTAAGTGA GCGCTCACTTACGT-3′ (Cho et al., 2011)) using a thermocycler as explained (Cabré et al., 2015).
- Hybridize the fluorescently labeled oligonucleotides (tagged in 5′ with FITC or Alexa 647) with a 10% excess of the complementary unlabeled one, following the same procedure.

Tip: oligonucleotides, including the ones labeled with fluorescent dyes, can be purchased from different companies.

2.1.3 Purification of DNA crowder

- Sonicate DNA (from salmon sperm, Wako Pure Chemical Industries, Osaka, Japan) for 15 min in a bath.
- Purify the DNA by the phenol:chloroform:isoamyl alcohol extraction method followed by ethanol precipitation, as described in Biswas et al. (2012).
- Resuspend the dried pellet in 50 mM Tris-HCl, 300 mM KCl, pH 7.5, and store at −20 °C until use.
- Estimate the concentration of DNA from its dry weight after purification (Monterroso, Zorrilla, Sobrinos-Sanguino, Keating, & Rivas, 2016).

2.1.4 Preparation of crowders and enriched phases

- Equilibrate the crowders (PEG 8, dextran T40 and 500, and Ficoll 70), without further purification, by extensive dialysis in 50 mM Tris-HCl, 300 mM KCl, pH 7.5.
- Determine their final concentrations from the refractive index increment as described (Monterroso, Reija, Jiménez, Zorrilla, & Rivas, 2016): dextran 500 and dextran T40, 0.142 and 0.147 mL/g respectively; PEG 8, 0.136 mL/g; Ficoll 70, 0.141 mL/g.
- Thoroughly mix defined volumes of binary mixtures of PEG 8 with a second crowder, rendering phase separation, to yield the following final nominal concentrations: 53 and 82 g/L (PEG 8 and dextran 500, respectively); 53 and 82–92 g/L (PEG 8 and dextran T40, respectively); 113 and

165 g/L (PEG 8 and Ficoll 70, respectively); and 92 and 183 g/L (PEG 8 and DNA (prepared as described in Section 2.1.3), respectively).
- Centrifuge the mixtures at 3000 rpm for 5 min in a bench centrifuge to favor phase separation. PEG phase remains always on top.
- Isolate the enriched phases by pipetting, discarding the material nearby the interface.

Tip: pipette first the phase on the bottom, depending on the mixture liquids seem to redistribute when removing the top phase first.

2.1.5 Preparation of a labeled PEG phase

- Follow the procedure described in Monterroso, Zorrilla, et al. (2016) for labeling of PEG with Alexa Fluor 488 or Alexa Fluor 647 carboxylic acid succinimidyl ester dyes (covalently linked to an amino derivative of the polymer, NH_2-PEG-NH_2, from Nanocs).
- After removal of the unreacted dye by size exclusion chromatography, calculate the concentration of dye attached to PEG from its molar absorption coefficient.
- When needed, add labeled PEG to the PEG phase so that the final concentration of the probe in the LLPS system is 1 μM (bulk systems, see Section 2.2, and manually encapsulated systems, see Section 2.3.1) or 2 μM (microfluidic encapsulation, see Section 2.3.2.2).

2.2 Bulk systems

2.2.1 Analysis of FtsZ·SlmA·SBS biomolecular condensates

- Prepare 50 mM Tris-HCl, 300 mM KCl, 1 mM $MgCl_2$, pH 7.5 solutions with a crowding agent (dextran 500, Ficoll 70, PEG 8 or DNA) at the required concentration (usually 150 g/L).
- Add FtsZ, SlmA and the SBS double-stranded oligonucleotide on these solutions. Their typical concentrations are 12, 5 and 1 μM, respectively.
- For visualization, add tracer amounts (usually 1 μM) of one or two of these elements labeled with fluorescent dyes.

2.2.2 Dynamics of FtsZ·SlmA·SBS biomolecular condensates and functionality of FtsZ within them

- Prepare samples of the condensates with the FtsZ, SlmA and SBS concentrations indicated in Section 2.2.1, with 1 μM labeled FtsZ. Visualize them after 30 min incubation by confocal microscopy (see Section 2.4) as your initial state.

- For dynamics experiments, add 0.5–1 μM FtsZ labeled with a spectrally different dye and follow its diffusion into the originally formed condensates with time.
- For experiments on FtsZ functionality, add 1 mM GTP and follow the conversion of condensates into fibers with time.

2.2.3 Turbidity measurements of biomolecular condensates

- Prepare samples of the condensates as explained in Section 2.2.1 and incubate 30 min.
- Measure the turbidity at 350, 400 and 450 nm using a plate reader and subtract background signal.
- For time dependence measurements, record data every 5 min for 90 min.

2.2.4 Preparation of bulk LLPS systems for imaging of FtsZ distribution

- Form simple emulsions by thoroughly mixing the corresponding enriched phases, prepared as described in Section 2.1.4, at a 3:1 (PEG/second crowder) ratio (Fig. 2). Identify the phases by using one of them labeled (PEG, labeled according to Section 2.1.5, or fluorescein labeled dextran 500, Molecular Probes/Invitrogen) with a final 1 μM dye concentration.
- Add to this mixture FtsZ containing a tracer amount (1 μM) of FtsZ labeled with a dye spectrally different from that of the labeled phase. Adjust the concentration of $MgCl_2$ to reach 1 mM, and that of KCl to 300 mM.
- When required, trigger formation of FtsZ fibers by adding GTP.

Tip: the total volume of solutions added to the LLPS system (*i.e.*, protein, nucleotide and any other component) should not constitute a big part of the total volume of sample, as an excessive dilution of the phases can cause the loss of their physical separation. Compatible percentage depends on the particular mixture.

- Image the samples as described below.

2.2.5 Calculation of FtsZ partition coefficients

- Mix the two crowders for a particular LLPS system in a 1:1 volume ratio (Fig. 2).
- Add FtsZ labeled with Alexa 488 (0.5 μM) and unlabeled FtsZ up to the total concentration required in each sample and mix gently. Adjust the concentrations of $MgCl_2$ and KCl to 1 and 300 mM, respectively. When required, add GTP to the samples.

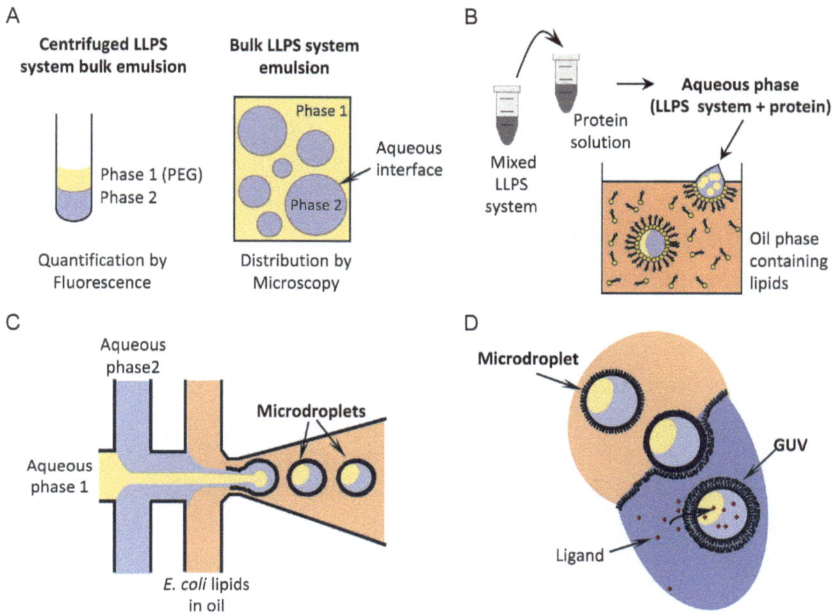

Fig. 2 Overview of experimental approaches. (A) Bulk LLPS systems emulsions. Left, evaluation of protein partition within the phases is determined by fluorescence measurements of the protein. Right, assessment of protein distribution and organization through confocal microscopy imaging. (B) Manual encapsulation of LLPS systems in microdroplets stabilized by *E. coli* lipids. (C) Scheme of the microdroplets production by microfluidics. (D) Scheme of the generation of GUVs from collected microdroplets. *Panels (A and B) modified from Monterroso, B., Zorrilla, S., Sobrinos-Sanguino, M., Keating, C. D. & Rivas, G. (2016). Microenvironments created by liquid-liquid phase transition control the dynamic distribution of bacterial division FtsZ protein.* Scientific Reports, 6, *35140.*

- Allow to equilibrate for 30 min until reaching phase separation. PEG phase remains always on top.
- Centrifuge at 3000 rpm for 5 min in a bench centrifuge.
- Measure the fluorescence emission intensity of an aliquot of each of the phases in a plate reader. Evaluate the background and subtract it from the measured intensities.

Tip: transparent glass tubes were used for large volumes to easily identify the interface. For small volumes, PCR tubes, more transparent than the regular plastic tubes, were employed.

- Calculate the concentration of protein according to a calibration curve, obtained from samples containing known amounts of FtsZ-Alexa 488 diluted in the corresponding phase. Previously, perform control measurements for linearity of the signal with FtsZ concentration.

- Calculate partition coefficients, *K*, within the binary mixture as the ratio of the concentration of FtsZ in the PEG-rich phase to that in the other phase.
- Calculate fraction of FtsZ at the interface as the difference between the total amount of protein added and the sum of those in each phase.

2.3 Confined systems

2.3.1 Encapsulation of FtsZ in LLPS systems encircled by a lipid monolayer through manual emulsion

- Thoroughly dry the required amount of a chloroform solution of phospholipids from *E. coli* (Polar extract; Avanti Polar Lipids, Alabama, USA) in a Speed-Vac device or under nitrogen flow, until obtaining a film.

Tip: when drying the lipids by nitrogen stream, leave under vacuum for an hour to ensure complete removal of any residual chloroform.

- Resuspend the film in a volume of mineral oil to get a final concentration of 20 g/L lipids by two cycles of vortex and 15 min sonication in a bath.
- Thoroughly mix the two enriched aqueous phases of choice (purified from each other as explained in Section 2.1.4) by vortexing to enhance the formation of very small droplets of one in the other (Fig. 2). 3:1 or 1:1 volume ratios of PEG/second crowder were employed with similar results.
- Add on the phases thus mixed FtsZ containing a tracer amount of labeled FtsZ (1 μM) spectrally different from that of the labeled phase. Adjust the concentration of $MgCl_2$ and that of KCl to 1 and 300 mM, respectively. When required, include GTP.
- Add this mixture on the oil solution containing the dispersed lipids.
- Gently mix by pipetting up and down to render water in oil emulsion droplets.
- Visualize the samples in the microscope as described below.

2.3.2 Microfluidics encapsulation of FtsZ in a compartmentalized cytoplasm mimic within a lipid membrane

2.3.2.1 Microfluidic chip fabrication

- PDMS microfluidic devices are constructed by conventional soft lithographic techniques from silicon masters. Fabrication details of the masters can be found in Mellouli et al. (2013).
- Mix PDMS base Sylgard™ 184 and curing agent (Dow Corning GmbH, Germany) in a 10:1 (*w*/w) ratio, stirring vigorously until whitish foamy appearance.

- Eliminate the air bubbles produced upon mixing of the PDMS by degassing in a vacuum chamber. Keep under vacuum until bubble free, transparent appearance.
- Rinse the master with ethanol and promptly dry under nitrogen stream.
- Decant the PDMS onto the master and further eliminate any possible remaining air in the vacuum chamber.
- Polymerize the PDMS by heating the master in an oven at 65 °C for at least 4 h.
- Cut the PDMS comprising the device(s) with a scalpel and peel off from the master, placing it upwards in a mat.
- Punch the inlet and outlet holes of the device(s) using a biopsy punch of appropriate diameter regarding dimensions of the inlets in the device and of the injection tubing.
- Rinse the PDMS (side with the channels) and a coverslip with ethanol and promptly dry under nitrogen stream.

- Activate the surfaces of the device and the coverslip by oxygen plasma (Diener electronic GmbH, Germany).

Tip: determine power and time settings depending on the plasma instrument you have.

- Seal the channels by placing the PDMS on the coverslip, face down, and gently pressing from one extreme to the other to remove any possible air between both surfaces.
- Keep the device in an oven at 65 °C for at least 6 h to ensure proper bonding.
- Coat the channels at room temperature by flushing through them a hydrophobic solution, for example, Aquapel (Pittsburgh Glass Works, LLC), and allow it to dry overnight at 65 °C.

2.3.2.2 Encapsulation of FtsZ in PEG/dextran solutions by microfluidics

The procedure here described (Fig. 2) was developed for a microfluidic device consisting of two inlets for internal phases and one inlet for continuous phase, kindly provided by W. Huck's lab (Radboud University of Nijmegen). Details on the design can be found in Mellouli et al. (2013).

- Prepare the following aqueous solutions in 50 mM Tris-HCl, 300 mM KCl, 1 mM $MgCl_2$, pH 7.5 (phases purified from each other as described in Section 2.1.4).
- Aqueous solution 1. PEG solution with PEG-Alexa 647 (2 μM) and FtsZ (12 μM) with a tracer amount labeled with Alexa 488 (1 μM FtsZ-Alexa 488).

- Aqueous solution 2. Dextran solution and FtsZ (12 μM) with a tracer amount labeled with Alexa 488 (1 μM FtsZ-Alexa 488).
- Aqueous solution 3. PEG solution with PEG-Alexa 647 (2 μM) and GTP.
- Aqueous solution 4. Dextran solution and FtsZ (24 μM) with a tracer amount labeled with Alexa 488 (2 μM FtsZ-Alexa 488).
- Prepare the following continuous phase solutions, in mineral oil.
- Oil solution 1. Follow procedure in Section 2.3.1.
- Oil solution 2. Take the required amount of a chloroform solution of phospholipids from *E. coli* and dope them with a 20% chloroform solution of 1,2-dimyristoyl-sn-glycero-3-phosphocholine (DMPC). Continue as in Section 2.3.1.
- For the encapsulation of the unassembled form of FtsZ, load syringes of internal phases with aqueous solutions 1 and 2, and syringe for continuous phase with oil solution 2.
- For the encapsulation of the FtsZ fibers, load syringes of internal phases with aqueous solutions 3 and 4, and syringe for continuous phase with oil solution 1.
- Adjust the tubing to the needle of each syringe, fill up the tubing with the solution and connect the end of the tubing to the corresponding inlet in the microfluidic device. For droplets collection, fit a piece of tubing to the outlet in the microfluidic device.
- Start flowing solutions by using automated syringe pumps (Cetoni GmbH, Germany), adjusting the flows of both aqueous solutions so that the dextran solution flow is around 70% that of the PEG solution, and ensure all solutions are moving toward the outlet.

Tip: start flowing at high speed, ~100 μL/h, until solutions reach the device. Then reduce the continuous phase flow to ~50 μL/h.

- Progressively reduce the flows of the aqueous solutions keeping the ~70% ratio between them until reaching a dextran phase flow around 50% the starting value.
- Progressively increase the flow of the oil phase until final flow value, 120 μL/h.
- Continue progressively reducing the flows of the aqueous solutions keeping the ~70% ratio between them until final flow values, 5 and 7 μL/h for dextran and PEG phases, respectively.

Tip: once at this final production flow, wait for 10 min to ensure homogeneous population of microdroplets.

- To analyze microdroplets off chip, collect them introducing the outlet tubing in at least 100 μL continuous phase, slightly below the solution surface and centered to avoid contact with the walls of the tube.

Tip: before collection, check for the homogeneous appearance of the flowing solution in the outlet tubing.

- Image the samples as described below.

2.3.2.3 Formation of Giant Unilamellar Vesicles from microdroplets and triggering of FtsZ polymerization

- Prepare the outer solution, that matches buffer composition inside the microdroplets supplemented with 117 mM sucrose.
- Measure the osmolarities of the solutions (Osmomat 3000. Gonotec GmbH, Germany). Osmolarity of the outer solution should be ~25 mOsmol/Kg above the highest osmolarity of the encapsulated solutions (PEG *ca.* 740 mOsmol/Kg, dextran *ca.* 710 mOsmol/Kg).

Tip: similar results are obtained using glucose instead of sucrose in the outer solution. Adjust the required concentration to meet the osmolarity requirements.

- Stabilize 700 μL of oil phase over 400 μL of outer solution for at least 1 h.
- Collect the microdroplets containing the unassembled form of FtsZ (aqueous solutions 1 and 2, oil solution 2) for 30 min, introducing the outlet tubing from the microfluidic chip slightly below the oil surface and centered to avoid contact with the walls of the tube.

Tip: positioning of the tubing slightly below the oil surface enables the migration of the microdroplets throughout the whole oil volume, probably helping further stabilization of the lipid monolayer, until reaching the oil/outer solution interface, where they accumulate before subsequent centrifugation into the aqueous solution.

- Centrifuge the solutions for 15 min at 1500 rpm in a bench centrifuge and remove the oil phase (Fig. 2).
- Wash the vesicles with outer solution. For triggering of FtsZ fibers formation, add 20 mM GTP to this solution.
- Centrifuge for 15 min at 2000 rpm.
- Visualize pelleted vesicles as described below.

Tip: cut the extreme of the tip you use to pipette the vesicles for visualization to reduce breakage.

2.3.2.4 Inclusion of SlmA and SBS on the encapsulated system

Follow the procedure as in Section 2.3.2.2 with the following aqueous solutions, in 50 mM Tris-HCl, 300 mM KCl, 1 mM $MgCl_2$, pH 7.5.

- Aqueous solution 1. PEG solution with FtsZ (25 μM) and a tracer amount labeled with Alexa 647 (2 μM FtsZ-Alexa 647).

- Aqueous solution 2. Dextran solution with SlmA (10 μM) and SBS-Fl (2 μM).
- Aqueous solution 3. PEG solution with FtsZ (25 μM) and a tracer amount labeled with Alexa 647 (2 μM FtsZ-Alexa 647).
- Aqueous solution 4. DNA solution with SlmA (10 μM) and SBS-Fl (2 μM).
- Optimal production of microdroplets containing PEG/dextran LLPS system (aqueous solutions 1 and 2) of around 16 μm is achieved with solutions delivered at 120 μL/h (oil phase) and 5 and 7 μL/h (dextran and PEG aqueous phases, respectively).
- Optimal production of microdroplets containing PEG/DNA LLPS system (aqueous solutions 3 and 4) of around 15–20 μm is achieved with solutions delivered at 120 μL/h (oil phase) and 6 μL/h (both DNA and PEG aqueous phases).

2.4 Fluorescence microscopy and image analysis

- For visualization by confocal microscopy of bulk solutions, microdroplets off chip and vesicles, place ~70 μL of sample in chambers fabricated by gluing a silicone isolator (Molecular probes/Invitrogen) to coverslips as previously described (Monterroso, Zorrilla, et al., 2016).

Tip: for visualization of microdroplets directly in the chip, simultaneously cut all inlet tubing before stopping the flow to keep them in the device reservoir.

- Prepare images and movies using Image J software (National Institutes of Health, USA). This software can be used to measure the distribution of diameters of the microdroplets or vesicles, and to obtain the intensity profiles in the red and green channels within them, applying the line tool through their equatorial section.

3. Summary of experimental results

3.1 Crowding-driven phase transitions control FtsZ spatial organization

The bacterial division ring is built on GTP-triggered dynamic FtsZ filaments that disband when GTP is hydrolyzed by the protein, because of GDP accumulation. Profuse work has been devoted to the characterization of the mechanisms ruling the formation of these filaments in dilute solution (Haeusser & Margolin, 2016). Fewer studies have dealt with the effect of crowding in FtsZ (Rivas, Alfonso, Jiménez, Monterroso, & Zorrilla,

2013), showing certain crowding conditions favor lateral interactions of the fibers, giving rise to bundles. The GDP bound FtsZ, on the other hand, forms short oligomers promoted by low salt, high magnesium and crowding (Rivas, Fernández, & Minton, 2001; Rivas et al., 2000). Little is known, however, about the possible influence of microenvironments and compartments within the bacterial cytoplasm in the spatial distribution and reactivity of this protein. The following results describe recent findings in this regard, from studies in which model systems displaying phase separation have been used to mimic these microenvironments (Monterroso, Zorrilla, et al., 2016; Sobrinos-Sanguino, Zorrilla, Keating, Monterroso, & Rivas, 2017).

3.1.1 FtsZ in bulk LLPS systems

We have used different LLPS systems consisting of two macromolecules, at concentrations beyond their phase separation threshold, to evaluate the behavior of unassembled FtsZ and of FtsZ fibers (Monterroso, Zorrilla, et al., 2016). In these systems, one of the phases was always PEG 8, the other one being either a polysaccharide (Ficoll 70, dextran 500, or dextran T40) or DNA at high concentration. Systems entailing PEG and a polysaccharide provided two neutral phases, the PEG-rich one being more hydrophobic (Keating, 2012), while that containing DNA offered a negatively charged phase mimicking a compartment enriched in nucleic acids, very abundant in the bacterial cytoplasm. FtsZ distribution in these model LLPS systems was studied through fluorescence imaging and protein partition assessment by fluorescence (Fig. 2), entailing the use of fluorescently labeled FtsZ. To preserve its polymerization activity, FtsZ was labeled in assembled form and depolymerized afterwards (González et al., 2003; Reija et al., 2011).

Imaging of emulsified LLPS systems, with FtsZ and one of the phases labeled with spectrally different dyes, allowed determining the protein distribution among the two phases and the interface between them (Figs. 3A and 4A). In these emulsions, 1:3 volume ratios of the purified phases were generally employed, usually rendering droplets of the lower volume phase surrounded by the other phase. In addition to the distribution, imaging experiments also allow detection and identification of FtsZ bundles (Fig. 4A). For quantitative determination of the partition coefficients, the two phases were allowed to fully separate from each other, and the amount of FtsZ in each phase was subsequently determined from fluorescence intensity measurements (Figs. 3B and 4B). Determination of the concentration of FtsZ in the upper phase, in our case PEG, was simple as an aliquot could be easily extracted and measured. Careful pipetting was important in the

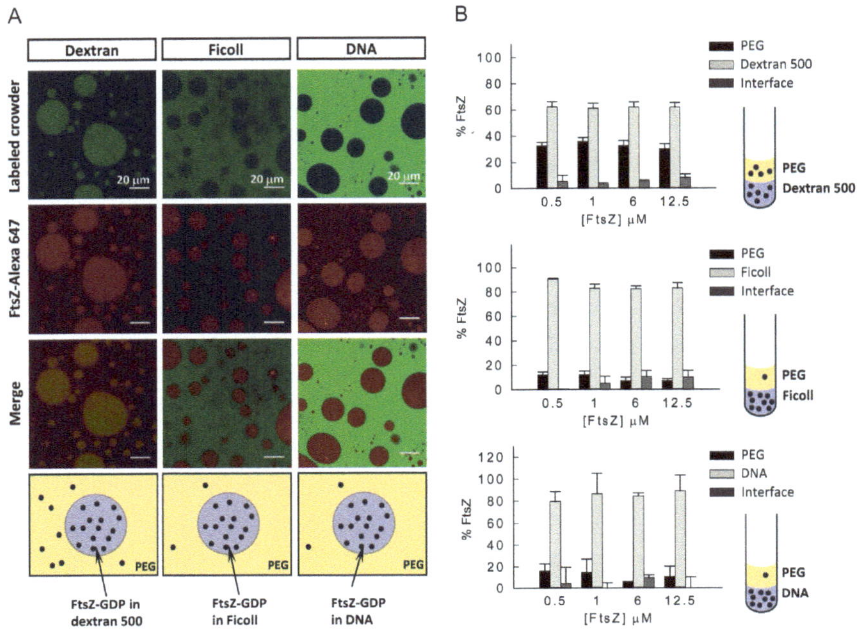

Fig. 3 Distribution of unassembled FtsZ in different model LLPS systems. (A) Confocal images of FtsZ-GDP in binary mixtures of PEG with the specified crowder 2 and illustration of FtsZ location at the bottom. (B) Concentration dependence of the distribution of FtsZ-GDP within these LLPS systems, determined by fluorescence, and illustration on the right. *Figure modified from Monterroso, B., Zorrilla, S., Sobrinos-Sanguino, M., Keating, C. D. & Rivas, G. (2016). Microenvironments created by liquid-liquid phase transition control the dynamic distribution of bacterial division FtsZ protein.* Scientific Reports, 6, *35140.*

evaluation of FtsZ in the bottom phase, to avoid contaminations with material from the interface or from the upper phase. Relevant to the FtsZ system was the calculation of the amount of protein at the interface, location at which FtsZ fibers often tended to accumulate. This is calculated as the difference between the total amount of protein initially loaded in the sample and the amount of protein in each phase. Consistence between these calculations and the corresponding images was always verified.

Data showed that the distribution of FtsZ-GDP was uneven in all the LLPS systems studied, with preferential partition into the non-PEG phase (Fig. 3), (Monterroso, Zorrilla, et al., 2016). Partition was more marked in the case of the PEG/DNA and PEG/Ficoll systems, in which the fraction remaining in the PEG phase was negligible. In the PEG/dextran systems, in contrast, around one third of the protein was detected in the PEG phase.

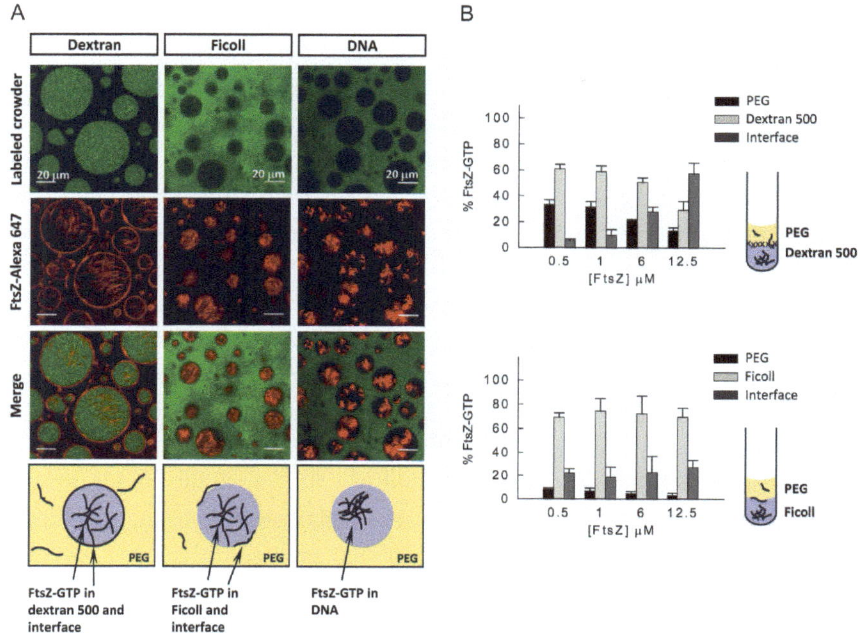

Fig. 4 Distribution of FtsZ fibers in different model LLPS systems. (A) Confocal images of FtsZ-GTP in bulk emulsions of PEG with the specified crowder 2 and illustration of fibers location at the bottom. (B) Concentration dependence of the distribution of FtsZ-GTP in the phases within the LLPS systems, determined by fluorescence, and illustration on the right. *Figure modified from Monterroso, B., Zorrilla, S., Sobrinos-Sanguino, M., Keating, C. D. & Rivas, G. (2016). Microenvironments created by liquid-liquid phase transition control the dynamic distribution of bacterial division FtsZ protein.* Scientific Reports, 6, *35140.*

The amount of protein at the interface was very low in all cases, below 10% (Fig. 3). This is probably reflecting the moderate size of the FtsZ-GDP oligomers under the experimental conditions used in this study, as interfacial localization is usually found for large entities (Cabral, 2007).

The behavior of FtsZ upon addition of GTP in these LLPS systems was also analyzed (Fig. 4), (Monterroso, Zorrilla, et al., 2016). Confocal microscopy images and fluorescence intensity measurements showed that FtsZ was not homogeneously distributed, and large structures reflecting the formation of fibers were detected in all cases (Fig. 4A). However, the distribution and topology of these fibers was different depending on the particular LLPS system. Thus, in PEG/dextran 500, elongated fibers were observed, decorating all the interfacial space in the confocal images (Fig. 4A). Accordingly, a large fraction of the total protein, reaching almost 60% at the highest concentration of FtsZ tested, was estimated to be at the liquid/liquid interface through the quantitative fluorescence determinations (Fig. 4B). Interfacial localization of large structures, previously described for cells and nanoparticles, seems to be driven by the concomitant reduction in the overall surface tension in the system (Keating, 2012). The scenario was different in the PEG/Ficoll 70 system, in which elongated fibers could also be distinguished, homogeneously distributed within the Ficoll phase (Fig. 4A). In this case, less than 30% of the total protein was found at the interface through the quantitative fluorescence measurements (Fig. 4B). It was initially hypothesized that the disparity in sizes of the two crowders, dextran and Ficoll (500 *vs* 70 KDa), might explain the different final distribution of the FtsZ fibers in the two LLPS systems. The results obtained in PEG/dextran T40, similar to those in PEG/dextran 500 despite the closer sizes of dextran T40 and Ficoll 70, ruled out this possibility (Monterroso, Zorrilla, et al., 2016). This pointed toward the nature of dextran, a randomly coiled glucose polymer, and Ficoll, a globular sucrose polymer (Laurent & Granath, 1967) as a possible source of the observed differences.

A markedly distinct organization of the FtsZ fibers triggered by GTP was found in the PEG/DNA system (Fig. 4A), probably related with the charged nature of the DNA-rich phase in which they preferentially partitioned. These fibers were not homogeneously dispersed as in the Ficoll or dextran phases, but forming very large irregular arrangements maybe because of the repulsion between the negatively charged DNA molecules and FtsZ, also negatively charged at the working pH. The heterogeneity in this system precluded reliable quantification of partition and amount of protein at the interface by fluorescence (Monterroso, Zorrilla, et al., 2016). Nevertheless, it is obvious from the images that FtsZ structures are not prone to locate at the interface (Fig. 4A), in contrast to that found in the other LLPS systems to a higher or lower extent. As in the case of FtsZ fibers, variations in the tendency of other large structures to populate the interface with the type and concentrations of crowders conforming the phases have been reported (Atefi, Joshi, Mann, & Tavana, 2015; Helfrich, El-Kouedi, Etherton, & Keating, 2005). Therefore, the distribution of FtsZ seems to depend both on the nature of the LLPS system and on the protein association state, regulated by GTP, which largely determines interfacial localization.

3.1.2 FtsZ in cell mimetic systems displaying a compartmentalized cytoplasm

The impact of compartmentalization on the distribution and localization of FtsZ was also studied through its encapsulation in containers of picoliter size surrounded by a lipid boundary, enclosing the above described LLPS systems (Monterroso, Zorrilla, et al., 2016; Sobrinos-Sanguino et al., 2017). This kind of platform better captures many of the key features of a bacterial cell such as a crowded and compartmentalized cytoplasm, microscale volume and a lipid boundary. From a practical point of view, and regarding the LLPS systems emulsified in bulk, these systems present the advantage that their composition remains stable, allowing monitoring of their temporal evolution by imaging. This kind of analysis is more limited in the bulk LLPS systems emulsions as, with time, the droplets of the phases tend to fuse to decrease surface tension.

Containers were initially generated by mechanical emulsion of an aqueous solution including the proteins and the LLPS system, previously emulsified, and an oil phase in which the lipids were dissolved (Fig. 2). Aqueous droplets stabilized by a lipid monolayer were obtained in this way, containing FtsZ and, most of them, the two aqueous phases. Occasionally, droplets with a single phase or with three phases, two of them equal, were found. The procedure was fast, easy to implement and the droplet yield was high. The main drawbacks were the wide size distribution of the droplets and the variability in the volume ratios of the two phases and in the concentration of FtsZ among the droplets. These limitations were overcome through the implementation of microfluidics based methods to encapsulate phase-separated systems within lipid membranes (Fig. 2). This procedure rendered hundreds of droplets with virtually monodisperse size and homogeneous composition, in terms of volume of the two phases and FtsZ concentration (Fig. 5).

Encapsulation of FtsZ-GDP and FtsZ-GTP in microdroplets containing the PEG/dextran LLPS system, either manually or by microfluidics, showed similar protein distribution among the phases and the interface as in the bulk systems. FtsZ-GDP partitioned preferentially into the dextran-rich phase, although some of the protein could be also found in the PEG (Fig. 5). The GTP-induced fibers displayed a clear tendency to accumulate at the interface of the two aqueous phases and at the lipid surface, those remaining being found in the dextran phase (Fig. 5). The temporal evolution of the system was analyzed through encapsulation of preformed FtsZ-GTP fibers and monitorization by time-lapse confocal microscopy, to obtain images at different times and the corresponding intensity profiles across the microdroplets, reflecting the distribution of the protein (Fig. 6). We found that as the FtsZ filaments visible by confocal disassembled, due to GTP depletion, the strong signal of protein at the interface between the two aqueous phases decreased and the protein adopted the distribution previously observed for

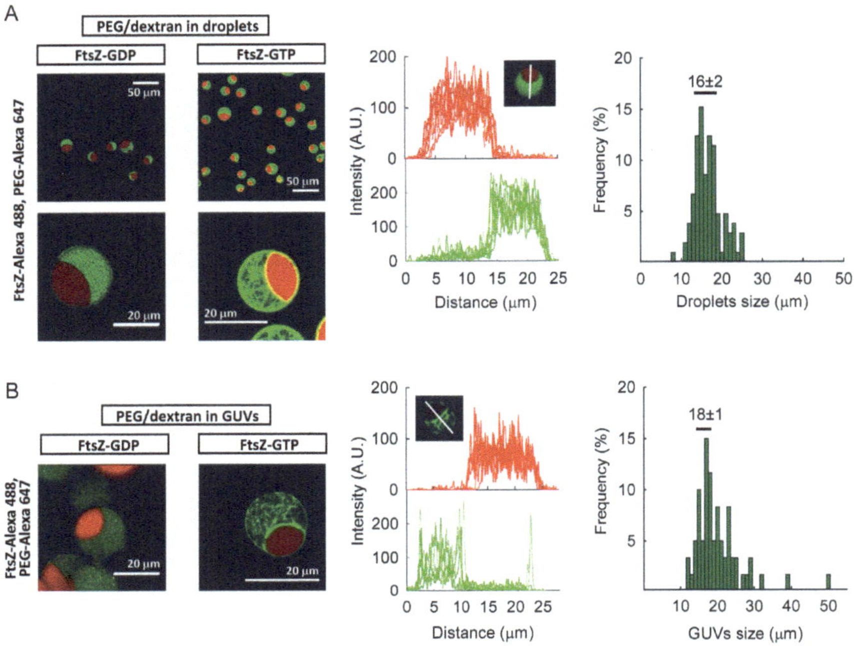

Fig. 5 Distribution of FtsZ within compartmentalized microfluidics microdroplets and GUVs. (A) Left, confocal images of microdroplets containing FtsZ-GDP or FtsZ-GTP. Middle, intensity profiles of the red and green channels of microdroplets containing FtsZ-GDP, obtained across lines as shown in the inset. Right, microdroplets size distribution. Average size is the mean value ± SD of the most populated sizes, indicated by the horizontal bar. (B) Same type of data than in A but in GUVs, that contain FtsZ-GDP or FtsZ-GTP (after external addition of GTP to the FtsZ-GDP containing vesicles). Intensity profiles correspond to vesicles containing either FtsZ-GDP or fibers. *Figure modified from Sobrinos-Sanguino, M., Zorrilla, S., Keating, C. D., Monterroso, B. & Rivas, G. (2017). Encapsulation of a compartmentalized cytoplasm mimic within a lipid membrane by microfluidics.* Chemical Communications (Cambridge), 53, *4775–4778.*

FtsZ-GDP. Therefore, the distribution of FtsZ in LLPS systems not only depends on its assembly state, but can also be reversibly modified by factors affecting it.

Likewise, induction of fiber formation in FtsZ previously encapsulated resulted in relocation of the protein within the phases, adopting the preferential distribution corresponding to encapsulated fibers (Fig. 5). This was concluded from experiments in which unassembled FtsZ was encapsulated in permeable vesicles and polymerized afterwards. For this, we went one step beyond in our development of cytomimetic containers displaying compartmentalization, by the adaptation of droplet transfer procedures (Carrara,

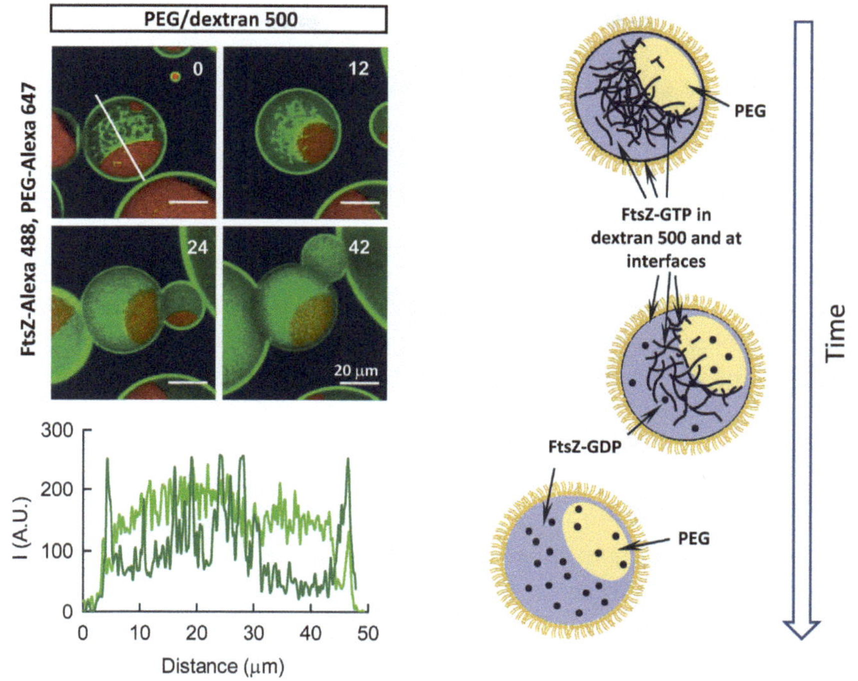

Fig. 6 Dynamic relocation of FtsZ in cellular mimicries displaying a compartmentalized cytoplasm. Confocal images of FtsZ fibers encapsulated by manual emulsion in lipid stabilized microdroplets encircling an LLPS system. Sequence shows the disassembly of the fibers with time, because of GTP depletion, and the concomitant redistribution of the protein (time in minutes). Below, intensity profiles of FtsZ-Alexa 488 along the line depicted in the image at initial (dark green) and final times (light green). Right, graphical representation of the process. *Figure modified from Monterroso, B., Zorrilla, S., Sobrinos-Sanguino, M., Keating, C. D. & Rivas, G. (2016). Microenvironments created by liquid-liquid phase transition control the dynamic distribution of bacterial division FtsZ protein.* Scientific Reports, 6, *35140.*

Stano, & Luisi, 2012) to convert the microdroplets generated by microfluidics into permeable giant unilamellar vesicles (GUVs) (Sobrinos-Sanguino et al., 2017). Although the production of GUVs was more time consuming and the yield was lower than in the case of the microdroplets, they present the advantage that, in addition of their bilayer better resembling the bacteria inner membrane, they allow for external addition of ligands into the system (Cisse, Okumus, Joo, & Ha, 2007). Along this line, we showed how by including the lipid DMPC in the lipid mixture, which did not introduce differences in the microdroplet formation step or in the GUV generation procedure, the formation of FtsZ fibers could be triggered by diffusion of externally added GTP through the packing defects it induces in the bilayer. Another strategy for membrane pore formation would entail the addition of hemolysin after GUV formation (Cabré et al., 2013; Cisse et al., 2007).

Encapsulation of FtsZ in the PEG/DNA phase allowed evaluating the dynamics of the GTP-induced fibers, which formed more irregular higher order arrangements in this LLPS system. It was found that the fibers disassembled with time, meaning that FtsZ within them was still able of hydrolyzing GTP (Monterroso, Zorrilla, et al., 2016). Longer time was required for disassembly compared with the fibers in PEG/dextran, though, probably related with their particular organization in the DNA phase.

3.2 FtsZ, together with the nucleoid-associated SlmA protein, forms dynamic phase-separated condensates in crowded cell-like media

Bacteria have developed different mechanisms that act in a coordinated fashion to control the formation and positioning of the FtsZ ring, one of them being nucleoid occlusion (Dewachter et al., 2018). Nucleoid occlusion is specifically aimed at antagonizing FtsZ assembly into a functional ring nearby the chromosome, to protect it from any damage inflicted by the division machinery. In *E. coli*, the nucleoid occlusion factor is SlmA, a protein that targets over 20 binding sites (SBS) spread through the chromosome macrodomains with the exception of the Ter (Cho et al., 2011). SlmA forms complexes with the SBS sites according to a 4:1 stoichiometry and it also binds directly to FtsZ, either in the absence or presence of GTP, shortening the lifetime of the GTP-induced fibers (Cabré et al., 2015; Tonthat et al., 2013).

The identification of condensates entailing the division proteins FtsZ and SlmA and the specific SBS sequences is an example of a serendipitous finding. After characterization of FtsZ in model LLPS systems, we were interested in the possible impact of known modulators of FtsZ assembly on its behavior in those systems. We started by SlmA, one of the division proteins we had previously studied in our laboratory. Addition of SlmA and the SBS over unassembled FtsZ in the PEG/dextran LLPS system led to the appearance of small droplets of a few micrometers in diameter, principally located within the dextran-rich phase where the three elements, FtsZ, SlmA and the SBS, accumulate. These small droplets resembled the biomolecular condensates described for eukaryotic proteins, of which there were no examples in bacteria at the time this study was conducted. This preliminary evidence prompted us to undertake an in-depth characterization of the structures, to determine the factors driving their formation, their possible implications in the context of division and, on a broader level, in the organization of the intracellular space in bacteria (Monterroso et al., 2019).

3.2.1 Identification of the factors determining condensation

Detection of condensate formation was conducted using two different techniques, confocal microscopy imaging and turbidity measurements. Imaging allowed detection of the round condensates and evaluation of their size and the colocalization of the elements by pairwise labeling with dyes of non-overlapping emission. Turbidity experiments with unlabeled proteins and DNA, performed in parallel, offered a quantitative measure, reflecting size and abundance of the condensates in each condition. As crowding is one of the factors known to promote biomolecular condensation, we first analyzed its impact on the appearance of FtsZ·SlmA·SBS droplets. We found that round structures containing the two proteins and the SBS were assembled in a variety of crowding agents (Fig. 7), while they were virtually absent from the diluted solutions tested (Monterroso et al., 2019).

Other factors known to lead to condensation are unstructured regions in proteins and multivalency, related with proteins harboring multiple domains of interaction and with nucleic acid sequences (Alberti, 2017; Banani et al., 2017; Shin & Brangwynne, 2017). These factors are somehow present in the FtsZ·SlmA·SBS complexes because the two proteins (aside of interacting with each other) form oligomers, the SBS oligonucleotide enhances the FtsZ-SlmA recognition, and FtsZ presents disordered regions. Accordingly, under the experimental conditions assayed, the condensates were largely favored by the presence of the three elements, although they could also be detected in the absence of the DNA. SlmA itself also seemed to be prone to form the structures, more obviously when bound to the SBS. Finally, condensate assembly was sensitive to the ionic strength of the solution (Fig. 7), the droplets becoming larger and more abundant at lower salt, probably reflecting the enhancement of electrostatic interactions among the elements.

3.2.2 Dynamics of the condensates

One of the hallmarks of biomolecular condensates is their dynamic nature. The dynamics of the FtsZ·SlmA·SBS condensates was first addressed by means of a capture assay (Fig. 8) previously described for other systems undergoing this kind of phase separation (Woodruff et al., 2017). Samples

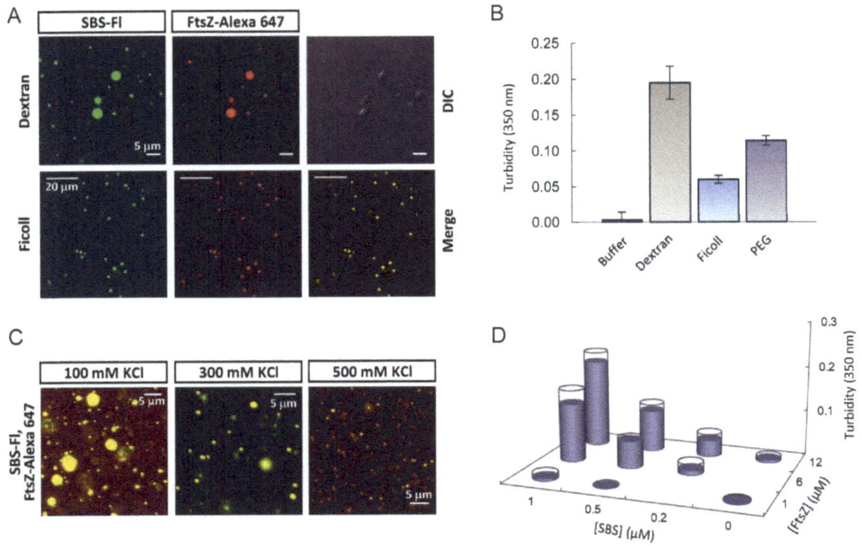

Fig. 7 Visualization and quantification of FtsZ·SlmA·SBS biomolecular condensates. (A) Confocal images of the condensates in dextran 500 or Ficoll 70 and (B) turbidity measurements in dilute conditions and in different crowders. (C) Effect of the ionic strength on the size and abundance of condensates. (D) Condensate formation as a function of FtsZ and SlmA·SBS concentration in dextran monitored by turbidity. SlmA concentration was five to sixfold that of the SBS. *Figure modified from Monterroso, B., Zorrilla, S., Sobrinos-Sanguino, M., Robles-Ramos, M. A., López-Álvarez, M., et al. (2019). Bacterial FtsZ protein forms phase-separated condensates with its nucleoid-associated inhibitor SlmA.* EMBO Reports, 20, *e45946.*

containing FtsZ·SlmA·SBS where only FtsZ was labeled were initially visualized in the confocal microscope to verify condensate formation. Then, FtsZ labeled with a non-overlapping dye was subsequently applied to the initial sample and its incorporation into the preformed condensates assessed by confocal imaging. The FtsZ·SlmA·SBS condensates incorporated the newly added FtsZ in the presence of all the crowders tested, including dextran, PEG or unspecific DNA. Time-lapse microscopy allowed monitoring of the stepwise incorporation of the freshly added protein inside the condensates.

Dynamism of the FtsZ·SlmA·SBS condensates and functionality of FtsZ within them was further assessed taking advantage of the ability of the protein to assemble into fibers triggered by GTP (Fig. 8). We found that the nucleotide elicited the assembly of fibers with the concomitant reduction in the number and size of the condensates, up to a point at which the fibers prevailed. These fibers were thinner than the usual ones observed under

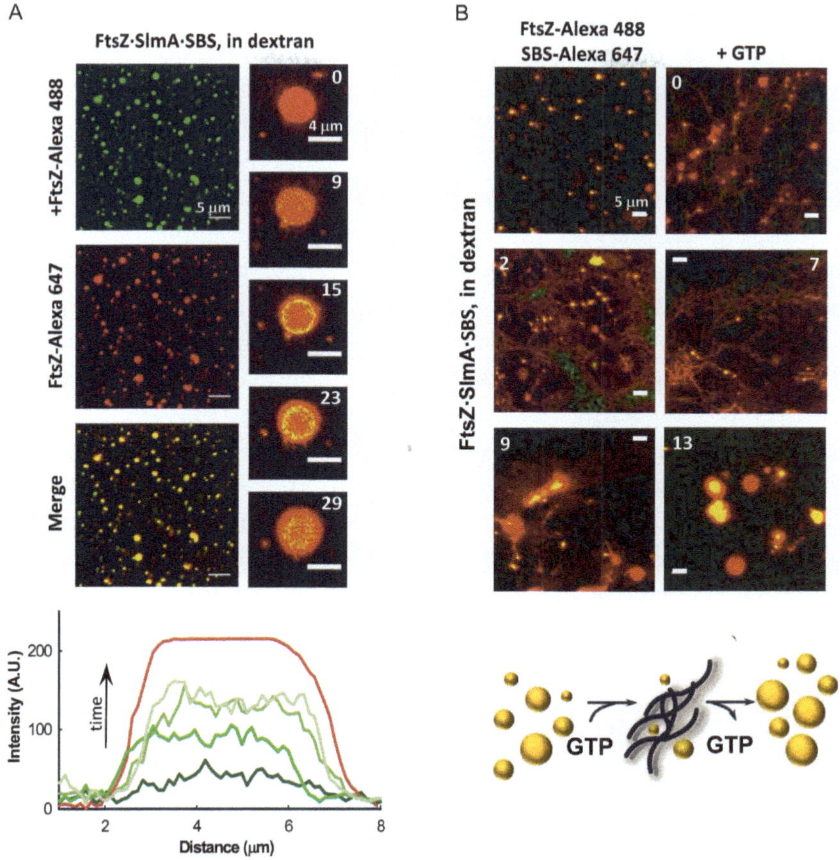

Fig. 8 Dynamics of FtsZ·SlmA·SBS biomolecular condensates. (A) Confocal microscopy images showing the final state of FtsZ·SlmA·SBS biomolecular condensates with FtsZ-Alexa 647 after addition of FtsZ-Alexa 488. On the right, temporal evolution (time specified in seconds) of a condensate upon supplementation with FtsZ-Alexa 488, and below the corresponding intensity profiles of the green channel. The red channel profile is shown as a reference. (B) Merged images showing fiber formation by FtsZ when GTP is added (time zero) to FtsZ·SlmA·SBS condensates and temporal evolution toward condensates formation after fiber disassembly due to GTP depletion (time in minutes). Below, scheme showing the interconversion of condensates and fibers. *Figure modified from Monterroso, B., Zorrilla, S., Sobrinos-Sanguino, M., Robles-Ramos, M. A., López-Álvarez, M., et al. (2019). Bacterial FtsZ protein forms phase-separated condensates with its nucleoid-associated inhibitor SlmA.* EMBO Reports, 20, *e45946.*

crowding conditions and they disassembled faster, due to the action of SlmA, with the subsequent reappearance of the condensates. The conversion between condensates and fibers regulated by GTP availability and depletion supports the dynamic nature of the biomolecular condensates.

3.2.3 Formation of condensates in cytomimetic systems

The formation of condensates in lipid stabilized microdroplets displaying compartmentalization and crowding as cell mimicries was also studied (Monterroso et al., 2019). Encapsulation of FtsZ·SlmA·SBS inside microdroplets containing PEG/dextran or PEG/DNA as model compartments showed again the formation of the condensates in these confined systems (Fig. 9). They accumulated in the dextran and DNA phases rather than in the PEG (probably because of the larger tendency of the conforming proteins toward those phases) and, interestingly, also at the lipid boundary. The conversion between FtsZ filaments and condensates was also assessed inside the microdroplets by encapsulating FtsZ, SlmA and the SBS and triggering fiber formation by adding GTP shortly before encapsulation. After disassembly of the filaments, FtsZ·SlmA·SBS condensates were found in the images (Monterroso et al., 2019). This demonstrated the formation of dynamic condensates occurs also in these kinds of systems, closer to cellular models, and their particular accumulation at the lipids interfaces, which may be important for the function of the proteins.

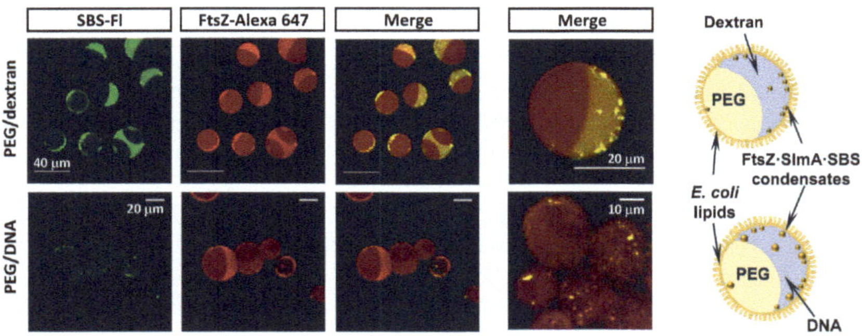

Fig. 9 Microfluidic encapsulation of FtsZ·SlmA·SBS biomolecular condensates inside lipid stabilized compartmentalized microdroplets. Confocal microscopy images of microdroplets containing FtsZ, SlmA and SBS and illustration of the elements distribution on the right. *Figure modified from Monterroso, B., Zorrilla, S., Sobrinos-Sanguino, M., Robles-Ramos, M. A., López-Álvarez, M., et al. (2019). Bacterial FtsZ protein forms phase-separated condensates with its nucleoid-associated inhibitor SlmA.* EMBO Reports, 20, *e45946.*

4. Concluding remarks and future prospects

To shed light on the mechanisms governing cell division, one of the research lines toward the bottom-up reconstitution of bacterial cell division elements focuses on their assembly in containers reproducing the crowded and phase-separated cell interior. In this work, we have analyzed two examples of this approach. In the first one, the partitioning of FtsZ in aqueous two-phase systems mediated by non-associative liquid-liquid phase separation (Crowe & Keating, 2018) has revealed a differential partition pattern depending on the state of assembly of FtsZ linked to GTP binding and hydrolysis. In the second example, the FtsZ protein underwent associative LLPS when mixed with the DNA-binding nucleoid-associated SlmA protein under crowding conditions, producing condensates capable of reversible evolution to fibers in the presence of GTP (Fig. 10).

The fact that central elements of the bacterial division machinery can assemble into dynamic condensates by liquid-liquid phase separation in crowded cell-like environments constitutes a compelling suggestion that biomolecular condensation may also help to organize the intracellular space in bacteria. These findings provide a novel mechanism to modulate, in time and space, the functional reactivity of FtsZ during the cell cycle. For example, the formation of FtsZ·SlmA·SBS condensates may have a potential impact on the action of SlmA over FtsZ in bacterial division. Our observations show the shuttling of "monomeric" FtsZ between the condensed

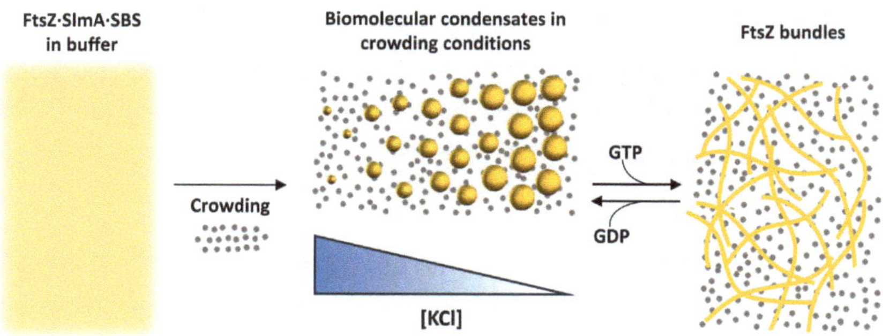

Fig. 10 Scheme of FtsZ·SlmA·SBS condensates formation. In crowding conditions, FtsZ·SlmA·SBS forms condensates of size and abundance determined, among other things, by KCl concentration. Addition of GTP to the condensates triggers FtsZ polymerization into fibers. FtsZ incorporates back into condensates after GTP exhaustion.

phase and the fibers in crowding media. Which state predominates under a particular set of conditions depends on GTP concentration, which promotes fiber formation. Before division, SlmA·SBS may partially recruit FtsZ forming FtsZ·SlmA·SBS condensates, maybe aiding in the localization of these structures nearby the membrane. Under division conditions, SlmA at midcell decreases, and other elements would compete for the interaction with FtsZ that would leave the condensates. In the presence of GTP, it would form a membrane-anchored FtsZ ring. In non-central regions, FtsZ would still be under the control of SlmA, which protects the chromosomes from scission by aberrant division ring formation.

Several relevant pending questions need to be answered, being the detection of FtsZ condensates *in vivo* using super-resolution technologies, one of the most demanding ones. Research on the use of mutants to define the structural nature of the elements involved in the formation of the condensates is underway. To unravel the mechanism of FtsZ condensation is required a precise quantitative understanding of the parameters that control the formation of these dynamic structures. The latter is not a trivial task, as several physicochemical factors (non-associative and associative LLPS, protein solubility and micelle formation, etc.) may cooperate synergistically in ways poorly known. Among other things, to measure precisely the composition-dependence of the phase diagrams using microspectroscopy and imaging analysis will be crucial.

Besides biomolecular condensates, other related structures as hydrogels and fibrils can be found under certain circumstances, which are especially important in, but not restricted to, disease states. Because of these remarkable cases, the study of the final state of matter resulting from the condensation processes has been a focus of attention. Complementary efforts to study the phase separation processes themselves will be relevant. For example, in the case of bacterial division, phase separation could be an active process that may be modulated non-specifically by crowding and specifically by proteins (*i.e.*, those regulating divisome/nucleoid structure and dynamics), and that dynamically acts on the membrane by itself. In this regard, a few studies started to investigate how the local accumulation of polymerizable FtsZ protein could produce droplet deformations and, ultimately, division. Interestingly, a recent study has demonstrated that the dissipative self-assembly of FtsZ filaments in the presence of GTP in RNA/polypeptide complex coacervates resulted in their elongation and eventual fission (Te Brinke et al., 2018). This study suggests that energy-consuming processes might lead to active fission of coacervate droplets by the action of FtsZ assembly.

Acknowledgments

This work was supported by the Spanish government (BFU2014-52070-C2-2-P, BFU2016-75471-C2-1-P, 2019AEP088, and PID2019-104544GB-I00). M.A.R.-R. was supported by the Agencia Estatal de Investigación and the European Social Fund through grant BES-2017-082003. This article is dedicated to the memory of all the victims of the coronavirus pandemic.

References

Abbondanzieri, E. A., & Meyer, A. S. (2019). More than just a phase: The search for membraneless organelles in the bacterial cytoplasm. *Current Genetics*, *65*, 691–694.

Alberti, S. (2017). Phase separation in biology. *Current Biology*, *27*, R1097–R1102.

Atefi, E., Joshi, R., Mann, J. A., Jr., & Tavana, H. (2015). Interfacial tension effect on cell partition in aqueous two-phase systems. *ACS Applied Materials & Interfaces*, *7*, 21305–21314.

Ausländer, S., Ausländer, D., & Fussenegger, M. (2017). Synthetic biology—The synthesis of biology. *Angewandte Chemie (International Ed. in English)*, *56*, 6396–6419.

Banani, S. F., Lee, H. O., Hyman, A. A., & Rosen, M. K. (2017). Biomolecular condensates: Organizers of cellular biochemistry. *Nature Reviews. Molecular Cell Biology*, *18*, 285–298.

Biswas, N., Ichikawa, M., Datta, A., Sato, Y. T., Yanagisawa, M., et al. (2012). Phase separation in crowded micro-spheroids: DNA–PEG system. *Chemical Physics Letters*, *539–540*, 157–162.

Cabral, J. M. (2007). Cell partitioning in aqueous two-phase polymer systems. *Advances in Biochemical Engineering/Biotechnology*, *106*, 151–171.

Cabré, E. J., Monterroso, B., Alfonso, C., Sánchez-Gorostiaga, A., Reija, B., et al. (2015). The nucleoid occlusion SlmA protein accelerates the disassembly of the FtsZ protein polymers without affecting their GTPase activity. *PLoS One*, *10*, e0126434.

Cabré, E. J., Sánchez-Gorostiaga, A., Carrara, P., Ropero, N., Casanova, M., et al. (2013). Bacterial division proteins FtsZ and ZipA induce vesicle shrinkage and cell membrane invagination. *The Journal of Biological Chemistry*, *288*, 26625–26634.

Carrara, P., Stano, P., & Luisi, P. L. (2012). Giant vesicles "colonies": A model for primitive cell communities. *Chembiochem*, *13*, 1497–1502.

Cho, H., McManus, H. R., Dove, S. L., & Bernhardt, T. G. (2011). Nucleoid occlusion factor SlmA is a DNA-activated FtsZ polymerization antagonist. *Proceedings of the National Academy of Sciences of the United States of America*, *108*, 3773–3778.

Cisse, I., Okumus, B., Joo, C., & Ha, T. (2007). Fueling protein DNA interactions inside porous nanocontainers. *Proceedings of the National Academy of Sciences of the United States of America*, *104*, 12646–12650.

Crowe, C. D., & Keating, C. D. (2018). Liquid-liquid phase separation in artificial cells. *Interface Focus*, *8*, 20180032.

Dewachter, L., Verstraeten, N., Fauvart, M., & Michiels, J. (2018). An integrative view of cell cycle control in Escherichia coli. *FEMS Microbiology Reviews*, *42*, 116–136.

González, J. M., Jiménez, M., Vélez, M., Mingorance, J., Andreu, J. M., et al. (2003). Essential cell division protein FtsZ assembles into one monomer-thick ribbons under conditions resembling the crowded intracellular environment. *The Journal of Biological Chemistry*, *278*, 37664–37671.

Haeusser, D. P., & Margolin, W. (2016). Splitsville: Structural and functional insights into the dynamic bacterial Z ring. *Nature Reviews. Microbiology*, *14*, 305–319.

Helfrich, M. R., El-Kouedi, M., Etherton, M. R., & Keating, C. D. (2005). Partitioning and assembly of metal particles and their bioconjugates in aqueous two-phase systems. *Langmuir*, *21*, 8478–8486.

Keating, C. D. (2012). Aqueous phase separation as a possible route to compartmentalization of biological molecules. *Accounts of Chemical Research*, *45*, 2114–2124.
Laurent, T. C., & Granath, K. A. (1967). Fractionation of dextran and Ficoll by chromatography on Sephadex G-200. *Biochimica et Biophysica Acta*, *136*, 191–198.
Martos, A., Jiménez, M., Rivas, G., & Schwille, P. (2012). Towards a bottom-up reconstitution of bacterial cell division. *Trends in Cell Biology*, *22*, 634–643.
Mellouli, S., Monterroso, B., Vutukuri, H. R., te Brinke, E., Chokkalingam, V., et al. (2013). Self-organization of the bacterial cell-division protein FtsZ in confined environments. *Soft Matter*, *9*, 10493–10500.
Monterroso, B., Reija, B., Jiménez, M., Zorrilla, S., & Rivas, G. (2016). Charged molecules modulate the volume exclusion effects exerted by crowders on FtsZ polymerization. *PLoS One*, *11*, e0149060.
Monterroso, B., Zorrilla, S., Sobrinos-Sanguino, M., Keating, C. D., & Rivas, G. (2016). Microenvironments created by liquid-liquid phase transition control the dynamic distribution of bacterial division FtsZ protein. *Scientific Reports*, *6*, 35140.
Monterroso, B., Zorrilla, S., Sobrinos-Sanguino, M., Robles-Ramos, M. A., López-Álvarez, M., et al. (2019). Bacterial FtsZ protein forms phase-separated condensates with its nucleoid-associated inhibitor SlmA. *EMBO Reports*, *20*, e45946.
Ortiz, C., Natale, P., Cueto, L., & Vicente, M. (2016). The keepers of the ring: Regulators of FtsZ assembly. *FEMS Microbiology Reviews*, *40*, 57–67.
Osawa, M., & Erickson, H. P. (2013). Liposome division by a simple bacterial division machinery. *Proceedings of the National Academy of Sciences of the United States of America*, *110*, 11000–11004.
Ramm, B., Heermann, T., & Schwille, P. (2019). The E. coli MinCDE system in the regulation of protein patterns and gradients. *Cellular and Molecular Life Sciences*, *76*, 4245–4273.
Reija, B., Monterroso, B., Jímenez, M., Vicente, M., Rivas, G., et al. (2011). Development of a homogeneous fluorescence anisotropy assay to monitor and measure FtsZ assembly in solution. *Analytical Biochemistry*, *418*, 89–96.
Rivas, G., Alfonso, C., Jiménez, M., Monterroso, B., & Zorrilla, S. (2013). Macromolecular interactions of the bacterial division FtsZ protein: From quantitative biochemistry and crowding to reconstructing minimal divisomes in the test tube. *Biophysical Reviews*, *5*, 63–77.
Rivas, G., Fernández, J. A., & Minton, A. P. (2001). Direct observation of the enhancement of noncooperative protein self-assembly by macromolecular crowding: Indefinite linear self-association of bacterial cell division protein FtsZ. *Proceedings of the National Academy of Sciences of the United States of America*, *98*, 3150–3155.
Rivas, G., López, A., Mingorance, J., Ferrándiz, M. J., Zorrilla, S., et al. (2000). Magnesium-induced linear self-association of the FtsZ bacterial cell division protein monomer. The primary steps for FtsZ assembly. *The Journal of Biological Chemistry*, *275*, 11740–11749.
Rivas, G., & Minton, A. P. (2016). Macromolecular crowding in vitro, in vivo, and in between. *Trends in Biochemical Sciences*, *41*, 970–981.
Rivas, G., & Minton, A. P. (2018). Toward an understanding of biochemical equilibria within living cells. *Biophysical Reviews*, *10*, 241–253.
Rivas, G., Vogel, S. K., & Schwille, P. (2014). Reconstitution of cytoskeletal protein assemblies for large-scale membrane transformation. *Current Opinion in Chemical Biology*, *22*, 18–26.
Shin, Y., & Brangwynne, C. P. (2017). Liquid phase condensation in cell physiology and disease. *Science*, *357*, eaaf4382.
Sobrinos-Sanguino, M., Zorrilla, S., Keating, C. D., Monterroso, B., & Rivas, G. (2017). Encapsulation of a compartmentalized cytoplasm mimic within a lipid membrane by microfluidics. *Chemical Communications (Cambridge)*, *53*, 4775–4778.

Szwedziak, P., Wang, Q., Bharat, T. A., Tsim, M., & Löwe, J. (2014). Architecture of the ring formed by the tubulin homologue FtsZ in bacterial cell division. *eLife*, *3*, e04601.
Tarnopol, R. L., Bowden, S., Hinkle, K., Balakrishnan, K., Nishii, A., et al. (2019). Lessons from a minimal genome: What are the essential organizing principles of a cell built from scratch? *Chembiochem*, *20*, 2535–2545.
Te Brinke, E., Groen, J., Herrmann, A., Heus, H. A., Rivas, G., et al. (2018). Dissipative adaptation in driven self-assembly leading to self-dividing fibrils. *Nature Nanotechnology*, *13*, 849–855.
Tonthat, N. K., Milam, S. L., Chinnam, N., Whitfill, T., Margolin, W., et al. (2013). SlmA forms a higher-order structure on DNA that inhibits cytokinetic Z-ring formation over the nucleoid. *Proceedings of the National Academy of Sciences of the United States of America*, *110*, 10586–10591.
Vendeville, A., Larivière, D., & Fourmentin, E. (2011). An inventory of the bacterial macromolecular components and their spatial organization. *FEMS Microbiology Reviews*, *35*, 395–414.
Woodruff, J. B., Ferreira Gomes, B., Widlund, P. O., Mahamid, J., Honigmann, A., et al. (2017). The centrosome is a selective condensate that nucleates microtubules by concentrating tubulin. *Cell*, *169*, 1066–1077.

CHAPTER THREE

Terpolymer-stabilized complex coacervates: A robust and versatile synthetic cell platform

Alexander F. Mason[a,b,*], Wiggert J. Altenburg[a,b], Shidong Song[a,b], Marleen van Stevendaal[a,b], and Jan C.M. van Hest[a,b,*]

[a]Department of Biomedical Engineering, Institute for Complex Molecular Systems, Eindhoven University of Technology, Eindhoven, The Netherlands
[b]Department of Chemical Engineering and Chemistry, Institute for Complex Molecular Systems, Eindhoven University of Technology, Eindhoven, The Netherlands
[*]Corresponding authors: e-mail address: a.f.mason@tue.nl; j.c.m.v.hest@tue.nl

Contents

Abstract

The utilization of liquid-liquid phase separated systems has seen increased attention as synthetic cell platforms due to their innate ability to sequester interesting, functional, and biologically relevant materials. However, their applications are limited by the temporal stability of such condensed phases. While there are a number of strategies toward droplet stabilization, in our group we have developed a polymer-based approach to

Methods in Enzymology, Volume 646
ISSN 0076-6879
https://doi.org/10.1016/bs.mie.2020.06.008

stabilize complex coacervate microdroplets. These protocells are remarkably robust and have been utilized to support a number of new protocellular applications. Here, we describe in detail the methodologies we have developed for the synthesis of the starting components, their formation into stable, cargo-loaded protocells, and how these protocells are treated post-formation to purify and analyze the resultant functional self-assembled systems.

1. Introduction

The question of "What is Life?" has been posed for millennia, but advances in the natural sciences have recently enabled us to start to explore a vital avenue toward answering this question—the generation of living systems in the lab. There are two complementary approaches toward synthesizing life. The top-down approach, pioneered by Craig Venter (Hutchison et al., 2016), aims to minimize an organism's genome to the point where a set of essential genes are found. In the bottom-up approach, researchers utilize non-living materials to design and create self-assembled systems that exhibit increasingly complex structural and functional resemblance to living systems. If we can create structures in the lab that are able to recapitulate all of the most important features found in living systems, then we're not only closer to being able to chemically define life, but we will have an unprecedented level of understanding of biological processes leading to a wide array of potential medical, industrial, and environmental applications.

There are many excellent reviews that outline the core characteristics of living systems and the many varied ways in which researchers around the world are exploring the challenge of synthesizing life (Buddingh & van Hest, 2017; Yewdall, Mason, & van Hest, 2018). In the context of liquid-liquid phase coexistence, there has been a surge in their application as synthetic cell platforms due to their remarkable ability to strongly sequester functional materials and their resemblance to the crowded microenvironment found in the cell's cytoplasm. Indeed, one of the earliest origin of life theories proposed by Oparin has coacervation at its core, where colloidal aggregates concentrated prebiotic components (Oparin, 1953), and their role as RNA concentrators in the RNA-world hypothesis continues to be explored (Poudyal, Pir Cakmak, Keating, & Bevilacqua, 2018). In recent years, we have seen a variety of interesting cargoes incorporated into synthetic cells based on liquid-liquid phase coexistence, such as the machinery for in vitro transcription-translation (Sokolova et al., 2013; Tang, van Swaay, DeMello, Ross Anderson, & Mann, 2015), ribozymes for enhanced RNA processing activity, chloroplasts (Kumar et al., 2018), bacterial division proteins (Fanalista, Deshpande, Lau, Pawlik, & Dekker, 2018; te Brinke et al., 2018), and many others (Mason & van Hest, 2019).

One significant drawback of using liquid-liquid phase separated materials as synthetic cells is their innate lack of a boundary or interface separating the two phases, which results in rapid coalescence and poor temporal stability. In Nature, compartmentalization is one of the most important characteristics of life (Szostak, Bartel, & Luisi, 2001), as all cells have a membrane that provides a clear distinction between the cell and the outside world. Clearly, the functional and structural advantages provided by a membrane need to be incorporated into liquid-liquid phase separated protocells, and it is possible to achieve this with assemblies of phospholipids (Pir Cakmak, Grigas, & Keating, 2019; Tang et al., 2014) or alginate/gelatin-mediated in-situ hydrogelation (Douliez et al., 2017). In our group, we have pursued a different direction, where a bespoke, carefully-designed block copolymer has been synthesized to membranize a liquid-liquid phase separated material, amylose-based complex coacervates (Mason, Buddingh, Williams, & van Hest, 2017).

In this method, we describe how to synthesize and utilize terpolymer-stabilized coacervate protocells. Once the starting materials are synthesized, formulation of the protocells is straightforward, leading to a wide range of possible applications for this robust platform. For example, in recent years we have demonstrated the incorporation of enzymatic cascades, polymeric nanoreactors (Mason et al., 2019), and DNA-based supramolecular nanoscaffolds (Magdalena Estirado, Mason, Alemán García, van Hest, & Brunsveld, 2020). With the methods described herein, interested researchers will be able to not only reproduce our key findings, but also begin to introduce their own functional cargoes into this synthetic cell platform in order to examine the effect of crowding and confinement in a stable, persistent protocell model.

2. Synthesis of coacervate forming materials and terpolymer

This system is, at its heart, a totally synthetic bottom-up synthetic cell. There are two main components that comprise this robust protocellular framework, the coacervate core and the polymer membrane. In the following sections, the synthesis, purification and characterization of these materials will be described in detail.

2.1 Synthesis of modified amyloses

Amylose is a highly linear α-(1-4)-linked glucose polymer, often isolated from potato starch. In this form, amylose is sparingly soluble in water. Thus, for food and industrial applications, derivatization is often performed to improve solubility (Wulff, Steinert, & Höller, 1998). On each repeat glucose unit, there are 3 hydroxyl groups at C2, C3, and C6 that are free to react (Fig. 1). The degree of substitution (DS), which is defined as the average number of modified side groups per monomer, is thus limited to the range of 0–3, with 0 representing unmodified amylose and 3 representing fully modified amylose. In practice, the three derivatization sites do not display the same degree of reactivity, with modification at the C2 position occurring preferentially (Lindberg, Lindquist, & Stenberg, 1987). Modification of starches such as amylose is typically performed under basic conditions, and a number of different functional groups can be introduced. In this system, this approach is used to introduce functional groups bearing a quaternary amine or a carboxylic acid, resulting in amyloses with a positive or negative charge, respectively. These oppositely charged materials form complex coacervates when mixed, which is the core of this protocell system.

2.1.1 Quaternized Amylose (Q-Am)

Materials

- Heated stirrer plate equipped with an oil bath
- Round-bottomed flask equipped with a magnetic stirrer bar
- Dropping funnel
- Milli-Q water
- Sodium hydroxide
- Amylose (12–16 kDa, supplied by Carbosynth)
- Chloro-2-hydroxypropyltrimethyl ammonium chloride (65 wt% in water)
- Acetic acid
- pH paper

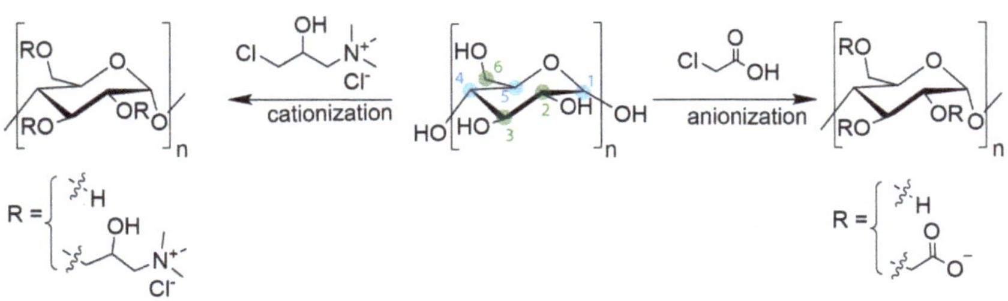

Fig. 1 Modification of amylose with either 3-chloro-2-hydroxypropyltrimethylammonium chloride solution or chloroacetic acid to form Q-Am and CM-Am, respectively.

- Ethanol
- 50 mL falcon tubes
- 1.5 mL Eppendorf tubes
- 3.5 kDa MWCO spin filters
- Centrifuge
- 3.5 kDa regenerated cellulose dialysis tubing
- Freeze drier

Method

1. Set up the heated stirrer plate to a temperature of 35 °C
2. Weigh out 2.8 g of NaOH directly into the round-bottomed flask. Add 14 mL of Milli-Q water and place it in the oil bath, stirring to dissolve
3. Weigh out 1.5 g of amylose, and slowly add it with a metal spatula to the stirred NaOH solution, waiting until each portion of amylose is dissolved before adding the next portion

Note: If the amylose is added all at once, the gummy solid can take a very long time to dissolve. To aid dissolution, use an ultrasonicator bath. The amylose is sparingly soluble in pH neutral water, thus it's important to add the NaOH first.

4. Once the amylose is fully dissolved, the solution may take on a yellow color. This is normal. Equip the round-bottomed flask with a dropping funnel, transfer 12 mL of chloro-2-hydroxypropyltrimethyl ammonium chloride (65 wt% in water), and start adding it dropwise to the amylose solution
5. Leave the reaction to stir overnight at 35 °C

Note: As the amylose modification reactions can be quite sluggish, an excess of reagent is used, resulting in a lack of control over the final DS. If a specific DS is required, it is necessary to monitor the reaction as it proceeds in order to quench it at the right point. Reaction monitoring is complicated by the fact that for both reagents, the NMR signals required for DS calculation overlap with the reagent peaks. It is thus necessary to perform a rapid purification to remove unreacted reagent from the sample in order to obtain a representative approximation of the DS. First, the polymer is separated from the bulk of reagent via precipitation in cold ethanol. As only a small amount of sample is needed for NMR analysis, this precipitation can be performed using only 1 mL of cold ethanol in a plastic Eppendorf tube and several drops of crude reaction mixture. The tube is vortexed, and a large amount of white precipitate is immediately noticed. The supernatant is discarded, pellet re-dissolved in Milli-Q, and purified further via spin filtration. This is repeated once more before the sample is diluted in D_2O and analyzed via ^1H NMR.

6. Once the desired degree of substitution is reached, the reaction is stopped via the addition of acetic acid. Add acetic acid until a neutral pH is obtained, as determined by frequent checks with pH paper
7. In a 500 mL conical flask equipped with a magnetic stirrer bar, dispense 200 mL of cold ethanol. Ethanol is cooled by storage in a −20 °C freezer for several hours before precipitation
8. Precipitate the crude reaction mixture into the rapidly stirred cold ethanol. A white precipitate is immediately visible, which is the desired product Q-Am
9. Discard the supernatant and collect the precipitate with Milli-Q water. Transfer the modified amylose solutions to 3.5 kDa regenerated cellulose dialysis tubing and dialyze extensively against Milli-Q (at least 2 days, with 3 changes of dialysate).
10. Transfer the dialyzed Q-Am to 50 mL falcon tubes and lyophilize for 48 h to obtain a white fluffy solid
11. To determine the degree of substitution (DS), dissolve a small amount of Q-Am in D_2O and analyze by ^1H NMR. The relevant peaks and method for DS calculation are shown in Fig. 2
12. Modified amyloses can be stored for extended periods (12–18 months) by placing them under argon at 4 °C

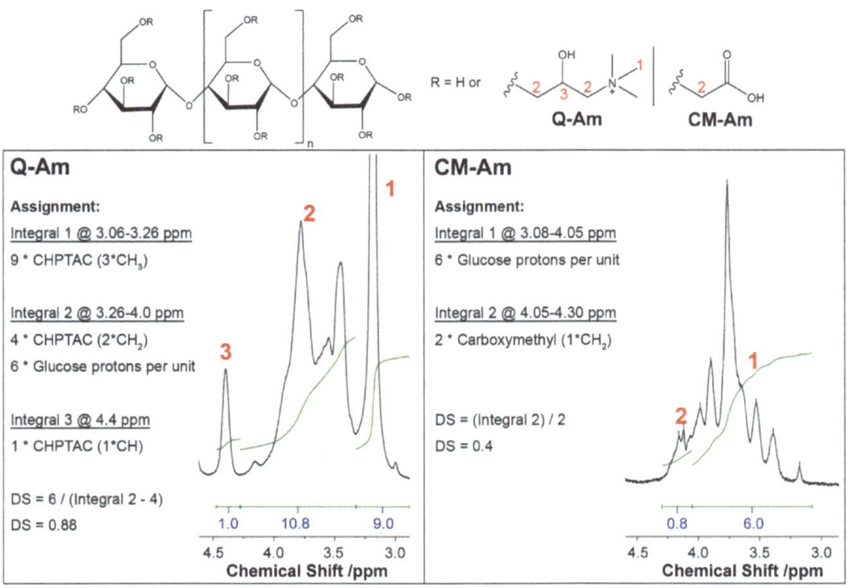

Fig. 2 ^1H NMR processing of highly charged amylose derivatives, including calculation of the degree of substitution (DS = number of modifying groups per glucose unit), of both quaternized (Q-Am) and carboxymethylated (CM-Am) products.

2.1.2 Carboxymethyl Amylose (CM-Am)

The synthesis of CM-Am mirrors that of Q-Am, except that chloroacetic acid is used in place of chloro-2-hydroxypropyltrimethyl ammonium chloride. As chloroacetic acid is a solid, a dropping funnel is not used. The ^1H NMR analysis is also different, with the relevant peaks and method for DS calculation shown below in Fig. 2.

2.2 Synthesis of terpolymer

The most vital component of this system is the carefully-designed triblock copolymer, as without this polymer, the coacervate microdroplets are not stabilized and lack utility in experiments requiring multiple populations or long experimental timeframes. The polymer consists of three main blocks: poly(ethylene glycol) (PEG), poly(ε-caprolactone-*gradient*-trimethylene carbonate) (PCLgPTMC), and poly(glutamic acid) (PGA). Each part of this multiblock structure is important to its final function. If the PEG moiety is too small, the polymer is internalized by the complex coacervate and does not assemble on the surface. The PCLgPTMC moiety is required for hydrophobic packing and the formation of a membrane. Furthermore, it gives the membrane semi-permeable characteristics, and is fluid at room temperature, enabling rapid and facile self-organization on the coacervate surface without needing to resort to complex methodologies or solvent mixtures. Finally, the PGA is essential for anchoring to the positively charged surface of the coacervate microdroplets. At neutral pH, the PGA displays a negative charge, which is electrostatically attracted to the coacervate core which has been engineered to contain an excess of positive polyelectrolytes.

The synthesis of this polymer proceeds in three main steps, depicted in Fig. 3. First, a diblock copolymer is prepared according to a modified literature procedure (Couffin, Delcroix, Martín-Vaca, Bourissou, & Navarro, 2013), where PEG is used in the organocatalyzed ring-opening of ε-caprolactone and trimethylene carbonate. In the second step, the terminal alcohol group is converted to a terminal amine, which is subsequently used in the third step to initiate the ring-opening polymerization of *N*-carboxyanhydride γ-benzyl-L-glutamate. The final terpolymer is obtained after deprotection of the benzyl protecting group.

2.2.1 Synthesis of poly(ethylene glycol)-block-poly(ε-caprolactone-gradient-trimethylene carbonate)

Materials

- Heated stirrer plate equipped with an oil bath
- NMR spectrometer

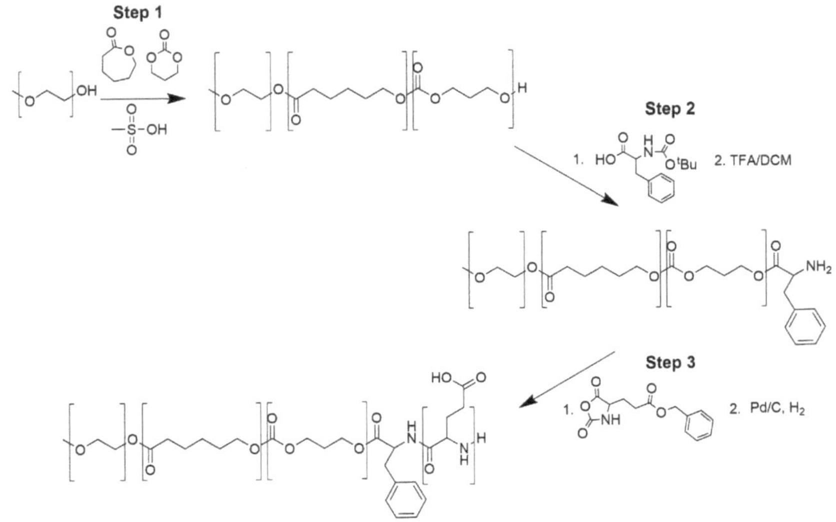

Fig. 3 Synthesis of PEG-*b*-PCLgPTMC-*b*-PGA terpolymer. Poly(ethylene glycol) monomethyl ether was used to initiate the ring-opening polymerization of ε-caprolactone and trimethylene carbonate (*step 1*). The terminal alcohol of this polymer was subsequently modified via a Steglich esterification with Boc-L-Phe-OH to yield a primary amine after TFA deprotection (*step 2*). The final poly(L-glutamic acid) block was introduced by the ring-opening polymerization of *N*-carboxyanhydride γ-benzyl-L-glutamate, followed by hydrogenation (*step 3*).

- Gel permeation chromatography (GPC) in THF
- Centrifuge
- 100 mL round-bottomed flask equipped with a magnetic stirrer bar (RBF)
- Positive displacement solvent pipettes (1000 + 100 μL)
- Rotavap
- Dry dichloromethane (DCM)
  Note: The reaction can be performed in DCM or toluene.
- Dry toluene
  Note: If you do not have access to dry toluene through a dry solvent system, toluene can be effectively dried by standing over activated 4 Å molecular sieves for 24 h.
- Poly(ethylene glycol) monomethyl ether, M_w 2000 g/mol (PEG2000)
  Note: PEG is hygroscopic and will contain a significant amount of water. Before use in this reaction, it is important to first dissolve a portion of PEG in dioxane, and freeze dry it. This effectively removes water from PEG. Store freeze-dried PEG under argon and re-purify often.
- Trimethylene carbonate (TMC)

- ε-caprolactone (CL)
  Note: CL can contain some water. If you are using an older, pre-opened bottle, it is wise to dry this monomer by stirring neat CL over CaH_2 for 24 h before distillation at reduced pressure.
- Methanesulfonic acid (MSA)
  Note: MSA is hygroscopic and difficult to dry. Purchase small bottles or sealed ampoules, and store MSA under argon immediately after opening.
- Methanol
- 500 mL conical flask equipped with a large flat magnetic stirrer bar
- Tetrahydrofuran (THF)
- Deuterated chloroform ($CDCl_3$)
- 50 mL falcon tubes

Method

1. Set up the heated stirrer plate to a temperature of 30 °C
2. Weigh out 400 mg of PEG2000 and 1121 mg of TMC in glass vials, and transfer to the RBF. Use small volumes of DCM to rinse the vials
3. Using a solvent pipette, transfer 1108 μL of CL to the RBF
4. Fill the RBF approximately half-full with toluene, and stir to dissolve
5. Use the rotavap to evaporate all of the toluene. Remove the RBF from the rotavap, and refill halfway with fresh dry toluene
6. Repeat step 5 a further two times, adding and removing toluene a total of three times

Note: This step is performed to remove trace amounts of water from the reaction mixture. Water and toluene form an azeotrope, and distillation of toluene will assist in the removal of water.

7. *Re*-dissolve dried reagents in 30 mL of dry toluene or DCM. Place the reaction mixture into the oil bath, and stir at 500 rpm
8. Using a solvent pipette, transfer 40 μL of MSA to the reaction mixture to initiate polymerization. Leave the mixture to react overnight under an inert atmosphere
9. The next day, transfer two small aliquots of crude reaction mixture to glass vials, and evaporate the majority of the solvent under a stream of nitrogen. Dissolve one portion in 2 mL of THF for GPC analysis, and 600 μL of $CDCl_3$ for ^1H NMR analysis. See Figs. 4 and 5 for reference data
10. If NMR indicates that the degree of polymerization is sufficient (50 TMC, 50 CL), and GPC indicates that the polymer is sufficiently monodisperse ($M_w/M_n < 1.2$), proceed with the following purification steps
11. Concentrate the crude reaction mixture via rotary evaporation

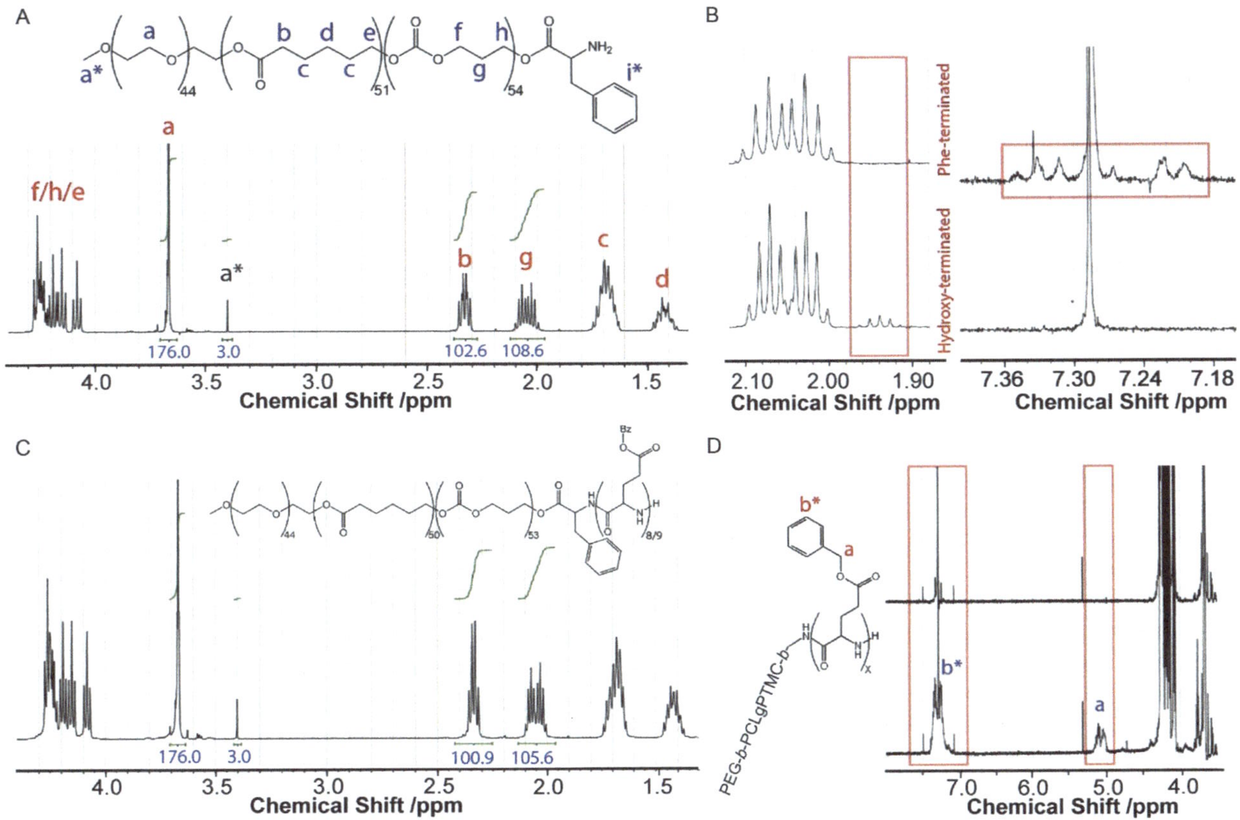

Fig. 4 ^{1}H NMR processing of terpolymer synthesis. (A) analysis of PEG-PCLgPTMC, (B) demonstration of the complete chain-end modification of PEG-PCLgPTMC with Boc-L-Phe, (C) analysis of the product of ring-opening polymerization of BLG-NCA and (D) specific evaluation of PBLG content of the resulting terpolymer using benzylic and aromatic protons at 5.0–5.2 and 7.1–7.4 ppm, with upper spectrum showing successful deprotection by hydrogenation.

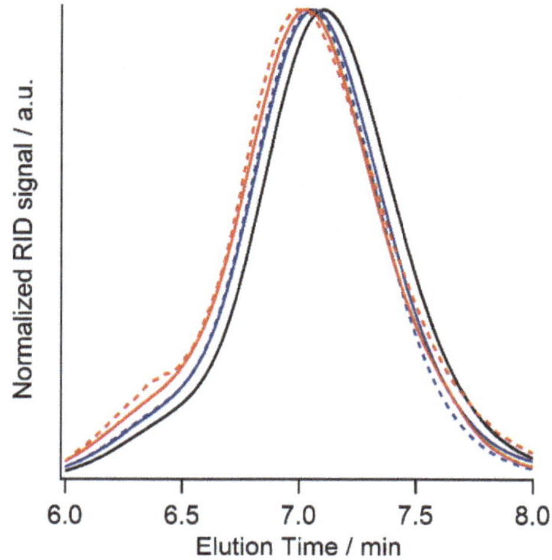

Fig. 5 GPC traces of PEG-PClgTMC-OH (black), PEG-PClgTMC-NH_2 (blue line with deprotected form as blue dashes) and PEG-PClgTMC-PBLG (red line with deprotected form as red dashes).

12. Precipitate the concentrated polymer solution into cold methanol, yielding a white precipitate

13. Centrifugation may be needed to collect precipitated polymer. In this case, transfer the supernatant containing pure polymer to 50 mL falcon tubes and centrifuge at 4000 × *g* for 5 min to collect the polymer

Note: If the methanol is not cold enough, or not used in enough excess, precipitation may be difficult. Use methanol that has been pre-cooled in a −30 °C freezer, and keep falcon tubes on ice while transferring to the centrifuge.

14. Collect pure polymer in a good solvent such as DCM, transfer to a tared container, and remove solvent in vacuo to calculate the polymerization yield

2.2.2 Chain-end modification with Boc-L-phenylalanine and deprotection

For the incorporation of a terminal amine onto the diblock copolymer prepared above, a carbodiimide-mediated esterification between PEG-PCLgTMC and Boc-L-phenylalanine is performed. Once this intermediate is obtained, the Boc group is removed via TFA mediated deprotection.

Supplementary materials:

- 25 mL RBF
- Separating funnel

- Acetonitrile
  Note: DCM can also be used as the solvent in the Steglich esterification, but the dicyclohexylurea (DCU) byproduct is less soluble in acetonitrile than DCM, facilitating its removal during purification.
- Boc-L-phenylalanine (Boc-L-Phe)
- *N,N'*-dicyclohexylcarbodiimide (DCC)
- 4-dimethylaminopyridine (DMAP)
- Trifluoroacetic acid (TFA)
- Saturated $NaHCO_3$
- 1 M NaCl
- Brine
- Magnesium sulfate

Method

1. Weigh out 1.2 g of PEG-PCLgPTMC copolymer into a 25 mL RBF and dissolve in 5 mL acetonitrile. Place the RBF in an ice bath and start stirring at 500 rpm
2. To this, add 42 mg of DCC, 23 mg of Boc-L-Phe, and 2.5 mg of DMAP. Use acetonitrile to rinse out glass vials, to a final volume of approximately 10 mL in the RBF
3. Leave the reaction to stir overnight, allowing it to warm to room temperature in the process
   Note: The reaction mixture may start becoming cloudy immediately, this is the insoluble DCU byproduct forming.
4. The next day, place the reaction mixture in the freezer for several hours to ensure full precipitation of DCU
5. Filter the crude reaction mixture rapidly, not allowing the apparatus to heat up too much
6. Concentrate the filtrate, and precipitate into cold methanol as described in Section 2.2.1 to purify
7. Analyze the pure product by GPC and NMR to ensure that the polymer has not degraded and that the terminal modification has gone to completion. See Figs. 4 and 5 for reference data
8. In a 25 mL RBF, dissolve the polymer in 5 mL of DCM, place it in an ice bath, and add 5 mL of TFA
9. Continue to stir the reaction mixture for 2 h, allowing it to warm to room temperature
10. Concentrate the crude reaction mixture in a rotavap
    Note: TFA fumes are extremely corrosive. Perform step 10 using a rotavap inside a fumehood and clean it thoroughly after use.

11. Dilute the polymer in 20 mL of DCM, and wash the organic phase with 20 mL saturated $NaHCO_3$, 20 mL 1 M NaCl, and 20 mL of brine. Collect the organic phase, dry it over magnesium sulfate, filter and concentrate before precipitating once more into cold methanol
12. Confirm full Boc deprotection by ^{1}H NMR. The Boc signals at 1.45 ppm will disappear, resembling the initial PEG-PCLgPTMC spectrum in this region
13. Collect the precipitate with DCM and transfer to a tared 25 mL RBF before removing solvent in vacuo to yield PEG-PCLgPTMC-NH_2

2.2.3 Polymerization and deprotection of N-carboxyanhydride γ-benzyl-L-glutamate

The final PGA block is obtained by first using a modified literature procedure (Zou et al., 2013), where PEG-PCLgPTMC-NH_2 as a macroinitiator in the ring-opening polymerization of *N*-carboxyanhydride γ-benzyl-L-glutamate, which yields a block of benzyl-protected glutamate. This is subsequently deprotected via palladium catalyzed hydrogenation to yield the final PEG-PCLgPTMC-PGA terpolymer.

Supplementary materials:

- 25 mL Schlenk tube equipped with a side-arm
- Gas bubbler
- Dimethylformamide (DMF)
- Diethyl ether
- Dioxane
- *N*-carboxyanhydride γ-benzyl-L-glutamate (NCA-BLG)
- H-Cube apparatus
- Freeze drier

Method

1. Weigh out 1 g of PEG-PCLgPTMC-NH_2 copolymer into the 25 mL Schlenk flask and dissolve in 2.5 mL DMF. Place the flask in an ice bath and start stirring at 500 rpm
2. Weigh out 160 mg of NCA-BLG in a glass vial, and dissolve in 2.5 mL DMF. Transfer to the reaction mixture, rinsing with a small amount of DMF
3. Seal the Schlenk flask and connect a supply of nitrogen using a gas adapter. Connect a second outlet tube to the side-arm, leading to the gas bubbler. The aim here is to have a gentle flow of nitrogen over the reaction mixture, as the removal of carbon dioxide generated by the polymerization helps to push the reaction to completion

4. Leave the reaction to stir overnight, allowing it to warm to room temperature in the process
5. To check the degree of polymerization before further purification, take a small aliquot of crude reaction mixture and dilute it directly into $CDCl_3$ for ^1H NMR analysis. Reference data can be found in Fig. 4
6. Precipitate the crude reaction mixture into diethyl ether to yield PEG-PCLgPTMC-PBLG as a white precipitate
7. Use THF GPC to confirm a slight increase in M_w from this polymerization
8. Collect the precipitate in 40 mL of THF and proceed to deprotect the benzyl group via hydrogenation. This is performed using a H-Cube apparatus, set to run at 1 mL/min, at 60 °C and 30 bar of H_2 using a 10% Pd/C cartridge

   Note: In our experience, using a traditional heterogeneous catalysis method (dispersed Pd/C and a hydrogen balloon or bomb) does effectively remove the benzyl groups, but the palladium binds to the polymer, leaving it with a gray coloration that is difficult to remove even via purification by dialysis. The terpolymer however retains its coacervate stabilizing properties.
9. Before purification, confirm full deprotection using ^1H NMR
10. Concentrate the hydrogenated product using a rotavap, and precipitate the polymer into cold methanol
11. Collect the polymer in 10 mL of dioxane, transfer to a tared container, and freeze dry for 48 h to yield PEG-PCLgPTMC-PGA as a soft white solid

3. Formation of terpolymer-stabilized coacervates

3.1 General methodologies

3.1.1 Stirrer plate

Coacervate protocells can reliably be formed using a stirrer plate. The advantage to this method is that if a small magnetic stirrer plate is utilized, the equipment needed for protocell formation can be easily transported, and set up next to analysis equipment if rapid prototyping or analysis is called for.

Materials

- Small magnetic stirrer plate (such as IKA Squid)
- 4 mL glass vials
- Small flat magnetic stirrer bar, length just smaller than vial diameter
- PBS buffer, pH 7.4

- Q-Am 1 mg/mL in PBS
- CM-Am, 1 mg/mL in PBS
- Terpolymer, 50 mg/mL in DMSO or PEG350

Note: A low molecular weight poly(ethylene glycol) such as PEG350 can be substituted with no observable detriment, which is useful if an encapsulated cargo or process is susceptible to organic solvents such as DMSO.

- Branson 2510 Ultrasonic Cleaner

Method

1. Pipette 200 μL of Q-Am into the glass vial equipped with a magnetic stirrer bar
2. Start stirring at 250 rpm, and add 100 μL of CM-Am
3. Directly add any cargo that you want to include, typically adding 1–5 μL of a concentrated solution
4. In the meantime, sonicate the terpolymer for 1 min (40 Hz, 185 W, degas mode)
5. After 5 min, add 5 μL of the terpolymer

Note: The amount of time elapsed before terpolymer addition directly affects protocell size (see following section). In typical experiments, where protocells on the order of 10–20 μm diameter are desired, a 5 min shaking time is used

6. Stop stirring and collect the sample

Note: Coacervate protocells sediment rapidly due to their density. When collecting, ensure that any sedimented protocells are suspended via gentle pipetting up and down before transfer.

3.1.2 Eppendorf shaker

An alternative approach to the stirrer plate is to make use of a shaker block, where multiple tubes can be shaken at the same time at a high speed. The main advantage of this method over the stirrer plate is that all tubes are close together, thus making it easier to do multiple samples. Furthermore, since stir bars are not required, contamination from previously made samples is not possible and time is saved in washing stir bars between samples. The shaker method is used preferentially in our lab, and a short video demonstrating the formulation can be found online in Supplementary video 1 in the online version at https://doi.org/10.1016/bs.mie.2020.06.008.

Materials

- MixMate (Eppendorf) or any other 1.5 mL tube shaker capable of 1500 rpm
- 1.5 mL Eppendorf tubes
- PBS buffer, pH 7.4

- Q-Am 1 mg/mL in PBS
- CM-Am, 1 mg/mL in PBS
- Terpolymer, 50 mg/mL in DMSO or PEG350
- Branson 2510 Ultrasonic Cleaner

Method

1. Set the shaker to 1500 rpm
2. Add 75 μL PBS and 25 μL of CM-Am in a fresh Eppendorf tube
3. Start shaking and add 50 μL of Q-Am

Note: The concentration of amylose stocks and the ratio of Q-Am:CM-Am can be varied significantly. At lower concentrations, the ripening process is slower, resulting in a decreased number of smaller protocells. When using multivalent cations in the buffer such as Mg^{2+} or Zn^{2+}, we find that lower ratios of Q-Am:CM-Am are needed to form large protocells

4. Directly add any cargo that you want to include, typically adding 1 μL of a concentrated solution
5. In the meantime, sonicate the terpolymer for 1 min (40 Hz, 185 W, degas mode).
6. After 5 min, add 5 μL of the terpolymer

Note: The amount of time elapsed before terpolymer addition directly affects protocell size (see following section). In typical experiments, where protocells on the order of 10–20 μm diameter are desired, a 5 min shaking time is used.

7. Stop shaking and collect the sample

3.2 Size control

3.2.1 Terpolymer addition time

As coacervate droplets gradually coalesce and ripen, it is possible to control the size of these protocells by adding stabilizing terpolymer at different times (Mason et al., 2017). For example, it is possible to generate coacervate protocells ranging from 5.23 ± 0.95, 22.7 ± 5.9 to 39.3 ± 9.6 μm by addition of terpolymer after 15 s, 2 min and 5 min, respectively (Fig. 6). However, the size distribution becomes more polydisperse as the size increases. This method can be conducted via the stirrer plate or Eppendorf shaker methodologies. The materials and methods are the same as in Section 3.1; only the mixing time before terpolymer introduction is changed.

3.2.2 Volume

Alternatively, the volume of amylose solutions also contributes to variation in size. We observed a non-linear relationship between volume of amylose solutions and protocellular size with an overall decreasing tendency in

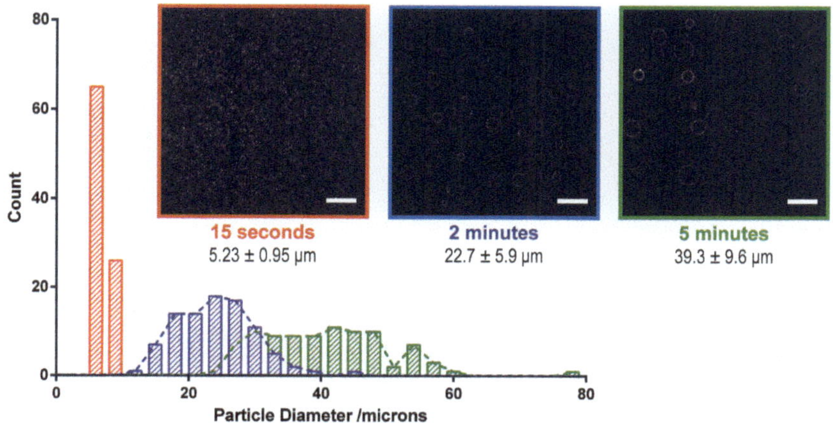

Fig. 6 The size of the coacervate protocells can be easily controlled by adding the stabilizing terpolymer at different time points. To obtain the smallest possible population, polymer is added immediately after coacervation is started (15 s, ▨). At 2 (-▨-) and 5 (-▨-) minutes post coacervation, the populations are larger. Scale bar represents 50 μm, histogram constructed by counting 91 particles for each time point.

size as volume decreases. One unit volume (1 V) denotes 200 μL Q-Am and 100 μL CM-Am. As volume decreased from 1 to 0.2 V (40 μL Q-Am and 20 μL CM-Am), diameter of coacervate protocells decreased from 10.3 ± 3.4 to 2.57 ± 0.71 μm, shown in Fig. 7. A significant drop in size was seen from 0.5 to 0.4 V where size declines from 9.34 ± 2.57 to 3.10 ± 1.29 μm. This could be interpreted as a threshold amount of amylose derivatives which is needed to generate large coacervate protocells. The materials and methods are the same as in Section 3.1; only the total volume of modified amyloses is changed, and terpolymer is added after 1 min of growth time instead of 5.

3.2.3 Ultrasonication

In order to access very small (<2 μm) monodisperse protocells, the following ultrasonication method has been developed. With such method, diameter of coacervate protocells can be reduced to around 1.2 μm. Interestingly, the coacervate growth time has a minor influence on the size, as only slight differences in size were observed when terpolymer was added after 1.5, 3, 5 min (Fig. 8).

Materials

- PBS buffer, pH 7.4
- Q-Am 1 mg/mL in PBS

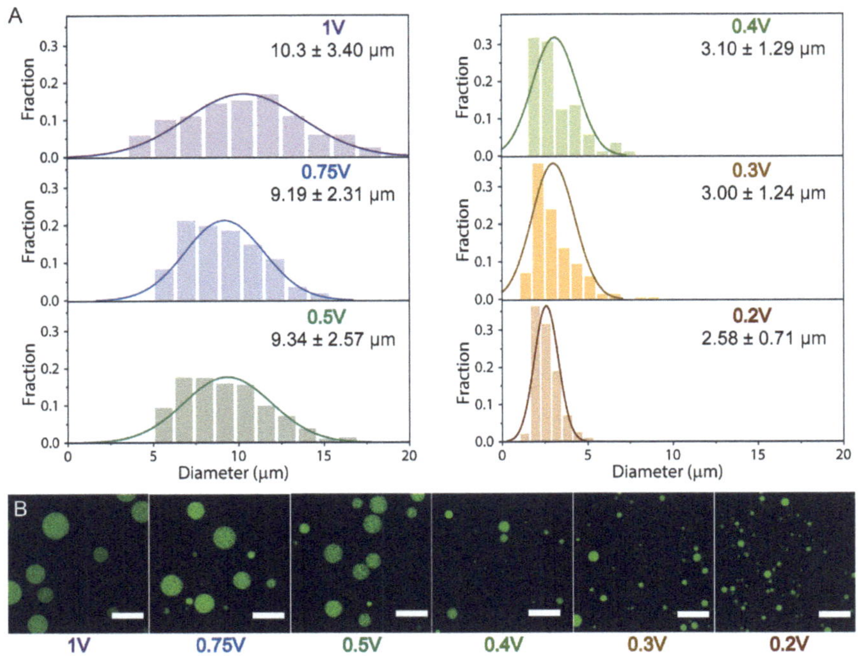

Fig. 7 Size of coacervate protocells is dependent on volume of amylose derivative solutions. 1 V denotes 200 μL Q-Am and 100 μL CM-Am. (A) Histogram charts were obtained by analyzing at least 100 protocells for each volume using confocal microscopy (B) Scale bars represent 20 μm.

- CM-Am, 1 mg/mL in PBS
- Terpolymer, 50 mg/mL in DMSO or PEG350
- Succinylated Atto-488 modified BSA (scBSA-A488, 1.2 mg/mL)
- Branson 2510 Ultrasonic Cleaner

Method

1. Add 200 μL Q-Am solution in a 2 mL Eppendorf tube
2. Place the tube in the sonication bath and start ultrasonic treatment (normal mode, 40 Hz, 185 W).
3. Add 100 μL CM-Am solution into the tube
4. Directly add 10 μL 10 × diluted mBSA
5. Wait for 1.5 min, then add 10 μL polymer solution into the tube

Note: As seen below, the amount of time before terpolymer is added doesn't have a significant impact on the final size of the protocells.

6. Leave the tube in sonication bath for another 30 s before removing the sample

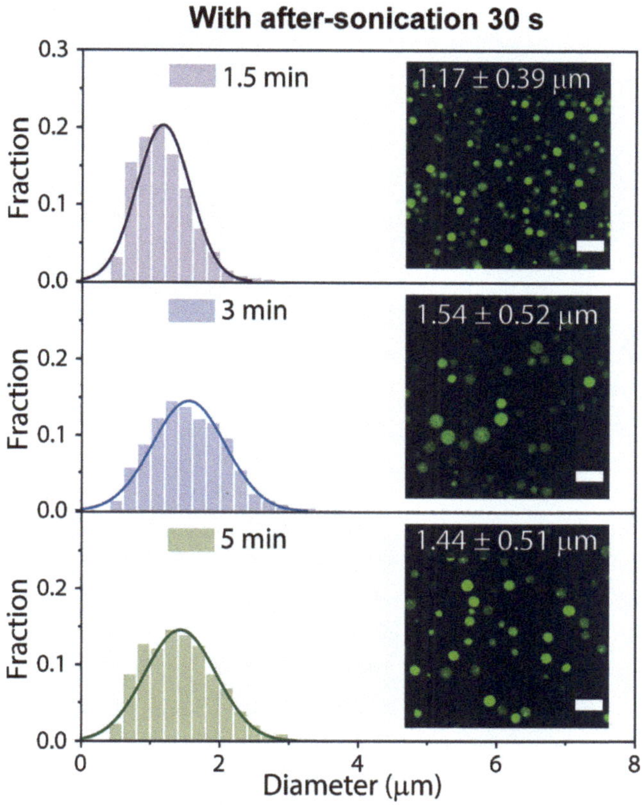

Fig. 8 Size of coacervate protocells is significantly reduced with the sonication bath method. Coacervate growth time doesn't have a significant effect on the size. Scale bar represents 5 μm. Histogram charts were obtained by including at least 280 protocells each condition.

3.3 Encapsulation of cargo

An inherent drawback of the liquid-liquid phase separated droplets is that cargo encapsulation is often biased due to segregative mechanisms, often electrostatics. In the case of the polymer stabilized coacervates, there is a positively charged core, therefore the uptake of negatively charged components is easily achieved (Fig. 9A). This is exemplified by the uptake of different DNA species, dextran of different molecular weight or succinylated BSA (Fig. 9B) (Yewdall et al., 2019). The charge based uptake is tunable by varying the ratio between the positively charged (Q-Am) and negatively charged (CM-Am) component. All previous examples shown had a 2:1 ratio of Q-Am to CM-Am, resulting in efficient partitioning inside the coacervate of the negatively charged cargo. Upon changing to either a lower (1:1) or

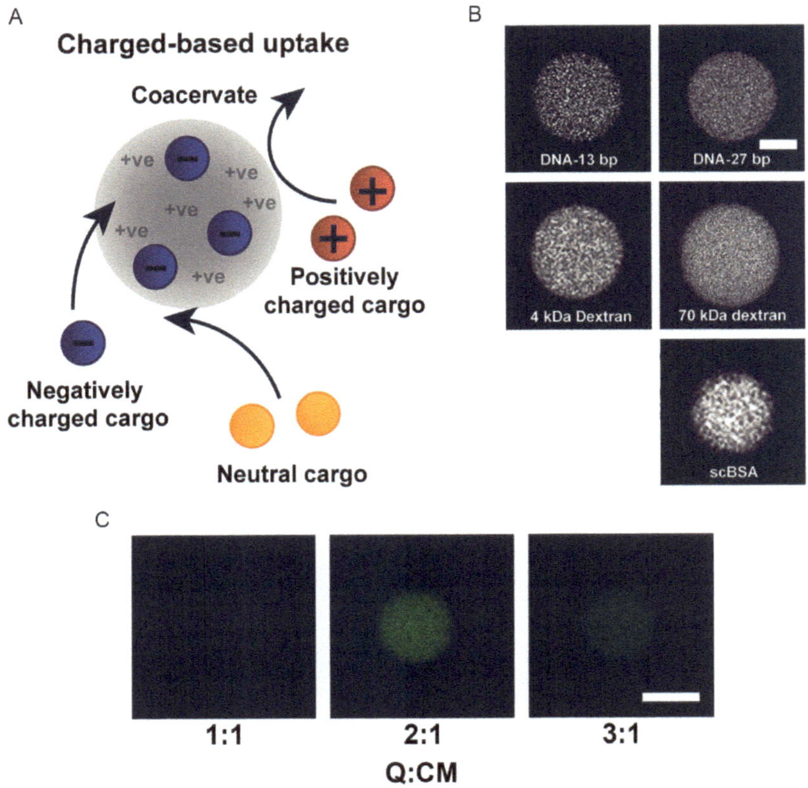

Fig. 9 (A) Schematic representation of cargo uptake into the coacervate-based protocells. Negatively charged components are sequestered, meanwhile species with a neutral or positive charge are excluded. (B) Confocal micrographs of coacervates loaded with different negatively charged components. Scale bar: 5 μm. (C) Confocal micrographs of coacervate sample with different charge ratios with negatively charged-30GFP.

higher (3:1) ratio, the uptake of a negatively charged cargo (-30GFPP) is clearly influenced (Fig. 9C). Alternatively, the cargo of interest can be modified as well. In the case of proteins, different strategies are known to modify the surface, e.g., succinylation (Mason et al., 2017) or mutagenesis (Lawrence, Phillips, & Liu, 2007; Spector et al., 2000). Although these approaches have been used with success, these altered species can be labor intensive to make and often suffer from a loss of function (Zhang et al., 2011). This limits the general applicability of electrostatic-based approaches, and when they cannot be performed due to the presence of a neutral or positive surface charge on the cargo of interest, a more specific methodology must be employed.

All in all, the uptake of proteins and other biomolecular cargo is tunable depending on the interaction with the coacervate core. In general, as long as there is enough driving force, e.g., charge, a macromolecule can be effectively sequestered into the coacervate core.

4. Purification strategies

The terpolymer membrane imparts the coacervate-based protocells with remarkable mechanical stability. Protocells can be easily manipulated and transferred without losing their structure or their contents. This enabled us to develop two methods to purify coacervates. Purification can be performed by means of dialysis and centrifugation as previously described. These methods can for example be adopted for the removal free cargo or excessive polymer (Section 4.1) or to buffer exchange coacervates into cell culture medium (Section 4.2) without loss of protocell structure (Fig. 10).

Materials

- Cell culture medium; vascular cell basal medium + Endothelial Cell Growth Kit-VEGF (VCBM)
- Note: These experiments have also been done with Dulbecco's Modified Eagle Medium + Fetal bovine serum
- Q-Am and CM-Am in cell culture medium (0.5 mg/mL)
- Terpolymer in PEG350 (containing 1 v/v% 1 mg/mL Nile red in DMSO)

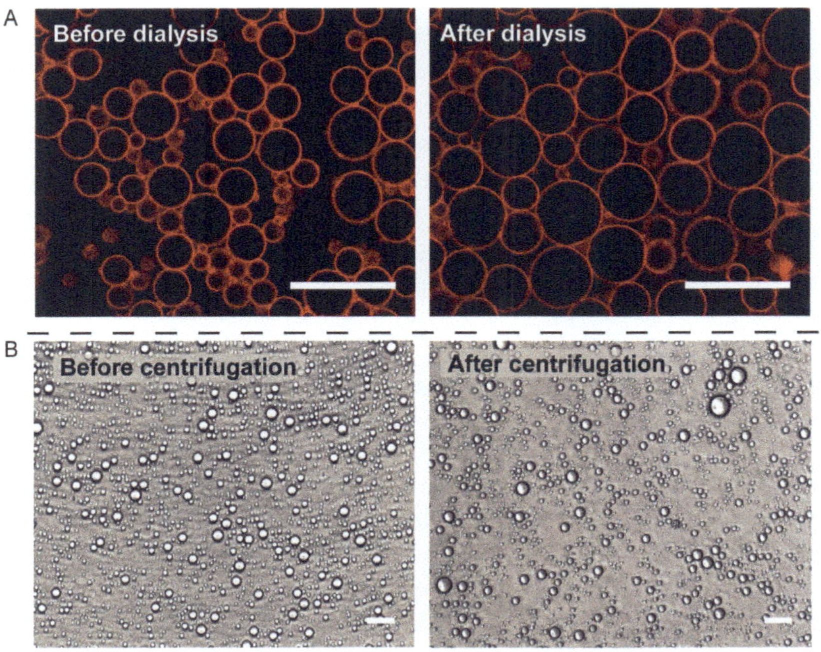

Fig. 10 Confocal microscopy image depicting the stability of coacervates in cell culture medium following mechanical disruption like (A) extensive dialysis for 24 h and (B) centrifugation. Terpolymer is stained with Nile red. Scale bars represent 50 μm.

- Succinylated BSA conjugated to Cy5 (scBSA-Cy5)
- Float-A-lyzer G2 dialysis device MWCO 1000 kDa
- Centrifuge tube filters (cellulose acetate membrane, pore size 0.22 μm, sterile)

4.1 Centrifugation

Method

1. Form coacervate protocells as per the method in Section 3.1, but use modified amyloses dissolved in cell culture medium instead of PBS
   Note: It is possible to use protocells formed in PBS, but if the solution environment changes radically, size changes may be observed.
2. Centrifuge stabilized coacervates into a pellet at 500 × *g* for 4 min
3. Gently remove the supernatant without disturbing the pellet
4. Resuspend the pellet in the same volume of cell culture medium

4.2 Dialysis

Method

1. Form coacervate protocells as per the method in Section 3.1, but use modified amyloses dissolved in cell culture medium instead of PBS
2. Pre-wet the membrane of the dialysis device according to the manufacturer's protocol using 70% ethanol and demineralized water
3. Condition the dialysis device by rinsing the inside of the membrane with VCBM
4. Dialyze terpolymer-stabilized coacervates overnight at 4 °C against 50 mL VCBM using a Float-A-lyzer G2 dialysis device with a MWCO of 1000 kDa
5. After dialysis, recover the coacervates from the bottom of the membrane by pipetting the solution up and down several times

5. Measurement considerations

5.1 Aqueous stability

In the above purification section, these terpolymer-stabilized protocells demonstrate remarkable chemical and mechanical stability, with protocells present after dialysis and centrifugation. However, for future biomedical applications, stability over even a longer time frame is required, as well as stability in challenging chemical environments (e.g., cell culture medium with high salt and protein concentration). To test this, protocells were incubated in a humidified incubator at physiological conditions (37 °C, 5% CO_2)

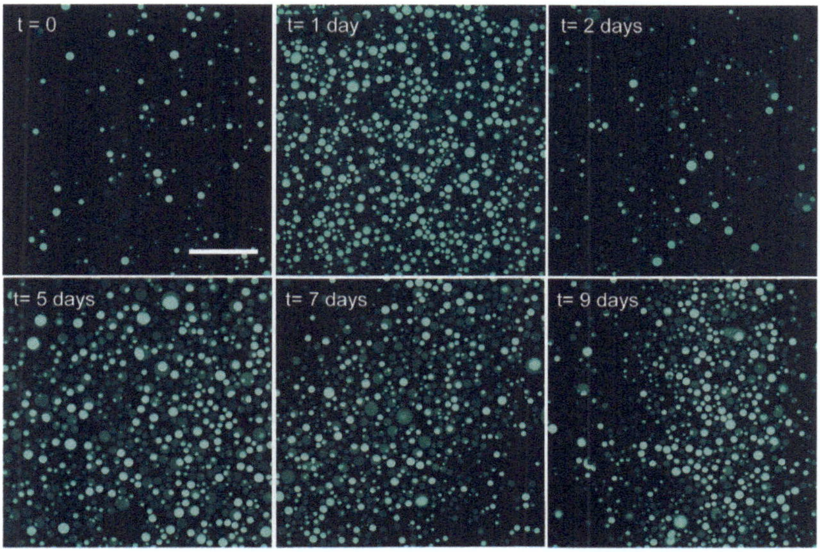

Fig. 11 Confocal microscopy image depicting the stability of coacervates loaded with scBSA-Cy5 in cell culture medium over a timeframe of 9 day incubation in a humidified incubator at physiological conditions. Scale bar represents 100 μm.

over a period of 9 days. The presence of fluorescent cargo and the retention of spherical morphology after this extended timeframe (Fig. 11) highlight the impressive capabilities of this platform to be used in a variety of biomedically relevant situations.

5.2 Confocal measurement

For routine analysis of protocell dispersions on the order of 100 μL, Ibidi 8-well slides are used. The coacervate-based protocells are significantly denser than water and sediment rapidly, thus protocells are gently pipetted up and down several times before sample transfer from Eppendorf tube to confocal slides. When working with small volumes, e.g., <50 μL, it is important to keep evaporation in mind. Especially in the case for reference curves of neat protein in solution, evaporation can have a dramatic effect on the protein concentration thus the amount of signal measured. Here, we provide a method to prepare slides as such that the evaporation of small volumes is minimized.

Materials

- Press-to-Seal™ Silicone Isolator with Adhesive, eight wells, 9 mm diameter, 0.5 mm deep
- Tweezers

- VWR Superfrost® microscope slide
- VWR rectangular cover slips, 24 × 50 mm
- 2-propanol
- Kimtech Tissue
- 20 μL of different dilutions for the protein of interest
- Styrofoam box

Method

1. Clean the microscope slide with 50 μL 2-propanol and wipe it dry with a tissue
2. Peel off the bottom plastic layer of the silicone isolator and place it on the slide and press it down firmly with a tissue
3. Pipette 20 μL of the protein solution in the middle of the well
   Note: Don't let any solution near the edges of the wells. If the solution comes in contact with the glue on the top layer the coverslip will not be able to seal properly.
4. Carefully remove the top plastic layer of the silicone isolator
5. Take a coverslip with the tweezers and place it on top of the isolator and press it down with a tissue
6. Place slide in a styrofoam box next to a wet tissue to prevent evaporation during transportation or waiting steps
7. Analyze the sample with microscopy at the same settings as the protocell sample of interest

5.3 Confocal analysis

Laser scanning confocal microscopy is often used for analysis of protocells. Although the technique allows for multiplex imaging of samples, obtaining quantitative data from these images is often very laborious. The amount of manual (non-automated) actions required for proper analysis can be daunting, especially if a time series is obtained, and any sample drift between frames complicates analysis further. Here, we provide a semi-automated approach for analyzing a time series of confocal micrographs. In addition to the analysis of drifting particles, the approach allows for easy analysis of particles containing a low signal/amount of cargo without boosting the contrast.

Materials

- ImageJ or FIJI
- UCB Vision Sciences plugin

Method

1. *Input*: stack of confocal micrographs where the membrane and the interior of the particle are imaged in different channels. Make sure to split the channels during import (Fig. 12A)
2. Transform the membrane image into a binary image (*Image > Adjust > Threshold*) (Fig. 12B)
   Note: When applying a threshold to an image, make sure the membranes are nicely saturated and the background is as clean as possible.
3. Open the Hough Circle Transform (*Plugins > UCB Vision Sciences > Hough Circle Transform*) and set it to advanced mode
4. Select the "Use local search space to speed up the search" option
   Note: Keep in mind that this plugin makes use of hard cut-off values, and requires optimization of parameters such as the minimum and maximum search radius, radius increment, and the number of circles found. The first parameter is dependent on pixel size, e.g., for 0.18 μm/pixel a minimum and maximum radius of 30 and 100 pixels results in a radius of 5.4–18 μm respectively (typical radius for these membrane bound coacervates).

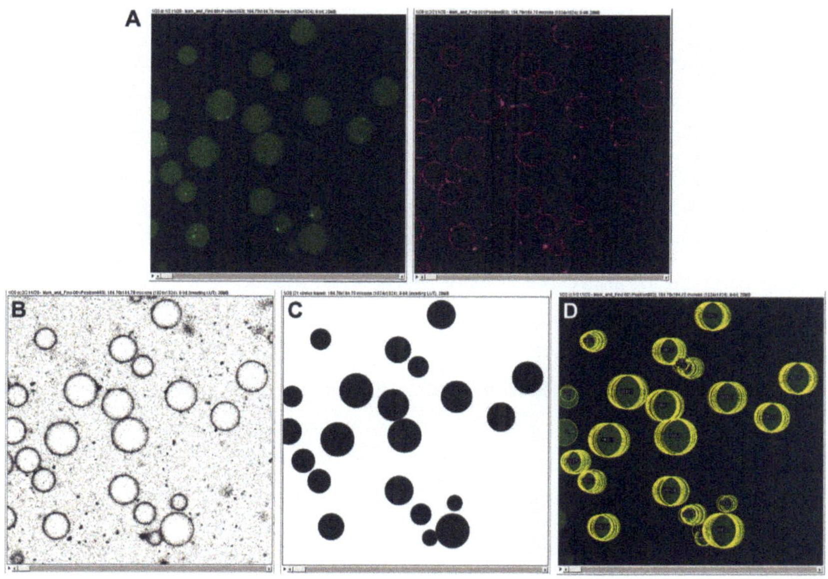

Fig. 12 Process of analyzing a population of protocells. (A) Two channel LSCM micrograph of a population of coacervates. Left: channel showing the cargo to be analyzed. Right: channel showing the protocell membrane. (B) Membrane image turned into a binary image by the threshold function. (C) Image showing all the particles found by the script. (D) Identified particles over time overlaid with the cargo image.

5. Select the "Circle outline overlaid on the original mask image" and press ok to run the plugin
Note: Depending on the background noise, the plugin might draw a circle on that noise and therefore miss a particle. An easy work around is to increase the maximum number of particles to be found.
6. Apply a threshold on the output image and select the overlaid circles only
7. Fill the circles by *Process > Binary > Fill Holes* and apply Watershed (*Process > Binary > Watershed*) in order to separate particles that are in contact with each other (Fig. 12C)
8. Use the analyze particles tool (*Analyze > Analyze Particles*) and select the "add to manager" and "include holes" options
Note: Set the size to the expected area of the particles based on the previously used radius.
9. Open the corresponding "cargo" image and *measure* in the manager (Fig. 12D)
10. The data in the second column (*mean*) can be used for further analysis
Note: ImageJ can output many other parameters if needed.

One must keep in mind this method is not completely foolproof and there are a few points the user should look into. First off, if particles surpass each other in Y-axis of the image this will lead to a different numbering of the particles. It is thus important to check if the overlay created by ImageJ is an accurate representation of your data. Additionally, over time particles might move out of focus or the field of view during the measurement, thus there is incomplete information of these particles over time. It is important to keep the type of analysis that is performed in mind, e.g., tracking a change inside an individual droplets or the entire population. Moreover, all the steps and/or plugin used to do this image analysis are quite common. Analysis with different programs such as Matlab, Mathematica or a Python script should be able to perform the same type of image analysis. Lastly, although this approach is written for a stack of images, it can also be applied to a single two channel confocal micrograph.

5.4 Fluorescence-activated cell sorting (FACS)

Flow cytometry is a method that is commonly used for counting and characterizing populations of cells (Givan, 2011). In this process, a sample (e.g., a suspension of cells) is hydrodynamically focused so preferably one cell at a time is shortly illuminated by a laser beam. The light that is scattered from these cells can be divided into forward scatter (FSC) and sideward scatter (SSC). FSC is generated when light diffracts around the cell and is detected along the path of the laser, which allows for the categorization of cells by size. In contrast, SSC is detected at a certain angle relative to the laser and provides information regarding the complexity, or granularity, of a cell. SSC is generated when light meets these intracellular structures and is refracted or reflected. In cell biology, FSC and SSC is mostly used for discrimination of different immune cells. In addition to FSC and SSC, fluorescently labeled cells can be categorized based on their fluorescent signal, which is used in a wide variety of cell based assays.

Similar to cells, synthetic micron-sized vesicles can also be characterized using flow cytometry. In these studies, mostly small or giant unilamellar vesicles (SUVs or GSUVs), were used. Childers et al. was the first to demonstrate that FSC analysis can be used to efficiently study the relative size and uniformity of SUVs (Childers et al., 1989). Over the last years, more advanced flow cytometry methods based on a fluorescence readout were developed to study processes in these liposome-based artificial cell populations (e.g., protein lipid binding (Temmerman & Nickel, 2009), reverse transcription polymerase chain reaction (Tsugane & Suzuki, 2018), and transmembrane signaling (Su et al., 2019)). In the latter, Su et al. used flow cytometry analysis to quantify enzyme activity in GUVs by measuring the catalytic breakdown of a membrane permeable fluorogenic substrate. The enzyme, β-glucuronidase (GUS), was activated following extracellular ligand binding and subsequent intracellular dimerization.

Contrary to these examples, coacervates-based artificial cells have, to the best of our knowledge, never been characterized using flow cytometry. The reason for this likely being their inherent instability, thereby not being able to withstand the high flow rates used in flow cytometry. However, flow cytometry has been proven to be a very high-throughput method for quantitative characterization of particles; while over 10,000 particles per minute can be easily analyzed using flow cytometry, it would be very laborious to obtain the same amount of data using confocal microscopy. Herein we developed a method to characterize terpolymer-stabilized coacervates-artificial cells using flow cytometry. We were able to study their size distribution using FSC (Fig. 13A and B), discriminate between coacervate populations based on fluorescence (Fig. 13C–E), and study protein uptake (Fig. 13F and G).

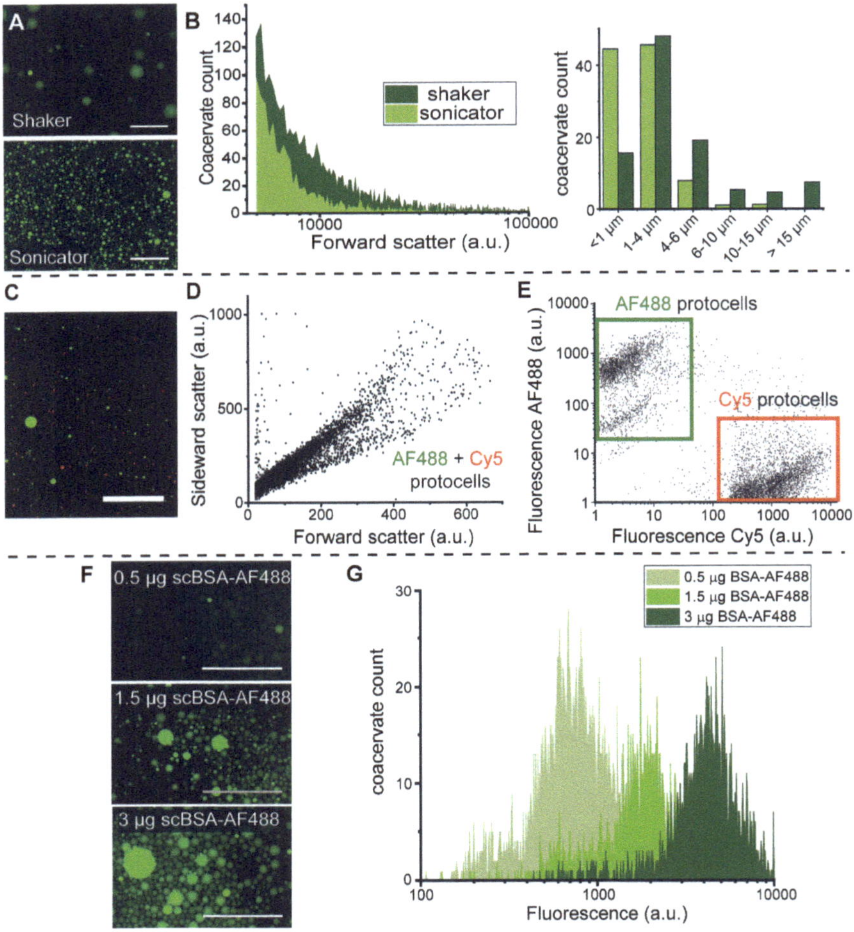

Fig. 13 Flow cytometry analysis of terpolymer-stabilized coacervates. (A) Microscopic image of coacervates loaded with scBSA-AF488 prepared either by shaking or sonicating. Scale bar represents 50 μm. (B) Left: histogram showing the FSC of coacervates prepared either by shaking (dark green) or sonicating (light green). Singlet coacervates were gated based on their FSC and fluorescence. Right: size distribution of coacervates prepared either by shaking (dark green) or sonicating (light green) relative to size calibration beads with size range (1–15 μm). (C) Microscopic image of coacervates loaded with either scBSA-AF488 or scBSA-Cy5 prepared by shaking. Scale bar represents 200 μm. (D) Dotplot depicting FSC and SSC for coacervates loaded with either scBSA-AF488 or scBSA-Cy5, highlighting the lack of discrimination. (E) Dotplot depicting Cy5 and AF488 fluorescence for coacervates loaded with either scBSA-AF488 or scBSA-Cy5, with clear separation observed. (F) Microscopic image of coacervates loaded with 0.5–3 μg scBSA-AF488. Scale bar represents 200 μm. (G) Histogram showing the intensity in fluorescent signal for coacervates loaded with 0.5–3 μg scBSA-AF488. For all measurements, a minimal of 1000 events were recorded.

Materials

- Q-Am 1 mg/mL in PBS
- CM-Am, 1 mg/mL in PBS
- Terpolymer, 50 mg/mL in PEG350
- Succinylated, Cy5 modified BSA (scBSA-Cy5)
- Succinylated, Alexa Fluor 488 modified BSA (scBSA-AF488)
- Flow Cytometry Size Calibration Kit (non-fluorescent polystyrene microspheres)
- 2 M sucrose solution

Method

1. Form coacervate protocells as per the method in Section 3.1, making two different populations, encapsulating either scBSA-Cy5 (10 μg) or scBSA-AF488 (0.5–3 μg).

   Note: In order to define background noise coming from possible aggregated protein or dust particles, a control is always prepared existing of PBS or a solution of succ-BSA-Cy5 or succ-BSA-AF488 in PBS
2. Dilute coacervate protocells in 2 M sucrose solution to prevent fast precipitation of the coacervates during flow cytometry analysis
3. Terpolymer-stabilized coacervates, controls, and size calibration beads were analyzed using a fluorescence activated cell sorter (FACS) ARIA III instrument (Becton Dickinson)
4. Data was processed using FACSDiva software (Becton Dickinson). In FACSDiva create the following dotplots:
   a) FSC-H vs. FSC-A and FSC-H vs. FSC-W to discriminate doublets and singlets
   b) FSC-H vs. SSC-H to discriminate background noise that has very few SSC from particles

   Note: Doublets are clustered particles that have a higher FSC-A and FSC-W values compared to FSC-H values due to their non-spherical morphology. Singlets are single spherical particles that have FSC-A and FSC-W values similar to FSC-H values. Exclusion of doublets is necessary as they can falsely give a higher fluorescent output per particle than expected.
5. In FACSDiva also create histograms for FSC-H and fluorescence channels. Cy5 was excited using a 633 nm laser and emission was detected between 650 and 670 nm. AF488 was excited using a 488 nm laser and emission was detected between 515 and 545 nm

6. Measure a coacervate sample and set the area scaling factor and the detector voltage of the FSC and SSC so that coacervates are visible within the dotplots. Measure also a control sample and use this to gate the singlet coacervates particles

   Note: The area scaling factor is introduced by Becton Dickson since the area of FSC and SSC is a calculated parameter. This is specific for FACSDiva software and should be adjusted dependent on the average size of the measured particles to ensure proper doublet discrimination
7. In order to determine the size distribution of the coacervates population measure size calibration beads with sizes ranging from 1 to 15 μm using the same FSC detector voltage

6. Conclusions and future perspectives

Over the last several years we have developed a new synthetic cell platform which offers researchers a unique combination of properties, where the advantages conferred by a liquid-liquid phase separated core are retained and discretized within a polymer-based membrane. With the methods described herein, other research groups should be able to reproduce this synthetic system in their own lab and begin to explore the many possibilities available by incorporating new and different kinds of functional cargo. Moreover, the techniques described here should be applicable to a range of other cell-sized stable protocellular platforms.

Acknowledgments

David S. Williams is acknowledged for his expertise in developing the original amylose modification and terpolymer synthesis protocols. The Dutch Ministry of Education, Culture and Science (Gravitation program 024.001.035) and the ERC Advanced grant Artisym 694120 are acknowledged for funding.

References

Buddingh, B. C., & van Hest, J. C. M. (2017). Artificial cells: Synthetic compartments with life-like functionality and adaptivity. *Accounts of Chemical Research*, *50*(4), 769–777. https://doi.org/10.1021/acs.accounts.6b00512.

Childers, N. K., Michalek, S. M., Eldridge, J. H., Denys, F. R., Berry, A. K., & McGhee, J. R. (1989). Characterization of liposome suspensions by flow cytometry. *Journal of Immunological Methods*, *119*(1), 135–143. https://doi.org/10.1016/0022-1759(89)90390-6.

Couffin, A., Delcroix, D., Martín-Vaca, B., Bourissou, D., & Navarro, C. (2013). Mild and efficient preparation of block and gradient copolymers by methanesulfonic acid catalyzed ring-opening polymerization of caprolactone and trimethylene carbonate. *Macromolecules*, *46*(11), 4354–4360. https://doi.org/10.1021/ma400916k.

Douliez, J.-P., Martin, N., Gaillard, C., Beneyton, T., Baret, J.-C., Mann, S., et al. (2017). Catanionic coacervate droplets as a surfactant-based membrane-free protocell model. *Angewandte Chemie, 129*(44), 13877–13881. https://doi.org/10.1002/ange.201707139.

Fanalista, F., Deshpande, S., Lau, A., Pawlik, G., & Dekker, C. (2018). FtsZ-induced shape transformation of coacervates. *Advanced Biosystems, 2*(9), 1800136. https://doi.org/10.1002/adbi.201800136.

Givan, A. L. (2011). Flow cytometry: An introduction BT. In T. S. Hawley & R. G. Hawley (Eds.), *Flow cytometry protocols* (pp. 1–29). Totowa, NJ: Humana Press. https://doi.org/10.1007/978-1-61737-950-5_1.

Hutchison, C. A., Chuang, R.-Y., Noskov, V. N., Assad-Garcia, N., Deerinck, T. J., Ellisman, M. H., et al. (2016). Design and synthesis of a minimal bacterial genome. *Science*, 351(6280), aad6253. https://doi.org/10.1126/science.aad6253.

Kumar, P. B. V. V. S., Fothergill, J., Bretherton, J., Tian, L., Patil, A. J., Davis, S. A., et al. (2018). Chloroplast-containing coacervate micro-droplets as a step towards photosynthetically active membrane-free protocells. *Chemical Communications, 54*(29), 3594–3597. https://doi.org/10.1039/C8CC01129J.

Lawrence, M. S., Phillips, K. J., & Liu, D. R. (2007). Supercharging proteins can impart unusual resilience. *Journal of the American Chemical Society, 129*(33), 10110–10112. https://doi.org/10.1021/ja071641y.

Lindberg, B., Lindquist, U., & Stenberg, O. (1987). Distribution of substituents in O-(2-hydroxyethyl)cellulose. *Carbohydrate Research, 170*, 207–214. https://doi.org/10.1016/S0008-6215(00)90905-4.

Magdalena Estirado, E., Mason, A. F., Alemán García, M. Á., van Hest, J. C. M., & Brunsveld, L. (2020). Supramolecular nanoscaffolds within cytomimetic protocells as signal localization hubs. *Journal of the American Chemical Society, 142*(20), 9106–9111. https://doi.org/10.1021/jacs.0c01732.

Mason, A. F., Buddingh, B. C., Williams, D. S., & van Hest, J. C. M. (2017). Hierarchical self-assembly of a copolymer-stabilized coacervate protocell. *Journal of the American Chemical Society, 139*(48), 17309–17312. https://doi.org/10.1021/jacs.7b10846.

Mason, A. F., & van Hest, J. C. M. (2019). Multifaceted cell mimicry in coacervate-based synthetic cells. *Emerging Topics in Life Sciences, 3*(5), 567–571. https://doi.org/10.1042/ETLS20190094.

Mason, A. F., Yewdall, N. A., Welzen, P. L. W., Shao, J., van Stevendaal, M., van Hest, J. C. M., et al. (2019). Mimicking cellular compartmentalization in a hierarchical protocell through spontaneous spatial organization. *ACS Central Science, 5*(8), 1360–1365. https://doi.org/10.1021/acscentsci.9b00345.

Oparin, A. I. (1953). *The origin of life* (2nd ed.). Mineola, NY: Dover Publications, Inc.

Pir Cakmak, F., Grigas, A. T., & Keating, C. D. (2019). Lipid vesicle-coated complex coacervates. *Langmuir*, 35(24), 7830–7840, Research-article. https://doi.org/10.1021/acs.langmuir.9b00213.

Poudyal, R. R., Pir Cakmak, F., Keating, C. D., & Bevilacqua, P. C. (2018). Physical principles and extant biology reveal roles for RNA-containing membraneless compartments in origins of life chemistry. *Biochemistry, 57*(17), 2509–2519. https://doi.org/10.1021/acs.biochem.8b00081.

Sokolova, E., Spruijt, E., Hansen, M. M. K., Dubuc, E., Groen, J., Chokkalingam, V., et al. (2013). Enhanced transcription rates in membrane-free protocells formed by coacervation of cell lysate. *Proceedings of the National Academy of Sciences of the United States of America, 110*(29), 11692–11697. https://doi.org/10.1073/pnas.1222321110.

Spector, S., Wang, M., Carp, S. A., Robblee, J., Hendsch, Z. S., Fairman, R., et al. (2000). Rational modification of protein stability by the mutation of charged surface residues. *Biochemistry, 39*(5), 872–879. https://doi.org/10.1021/bi992091m.

Su, J., Kitaguchi, T., Ohmuro-Matsuyama, Y., Seah, T., Ghadessy, F., Hoon, S., et al. (2019). Transmembrane signaling on a protocell: Creation of receptor-enzyme chimeras for immunodetection of specific antibodies and antigens. *Scientific Reports*, *9*, 18189. https://doi.org/10.1038/s41598-019-54539-7.
Szostak, J. W., Bartel, D. P., & Luisi, P. L. (2001). Synthesizing life. *Nature*, *409*(6818), 387–390. https://doi.org/10.1038/35053176.
Tang, T.-Y. D., Hak, C., Thompson, A. J., Kuimova, M. K., Williams, D. S., Perriman, A. W., et al. (2014). Fatty acid membrane assembly on coacervate micro-droplets as a step towards a hybrid protocell model. *Nature Chemistry*, *6*(6), 527–533. https://doi.org/10.1038/nchem.1921.
Tang, T.-Y. D., van Swaay, D., DeMello, A., Ross Anderson, J. L., & Mann, S. (2015). In vitro gene expression within membrane-free coacervate protocells. *Chemical Communications*, *51*(57), 11429–11432. https://doi.org/10.1039/C5CC04220H.
te Brinke, E., Groen, J., Herrmann, A., Heus, H. A., Rivas, G., Spruijt, E., et al. (2018). Dissipative adaptation in driven self-assembly leading to self-dividing fibrils. *Nature Nanotechnology*, *13*(9), 849–855. https://doi.org/10.1038/s41565-018-0192-1.
Temmerman, K., & Nickel, W. (2009). A novel flow cytometric assay to quantify interactions between proteins and membrane lipids. *Journal of Lipid Research*, *50*(6), 1245–1254. https://doi.org/10.1194/jlr.D800043-JLR200.
Tsugane, M., & Suzuki, H. (2018). Reverse transcription polymerase chain reaction in giant unilamellar vesicles. *Scientific Reports*, *8*(1), 9214. https://doi.org/10.1038/s41598-018-27547-2.
Wulff, G., Steinert, A., & Höller, O. (1998). Modification of amylose and investigation of its inclusion behavior. *Carbohydrate Research*, *307*(1–2), 19–31. https://doi.org/10.1016/S0008-6215(97)10097-0.
Yewdall, N. A., Buddingh, B. C., Altenburg, W. J., Timmermans, S. B. P. E., Vervoort, D. F. M., Abdelmohsen, L. K. E. A., et al. (2019). Physicochemical characterization of polymer-stabilized coacervate protocells. *Chembiochem*, *20*(20), 2643–2652. https://doi.org/10.1002/cbic.201900195.
Yewdall, N. A., Mason, A. F., & van Hest, J. C. M. (2018). The hallmarks of living systems: Towards creating artificial cells. *Interface Focus*, *8*(5), 20180023. https://doi.org/10.1098/rsfs.2018.0023.
Zhang, Z., Tan, M., Xie, Z., Dai, L., Chen, Y., & Zhao, Y. (2011). Identification of lysine succinylation as a new post-translational modification. *Nature Chemical Biology*, 7(1), 58–63. https://doi.org/10.1038/nchembio.495.
Zou, J., Fan, J., He, X., Zhang, S., Wang, H., & Wooley, K. L. (2013). A facile glovebox-free strategy to significantly accelerate the syntheses of well-defined polypeptides by N-carboxyanhydride (NCA) ring-opening polymerizations. *Macromolecules*, *46*(10), 4223–4226. https://doi.org/10.1021/ma4007939.

CHAPTER FOUR

Cell-free reconstitution of multi-condensate assemblies

Andrea Putnam* and Geraldine Seydoux
Department of Molecular Biology and Genetics, Howard Hughes Medical Institute, Johns Hopkins University, School of Medicine, Baltimore, MD, United States
*Corresponding author: e-mail address: aputnam5@jhmi.edu

Contents

Abstract

Biomolecular condensates (BCs) are intracellular condensates that form by phase separation of proteins and RNA from the nucleoplasm or cytoplasm. BCs often form complex assemblies where compositionally distinct condensates wet each other without mixing. In this chapter, we describe methods to reconstitute multi-condensate assemblies from purified components. We include protocols to express, purify, label, and analyze the dynamics of proteins and RNAs that drive multi-condensate assembly. Analysis of the condensation and wetting behaviors of condensates in cell-free reconstituted systems can be used to define the molecular interactions that regulate BCs in cells.

Methods in Enzymology, Volume 646
ISSN 0076-6879
https://doi.org/10.1016/bs.mie.2020.07.004

1. Introduction

Biomolecular condensates (BCs) are dynamic local concentrations of proteins and/or nucleic acids in cells (Boeynaems et al., 2018; Hyman, Weber, & Jülicher, 2014). BCs are thought to assemble by liquid-liquid phase separation driven by low affinity, multivalent interactions between BC components. BCs exhibit a range of material properties from liquid to solid (Alberti, Gladfelter, & Mittag, 2019; Boke et al., 2016; Jain et al., 2016; Rai, Chen, Selbach, & Pelkmans, 2018). Characterization of the proteomes and transcriptomes of BCs have revealed highly complex compositions with as many as 100–1000s of proteins and RNAs co-localized in BCs (Boisvert, van Koningsbruggen, Navascués, & Lamond, 2007; Ditlev, Case, & Rosen, 2018; Hubstenberger et al., 2017; Jain et al., 2016; Khong et al., 2017; Lee et al., 2020). In many cases, it is unclear which components drive BC assembly ("scaffolds") and which components are merely "clients" that localize to BCs (Banani et al., 2016). In cells, compositionally distinct BCs often wet each other's surfaces to form complex multi-condensate assemblies (Feric et al., 2016; Jain et al., 2016; Putnam, Cassani, Smith, & Seydoux, 2019; Wang et al., 2014; West et al., 2016). The complex composition and wetting behaviors of BCs make it difficult to derive principles of BC assembly from *in vivo* observations alone. *In vitro* reconstitution of minimal BCs from purified components can provide critical insights into assembly mechanisms. Reconstitution experiments are most informative when performed with proteins demonstrated to function as scaffolds for BC assembly *in vivo* (BC proteins). In this chapter, we provide methods for expressing and purifying BC proteins and for reconstitution and analysis of multi-condensate assemblies.

The methods presented in this chapter were developed to characterize the assembly of BC assemblies in the *C. elegans* germline called P granules. P granules are RNA-rich BC assemblies that segregate with the germ lineage during the first cell divisions of *C. elegans* embryos (Marnik & Updike, 2019; Seydoux, 2018). The function of P granules is to enrich mRNAs and epigenetic determinants in the germline founder cells. These two functions are mediated by two distinct condensates: MEG condensates, which enrich mRNAs, and PGL condensates, which enrich epigenetic determinants (Lee et al., 2020; Ouyang et al., 2019; Putnam et al., 2019; Wang et al., 2014). MEG condensates are assembled by the intrinsically disordered protein MEG-3 and its paralog MEG-4, and PGL condensates are assembled by

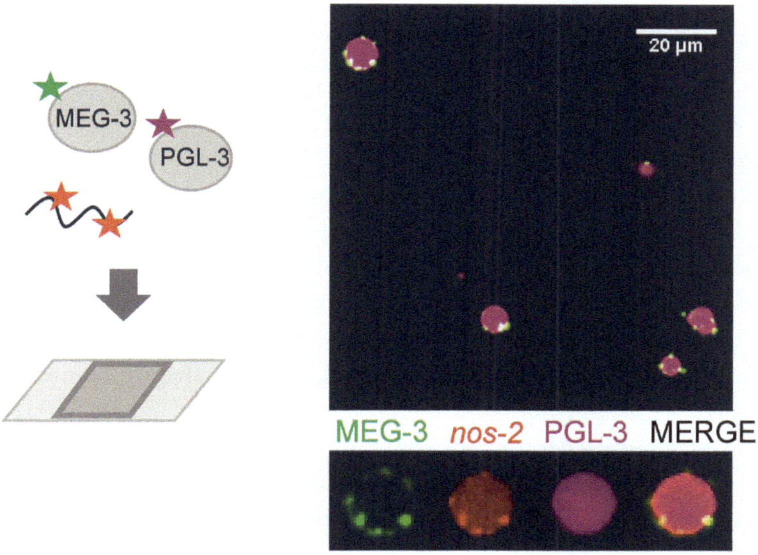

Fig. 1 Condensation reaction of fluorescently labeled MEG-3-Dylight488 (150 nM), PGL-3-Alexa647 (5000 nM), and *nos-2* RNA-Alexa546 (20 ng/μL) incubated for 30 min and visualized on a thin chambered glass slide with a 40 × objective. MEG-3 particles assemble on the surface of PGL-3 condensates, *nos-2* RNA is recruited to both phases, but is enriched in the MEG-3 phase.

the mostly globular proteins PGL-1 and its paralog PGL-3. In embryos, the two condensates form tight assemblies with a single PGL condensate at the center, and multiple MEG condensates at the periphery (Fig. 1). PGL condensates behave like liquid droplets, readily dissolving when exposed to high temperature or to a dilute environment (Brangwynne et al., 2009; Putnam et al., 2019; Saha et al., 2016). In contrast, MEG condensates are thermostable and resist dilution (Putnam et al., 2019). PGL and MEG condensates form independently of each other but require each other for normal dynamics and localization in embryos. Using purified PGL-3, MEG-3 and *in vitro* synthesized RNA, we have reconstituted a minimal P granule *in vitro* with the same core/shell architecture characteristic of P granules assembled *in vivo* (Putnam et al., 2019). Like other BC proteins, PGL-3 and MEG-3 present unique challenges for purification and handling, which we describe in Section 3. Protocols for purification and labeling of BC proteins are presented in Sections 3 and 4, protocols for *in vitro* synthesis and labeling of RNA in Section 4, and reconstitution assays in Section 5. Finally, we outline methods to visualize and analyze the dynamics of multi-condensate assemblies (Section 6).

2. Considerations when working with BC proteins

2.1 Protein expression

Because of their condensation properties, BC proteins frequently exhibit low solubility in aqueous buffers especially at high concentrations (Alberti et al., 2018). Many BC proteins are not expressed as soluble proteins in bacteria and instead become concentrated in inclusion bodies. Inclusion bodies, however, are easily purified away from other bacterial structures and can yield large concentrations of highly pure protein (Palmer & Wingfield, 2004). We provide methods for purifying proteins from inclusion bodies in Section 3.1. In this approach, the protein is denatured in urea before purification and returned to an aqueous buffer after purification. Intrinsically disordered proteins with a high tendency to form condensates can be stored in urea and diluted out of urea for condensation assays. Proteins with globular domains are refolded post-purification by step-dialysis out of urea as described in Section 3.1, but some proteins may be difficult to refold properly. Native purification may be preferable for those proteins.

Purification of BC proteins under native conditions is possible by attaching a solubilizing tag. Solubilizing tags can increase solubility of BC proteins both in bacteria and during purification. Maltose binding protein (MBP) is a particularly efficient solubilizing tag that has been used for the purification of several disordered proteins including PGL-3 (Alberti et al., 2018; Wang et al., 2018; Wang, Zhang, & Zhang, 2019). Because solubilizing tags affect condensation behavior, they must be removed before the purified protein is used in condensate assays. The sequence-specific TEV protease is a useful reagent for this step. In Section 3.2, we provide methods for native purification of MEG-3 and PGL-3 proteins expressed in *E. coli* as soluble MBP fusions. For highly condensing proteins, native purification may result in low protein yield and may be difficult to scale up. In those cases, native purification is still recommended as a secondary purification method to confirm results obtained with protein purified under denaturing conditions. Native purification of condensate proteins has also been reported using proteins expressed in insect cells (Alberti et al., 2018). This approach, however, presents two disadvantages. First, proteins purified from insect cells often contain post-translational modifications that may affect condensation behavior. Second, insect cells require equipment for tissue culture, which may not be accessible. All of the protocols presented in this chapter use proteins expressed in *E. coli*.

2.2 Protein purification

The second challenge when purifying BC proteins is to avoid condensation during purification. Condensation can cause the protein to exhibit anomalous behavior in purification columns and/or to bind non-specifically to tube and column surfaces (Alberti et al., 2018; Elbaum-Garfinkle et al., 2015). Condensation can be monitored by visual examination of the protein solution: solutions containing condensates appear turbid or cloudy. Protein concentration, buffer ionic strength, and temperature are all factors that influence protein solubility and that should be carefully controlled during purification. In general, we recommend conducting most purification steps under high salt conditions and at room temperature. Purification steps that require low salt such as binding to ion exchange columns or protease cleavage should be performed as quickly as possible. We have found that high concentrations of glycerol can delay or prevent phase separation during low salt steps. L-arginine has been reported to improve the solubility of arginine-rich proteins (Alberti et al., 2018). We find that addition of L-arginine to bacterial lysates also strongly improves binding of MBP-fusion proteins to the affinity matrix.

Maintaining protein concentration below the condensation threshold during purification is also important. Purification steps that increase protein concentration can cause the protein to phase separate either on the column or in the most concentrated fractions after elution. For proteins that form gels or aggregates, solutions may begin to solidify on the column or in tubes during storage. Proteins that have a strong tendency to condense, such as the mostly disordered protein MEG-3, may do so even when fused to MBP. To minimize this problem, we recommend performing batch purification using large resin volumes to maintain low protein concentration. Additionally, we recommend monitoring elution steps carefully. If a peak fraction begins to phase separate (become cloudy), condensation can sometimes be reversed by immediately diluting the fraction to a lower concentration or increasing the salt or temperature.

2.3 Labeling proteins and RNAs

Several techniques have been developed to monitor protein condensation *in vitro*, including turbidity and sedimentation assays, as well as direct imaging of droplets by light microscopy (Ceballos, McDonald, & Elbaum-Garfinkle, 2018; Elbaum-Garfinkle et al., 2015; Keenen, Larson, & Narlikar, 2018; Wang et al., 2019). When assembling condensates using multiple components, however, it is preferable to use methods that allow each component to be monitored independently. A convenient way to achieve this is to attach to a fluorescent tag with a unique emission spectrum to each component. Many condensation assays reported in the literature use fluorescent proteins as fluorescent tags for BC proteins. Although convenient, we do not recommend this approach since fluorescent proteins alter condensation behavior both *in vivo* and *in vitro* (Feric et al., 2016; Uebel & Phillips, 2019). GFP in particular, a well folded globular protein, can function as a solubilizing tag, which may be useful during protein purification, but is not suitable for condensation assays. As an alternative to protein tags, we recommend labeling proteins directly with fluorescent dyes. This approach provides several advantages: (1) dyes are attached covalently post-purification, (2) multiple color dyes are available for flexible labeling of multiple components, and (3) due to their high fluorescence yield, dyes need only be attached to a fraction of the molecules used in the condensation assay ("trace label"). We describe a facile protocol for labeling BC proteins with fluorescent dyes in Section 4.1. To label RNA, we recommend the use of fluorescent nucleotides for incorporation during the transcription reaction. An alternative approach is to use short labeled oligonucleotides as indirect probes for longer unlabeled RNAs (Tauber et al., 2020), but this approach is not recommended since oligonucleotides and RNA: RNA duplexes can alter the properties of condensates (Nott, Craggs, & Baldwin, 2016).

2.4 Assembling condensates

Designing condensation reactions for multicomponent systems can be challenging. Condensation of each component may be affected by salt, pH, temperature, and concentration (of itself and other components) in different ways. We recommend creating phase diagrams for each BC protein in isolation first by monitoring phase separation in varying concentrations of protein, salt and/or RNA (Ceballos et al., 2018). Keep in mind that phase separated systems can display re-entrant phase transitions where components like RNA can enhance phase separation at low concentration and dissolve droplets at high concentration (Banerjee, Milin, Moosa, Onuchic, & Deniz, 2017; Lee et al., 2020; Maharana et al., 2018). Because it is impractical when working with complex assemblies to vary all variables for all BC proteins systematically, we recommended starting with physiological concentrations of components. The relative molar ratio of each component is particularly important since relative concentration can dramatically alter the partitioning behavior of BC proteins and the wetting behavior of condensates (Banani et al., 2016; Feric et al., 2016). Choosing an appropriate condensate buffer is also critical. We recommended starting with a minimal buffer at physiological salt and pH. Additives that promote phase separation by inducing molecular crowding such as polyethylene glycol, ficoll, or dextran should be used with caution. Although the cell is a crowded environment, it is unclear how well commonly used crowding agents mimic crowding induced by biomolecules like protein and nucleic acid (Kuznetsova, Turoverov, & Uversky, 2014; Protter et al., 2018). In our experience, we have found that crowding agents are not needed when working with BC proteins demonstrated to drive condensation *in vivo*. When building multi-condensate assemblies, it is also critical to pay attention to the order of addition of components, since order can affect both condensation and wetting behavior. A good understanding of condensate wetting behavior in cells is critical to recreate physiologically relevant assemblies *in vitro*. In Section 5, we provide a protocol for the assembly of a three-component minimal P granule, which reproduces the core-shell architecture of P granules in embryos (Putnam et al., 2019; Wang et al., 2014).

2.5 Monitoring condensate dynamics

When BC proteins are first induced to condense (by buffer change or release from a solubilizing tag), phase separation initially results in the formation of many small condensates. If liquid, these condensates grow into fewer larger condensates through coalescence and Ostwald ripening, a thermodynamic process whereby small droplets lose material to larger droplets (Hyman et al., 2014). The material properties of liquid condensates can also evolve with some liquids hardening over time into gel-like droplets that stop growing (Boeynaems et al., 2018). Other BC proteins form condensates or aggregates that are immediately solid and do not grow (Boke et al., 2016; Putnam et al., 2019). Solid condensates can also evolve over time if allowed to interact and "stick" to each other to form larger assemblies. For these reasons, care should be taken to standardize observations with respect to time from the onset of the condensation reaction.

Observations of condensate dynamics include measurements of droplet number, size, wetting, and monomer exchange rate (frequently measured by FRAP). Material properties can also be measured by sensitivity to disruption. Liquid droplets, which are held together by networks of weak interactions, typically are disrupted by dilution, salt, pH, hexanediol, ATP and temperature changes (Kroschwald, Maharana, & Simon, 2017; Patel et al., 2017; Riback et al., 2017). Droplets that harden into gels are typically resistant to these factors but sensitive to SDS, whereas amyloid-like aggregates typically resist all treatments including SDS (Kryndushkin, Pripuzova, Burnett, & Shewmaker, 2013). Monitoring changes in condensate dynamics upon omission or mutation of specific components can provide key insights into the molecular interactions driving assembly. Insights gained from such reconstitution experiments can then be used to design cell-based experiments to further test hypotheses.

3. Preparation of recombinant MEG-3 and PGL-3 proteins

3.1 Purification under denaturing conditions

In this protocol we describe a method to purify proteins expressed in bacterial inclusion bodies. Native purification of highly condensing proteins is often not practical when the goal is to obtain microgram amounts of protein, especially for proteins that tend to condense even as MBP-fusions. For those proteins, we recommend non-native purification from protein recovered from bacterial inclusion bodies. The protein is tagged with a 6 × His-tag used for affinity purification from inclusion bodies dissolved in urea. The purified denatured protein can be used for condensation assays directly out of urea or refolded into its native configuration using a multistep dialysis procedure.

3.1.1 Materials and buffer recipes

Expression *E. coli*: Rosetta (DE3)
Antibiotics: Kanamycin
IPTG (GoldBio)
5 mL HisTRAP column (GE Healthcare)
Size exclusion column (GE Healthcare)
Terrific broth (see Section 3.1.3)
FPLC
Lysis: 20 mM HEPES pH 8, 1000 mM KCl, 10% glycerol, 0.5% Triton-X100, 0.4 mM PMSF, Roche Proteinase Inhibitors, 2 mM DTT
Solubilization: 20 mM HEPES pH 8, 1000 mM KCl, 10% glycerol, 0.5% Triton-X100, 20 mM imidazole, pH 7.5, 6 M urea 0.4 mM PMSF, Roche Proteinase Inhibitors, 2 mM DTT

Denaturing-His-Wash buffer: 20 mM HEPES pH 8, 1000 mM KCl, 10% glycerol, 25 mM imidazole, pH 7.5, 6 M urea, 2 mM DTT
Denaturing-His-Elution buffer: 20 mM HEPES pH 7.4, 1000 mM KCl, 10% glycerol, 250 mM imidazole, pH 7.5, 6 M urea, 2 mM DTT
Size Exclusion buffer: 20 mM HEPES pH 8, 1000 mM KCl, 10% glycerol, 6 M urea, 2 mM DTT
Dialysis buffer: 20 mM HEPES pH 8, 1000 mM KCl, 10% glycerol, 0 M urea, 2 mM DTT

3.1.2 Overexpression and purification

1. *Overexpression*: An expression plasmid (pET28a+) containing the 6×-His-MEG-3 protein is transformed into Rosetta (DE3) competent cells using recommended procedures. Transformed cells are inoculated in a ~10 mL overnight culture. The overnight stock is used to inoculate 1 L cultures of terrific broth (TB) which are then shaken at 37 °C for approximately 5 h until reaching an OD_{600} of ~1. Protein expression is induced by addition of 1 mM IPTG and shaking overnight at 16 °C for ~16 h.
2. *Harvest cells*: Centrifuge cultures at 4000 × *g* for 15 min @ 4 °C. Discard the supernatant. In general, we have found that for most proteins a 0.5 L of culture yields sufficient protein for purification. Use one pellet ~4 g (from 500 mL) for purification, and store a second pellet at −80 °C for future use.
3. *Cell lysis*: Resuspend cell pellets 1 × Lysis buffer (5 mL of Lysis buffer per gram of cells). Lyse the cell suspension on ice by sonication until cells are completely lysed. Transfer the lysed cells to 50 mL centrifuge tubes and centrifuge @ 25,000 × *g* for 30 min @ 4 °C. Remove the supernatant and keep the pellet. Resuspend the pellet in 20 mL of Lysis buffer. Repeat sonication and centrifugation steps. The pellet should be mostly white. If the pellet still appears brown/tan sonicate for additional cycles to fully lyse cells. Save samples of the crude lysate and the clarified supernatant to confirm by SDS-PAGE analysis that the expressed protein is in inclusion bodies. Pellets can be stored at −80 °C at this point.
4. *Inclusion body resuspension*: Resuspend pellet in 45 mL of solubilization buffer and transfer to a small beaker. Stir overnight at 4 °C. Transfer solubilized inclusion body pellets to 50 mL centrifuge tubes and centrifuge @ 25,000 × *g* for 30 min. Pass the solution through a 0.45 μm filter.
5. *His affinity*: The filtered supernatant is passed over a 5 mL HisTRAP column (GE Healthcare), and washed with five column volumes of

His-wash buffer. To elute the protein, wash with five column volumes of His-elution buffer. Pool elution fractions, and concentrate to ~2 mL.

6. *Size exclusion chromatography*: Inject the concentrated protein sample onto an S300 gel-filtration column. Examine the purity of each elution fraction by SDS-PAGE. Pool pure fractions and concentrate to ~2 mg/mL. Aliquot and store the protein at −80 °C. Protein in the urea storing buffer can be used directly for condensation experiments by dilution into condensation buffer. Alternately urea can be removed from the protein by dialysis before use.
7. *Dialysis*: Transfer 50–200 μL of protein to a mini-dialysis chamber. Mix size exclusion buffer with dialysis buffer to obtain dialysis buffers with decreasing concentrations of urea. Dialyze the protein solution at 4 °C in 4.5 M urea (2 h), 3 M urea (2 h), 1.5 M urea (2 h), 0 M urea (2 h). The first dialysis step (4.5 M urea) can be left overnight for convenience. However, it is critical that the protein is not left in 0 M Urea at 4 °C for prolonged periods of time as it will form a gel. To store the protein, aliquot, flash freeze, and store at −80 °C. Thaw aliquots at room temperature, not on ice. We have found that MEG-3 occasionally can aggregate or form a gel during dialysis, and as a result we recommend small volumes for dialysis.

3.1.3 Terrific broth recipe

1. Prepare 10 × phosphate buffer:
   23.1 g KH_2PO_4
   125.4 g K_2HPO_4
   Add water to 1 L
   Check pH with filter paper (~7.2–7.5)
   Filter sterilize
2. Add 24 g yeast extract, 12 g tryptone, 8 mL of 50% glycerol, and 900 mL of water in a 2 L flask
3. Autoclave, and cool to room temperature
4. Add 100 mL of 10 × phosphate buffer

3.2 Purification under native conditions

In this protocol, we describe a method to natively purify MBP-fusion proteins. First, the ORF coding for the protein of interest is cloned downstream of MBP (tagged with 6 ×-His) and a cleavage site for TEV protease in an *E. coli* expression vector. The MBP fusions are purified from *E. coli* lysates using amylose affinity chromatography in high salt buffer which removes most bacterial proteins. For proteins that can be maintained stably in high salt buffer without the MBP tag, such as PGL-3, the MBP tag is cleaved immediately after the MBP affinity purification using a TEV protease tagged with 6×-His (6×-His-TEV). The cleavage products, 6×-His-MBP and 6×-His-TEV, are then removed using a histidine affinity chromatography. For proteins that cannot be maintained stably under native conditions without MBP, such as MBP-MEG-3, TEV cleavage is omitted during the purification procedure and used to initiate condensation reactions.

We have found that most RNA-binding proteins bind efficiently to Heparin, an RNA-like polymer. Heparin resins functions as both nucleic acid affinity and cation exchange chromatography. We use a heparin affinity column as a final purification step. We have found that this step significantly increases protein purity and can be used to further concentrate the protein.

3.2.1 Materials and buffer recipes

Expression *E. coli*: Rosetta 2
Terrific broth (see Section 3.1.3)
Antibiotics: Ampicillin
IPTG (GoldBio)
Amylose resin (NEB)
5 mL HisTRAP column (GE Healthcare)
5 mL Heparin column (GE Healthcare)
TEV protease (Addgene: pRK793)
FPLC
Lysis buffer: 50 mM HEPES pH 7.5, 1000 mM NaCl, 40% glycerol, 400 mM L-arginine, 1% IGEPAL, 0.4 mM PMSF, Roche Proteinase Inhibitors, 2 mM DTT
MBP-Wash buffer: 50 mM HEPES pH 7.5, 1000 mM NaCl, 20% glycerol, 2 mM DTT
MBP-Elution buffer: 50 mM HEPES pH 7.5, 1000 mM NaCl, 20% glycerol, 20 mM maltose, 2 mM DTT
Dilution buffer: 25 mM HEPES pH 7.5, 0 mM NaCl, 40% glycerol, 2 mM DTT, 0.1% IGEPAL
Heparin-0 M NaCl buffer: 25 mM HEPES pH 7.5, 0 mM NaCl, 20% glycerol, 2 mM DTT, 0.1% IGEPAL
Heparin-1000 M NaCl buffer: 25 mM HEPES pH 7.5, 1000 mM NaCl, 20% glycerol, 2 mM DTT, 0.1% IGEPAL

3.2.2 Construct design

MBP fusions are created in an pMAL-c2X-derived expression vector (Addgene: pJW4.01HT) where the MBP coding sequence is followed by a 6 × His-tag, a TEV protease site, and cloning site for the protein of interest. Expression of the MBP fusion is driven by an IPTG-inducible promoter. TEV protease can be purchased (Promega #V6101) or purified as a 6 × His-tag fusion (Addgene: pRK793). We recommend purifying TEV protease following a protocol developed by the Waugh lab that yields concentrated TEV protease with sufficient activity for use in 500 mM NaCl (Tropea, Cherry, & Waugh, 2009). Inclusion of 6 × His-tags on both TEV and MBP facilitate removal of the protease and MBP cleavage product after the TEV reaction by washing the reaction products through a HisTrap column.

3.2.3 Overexpression and purification

1. *Overexpression*: An expression plasmid containing the MBP-fusion protein is transformed into Rosetta 2 competent cells using recommended procedures. Transformed cells are inoculated in a ~10 mL overnight (TB) culture. This overnight stock is used to inoculate a 1 L culture of TB which is grown at 37 °C in a shaker incubator for approximately 5 h until reaching an OD_{600} of ~1. The culture is induced by addition of with 1 mM IPTG and shaking overnight at 16 °C for ~16 h.
2. *Harvest cells*: Centrifuge the cultures at 4000 × *g* for 15 min @ 4 °C, and discard the supernatant. For most proteins 0.5 L of culture is sufficient to purify sufficient protein for purification. Use one pellet ~4 g (from 500 mL) for purification and store a second pellet at −80 °C for future use.
3. *Cell lysis*: Resuspend cell pellets in 1 × Lysis buffer (5 mL of Lysis buffer per gram of cells). Lyse the cell suspension on ice by sonication until cells are completely lysed. Transfer the lysed cells to 50 mL centrifuge tubes and centrifuge @ 25,000 × *g* for 30 min @ 4 °C. Clarified supernatants can be stored at −80 °C or used directly. Run an aliquot of the lysate on SDS-PAGE to verify induction of the protein of interest.
4. *Temperature considerations*: Although protein purifications are traditionally performed at 4 °C, we have found that low temperatures increase the propensity of PGL-3 and MEG-3 to phase separate. Therefore we recommend performing all following purification steps (in particular TEV cleavage) at room temperature.
5. *Amylose affinity*: The clarified supernatant is incubated with Amylose resin (~10 mL) pre-equilibrated with MBP-wash buffer while rotating

for 1 h. The suspension is poured into a gravity flow column and washed with five column volumes of MBP-wash buffer. To elute the protein, slowly wash with elution buffer and collect 2 mL elution fractions until the protein is completely eluted. For MBP-TEV-PGL-3, we use 500 mM NaCl instead of 1 M NaCl in the lysis, MBP-wash, and MBP-elution buffers. Pool protein-containing elution fractions. If MBP protein is to be used without further purification, fractions should be aliquoted, snap frozen and stored at −80 °C.

6. *MBP cleavage and reverse nickel*: MBP and the 6 × His-tag are cleaved from the protein of interest by addition of TEV protease. We recommend using high concentrations of TEV protease for ~4 h at 16 °C room temperature. Longer incubation times do not typically result in more efficient cleavage of protein. The cleaved solution is then passed over a 5 mL HisTRAP column (GE Healthcare) to remove the protease and the cleaved MBP, and washed with an additional five CV of MBP-wash buffer. Pool protein-containing fractions.
7. *Heparin affinity*: Pool fractions containing the cleaved protein are diluted into low salt high glycerol buffer (final salt concentration of ~100 mM NaCl) and loaded onto a 5 mL heparin affinity column (GE Healthcare). We recommend dilution into low salt and high glycerol buffer instead of dialysis. We have found that this approach minimizes condensation of the protein during the heparin purification step. The protein is step-eluted in heparin buffer with NaCl at 100, 250, and 500 mM for PGL-3 and up to 800 mM for MEG-3. The protein is recovered in the final high salt elution. We have found that a step elution results in much higher final protein concentrations.
8. *Storage conditions*: Peak fractions are pooled, aliquoted and flash-frozen, and stored at −80 °C. In general, we recommend avoiding centrifugal concentration as high concentrations can induce condensation and sedimentation of the protein. Limited freeze-thaw cycles do not alter the behavior of the protein. However, it is recommended to performed assays comparing the freeze/thawed protein and fresh protein side-by-side.

4. Fluorescent labeling of protein and RNA

4.1 Protein labeling with amine-reactive dyes

The following protocol uses amine-reactive NHS-ester dyes to trace label purified protein, remove unreacted dye, and prepare trace-labeled protein solutions. This method will label the N-terminal amine and lysine amino acids. If amine-labeling alters protein function or if MBP is proteolytically removed from an MBP-fusion protein to initiate a condensation reaction, labeling of reactive thiols (cysteine residues) is an alternative option (Nanda & Lorsch, 2014).

4.1.1 Materials, equipment and reagents

Purified protein of interest
Protein storage buffer
1 M sodium bicarbonate pH 8.0
1 μg/μL NHS-ester dye (amine reactive) in DMSO
  Alexa Fluor® 647 NHS Ester (Thermofisher: A37573)
  Alexa Fluor® 555 NHS Ester (Thermofisher: A37571)
  DyLight™ 488 NHS Ester (Thermofisher: 46403)
Three Zeba™ Spin Desalting Columns, 7K MWCO, 0.5 mL (Thermofisher: 89883)
Nanodrop
SDS-PAGE equipment

4.1.2 Labeling protocol

1. *Labeling reaction*: Protein storage buffers containing free amines are not recommended for use in labeling reactions, and can reduce labeling efficiency. However, we have found that a wide range of buffers, even sub-optimal, can be used in the labeling reaction including urea. If the protein storage buffer is not compatible, protein can be buffer exchanged into a compatible buffer, and then returned to the original buffer during removal of dye after the labeling reaction is completed.
2. Mix together the following components for a total volume of 70 μL **Reaction volume is not critical, but the load volume of Zeba spin columns should be between 70 and 150 μL.*
   50 μL of 25 μM protein.
   7 μL 1 M sodium bicarbonate (pH 8.0).
   **Addition of sodium bicarbonate is not critical; labeling occurs at pH > 7.5 and is most efficient at pH 8.*
   8 μL of protein storage buffer.
   5 μL of 5 μg NHS-ester dye (1 μg/μL in DMSO).
3. Incubate for 1–2 h at room temperature.
4. Pre-equilibrate three to four Zeba spin columns.
   1. Centrifuge each column for 1 min at 1500 × *g*, discard flow through.
   2. Add 300 μL of exchange buffer, spin column for 1 min at 1.5 × *g* and discard flow through. Repeat buffer exchange two additional times.

5. Add 70 μL labeling reaction to pre-equilibrated column and centrifuge for 2 min at 1500 × *g*. Collect the flow through, and pass the sample through two more columns. Three columns should be sufficient to remove free dye, but more can be used if the presence of free dye is suspected. Alternately, Pierce™ Dye Removal Columns (Thermofisher #22858) may be used for single step removal of dye.
6. Check concentration by monitoring the protein and fluorescent dye absorbances and calculated extinction coefficients using a NanoDrop. If the labeling efficiency is too high or low, both the concentration of the dye and protein can be varied.
7. We recommend visualizing the labeled protein by SDS-PAGE to validate the labeling reaction. Labeled protein can be visualized by fluorescence imaging of the gel using an imager such as a Typhoon system before staining with a protein dye. After fluorescence imaging, gels can be stained with a protein dye such as SimplyBlue™ SafeStain to visualize total protein levels (Fig. 2).
8. Aliquot the dye-labeled protein into small aliquots (~5–10 μL), snap freeze, and store at −80 °C.

4.1.3 Assembly of trace-labeled protein solutions for imaging

Only small amounts of fluorescently labeled protein are required for imaging. The concentration of fluorescently labeled protein will depend on the microscope and imaging conditions used. To make a trace-labeled protein solution, mix labeled protein with unlabeled protein. We find that ratios of 1/20 to 1/100 of labeled/unlabeled protein are good starting points for most applications. However, labeling ratios as high as 1/3 and as low as 1/1000 can also be effective. The final ratio should be kept constant between experiments where the fluorescence signal will be quantitatively compared. As the actual extent of protein labeling has varied efficiency, the concentration of both labeled and unlabeled protein must be considered when making fluorescently labeled solutions. For example if a protein is only 50% labeled, then adding 50 nM dye-labeled protein will also add an additional 50 nM of unlabeled protein. This additional unlabeled protein should be considered when preparing protein solutions.

4.2 RNA labeling with fluorescent nucleotides

The following protocol includes steps for generation, labeling, and purification of RNA transcription products (Fig. 3).

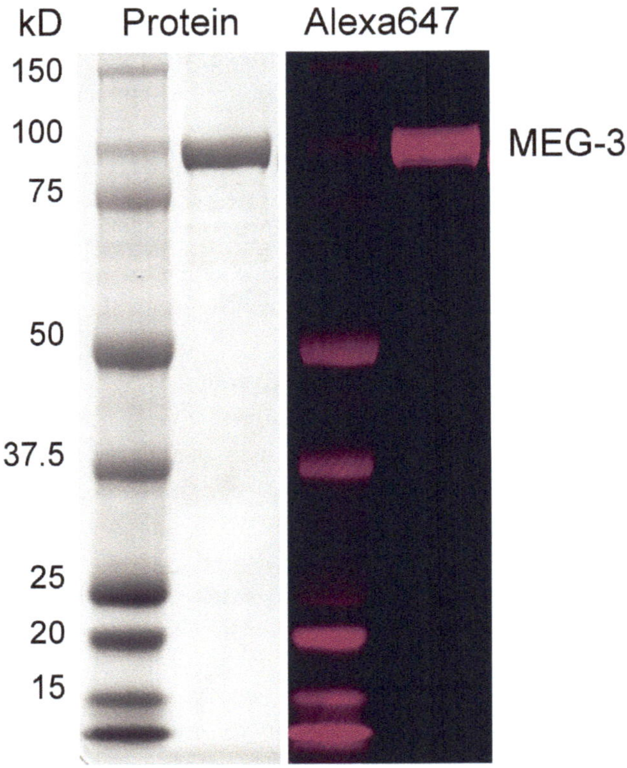

Fig. 2 Representative SDS-PAGE gel of MEG-3 protein (denaturing protocol) labeled with Alexa647 fluor and stained with SimplyBlue™ SafeStain (left panel) or fluorescently imaged before staining (right panel).

4.2.1 Materials, equipment and reagents

Micropipette and tips
totalRNA (or genomic cDNA)
PCR primers
Thin-walled PCR tubes
Thermal cycler
PCR reagents
Zero Blunt™ TOPO™ PCR Cloning Kit (Thermofisher: 451245)
Chemically competent cells
42 °C heat source
Agar plates with appropriate selection agent
37 °C incubator
Culture tubes
LB broth with appropriate selection agent

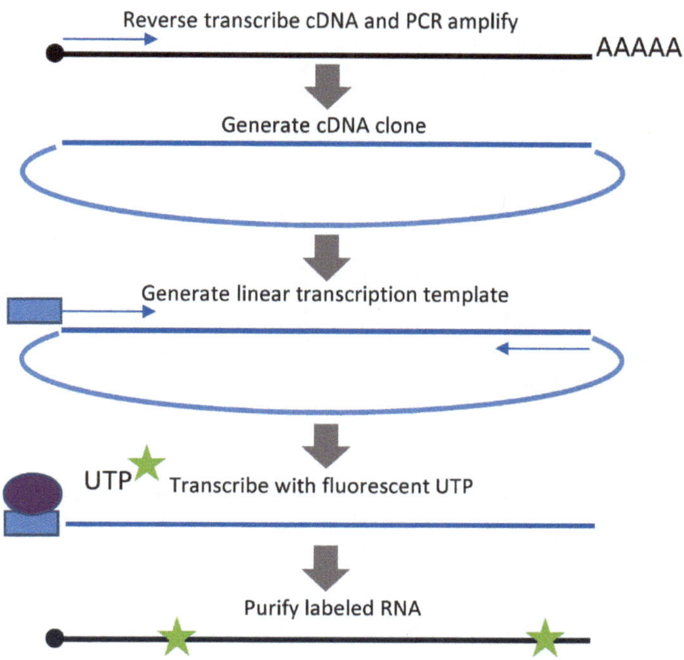

Fig. 3 Schematic of the RNA labeling transcription reaction. Black lines represent RNA, blue lines represent DNA, blue arrows represent PCR and reverse transcription primers, green stars represent fluorescent UTP, and the purple circle represents the RNA polymerase.

Miniprep kit (we recommend QIAprep Spin Miniprep kit; Qiagen cat. #27104)
PCR purification kit (optional—QIAquick PCR purification kit; Qiagen: 28104)
T7 mMessage Machine (Thermofisher: AM1344)
ChromaTide™ Alexa Fluor™ 488-5-UTP (Thermofisher: C11403), or 546-14-UTP (Thermofisher: C11404)
Agarose gel casting supplies
Gel box
Power supply
DNA size ladder, RNA size ladder

4.2.2 mRNA transcription and labeling protocol

1. Use reverse transcription (RT) to create a cDNA for the RNA of interest from total RNA (He, 2011) or use PCR to amplify the gene from an existing cDNA library. Ensure that the product is the correct size by agarose gel electrophoresis.

2. Follow the manufacturer's protocol to insert the RT or PCR product into the vector of choice using ZeroBlunt TOPO or a similar cloning strategy.
3. Screen transformants by sequencing to ensure that the accurate clone is selected.
4. Store reagents at 4 °C and use RNAse-free reagents for all subsequent steps to minimize any contamination that may degrade the RNA.
5. PCR amplify the linear transcription template. The polymerase start site sequence should be included in the primer 5′ of the gene specific primer. For T7 transcription we use: GAAAATTAATACGACTCACTATAG GGAGA. Follow the PCR reaction with a PCR clean up step.
6. Ensure that the product is the correct size by agarose gel electrophoresis. CRITICAL: this is the final template for transcription, if there are contaminating bands in this template, they will likely appear in the final transcription reaction. If the template is not pure, the correct band should be excised and gel purified. A second round of PCR can be performed on the gel purified product if the amount of material is insufficient for transcription.
7. Set up a transcription reaction according to the manufacturer's protocol. If significant smearing is observed in the final product, the reactions may have been contaminated by RNAse. We have used both SP6 and T7 mMessage machine kits successfully to generate capped mRNAs. To fluorescently label RNA, we use ChromaTide™ Alexa Fluor™ 488-5-UTP or 546-5-UTP Critical: while the polymerase can incorporate fluorescent nucleotides, these are not incorporated as efficiently as unlabeled nucleotides. Too much fluorescent UTP will cause incomplete transcription while too little results in insufficient labeling. We have found that a labeling ratio of 1:200 fluorescent UTP:unlabeled UTP results in high product yield with sufficient fluorescence for visualization. Transcription should be followed by a DNAse treatment to remove the original transcription template. *Optional: transcription can be followed by polyA tailing if desired.
8. Many options for purification are possible including phenol:chloroform extraction, spin columns, and gel purification. We have found that LiCl precipitation is fast and enables unincorporated fluorescent UTP to be efficiently eliminated from the reaction.
9. Precipitated RNA should be resuspended in water and stored at −80 °C. The concentration of the product can be measured by A260 absorbance using a nanodrop.

10. All transcription reactions should be run on a denaturing agarose gel to ensure the purity of the product. If additional bands are observed, we recommend optimizing the DNA template and transcription conditions to maximize the amount of RNA produced.

5. Condensation assays

5.1 Fluorescent multicomponent condensation assays

Condensation can be induced in several ways. Some of the most common are shifts from high to low salt, addition of a cofactor such as RNA, addition of a molecular crowding agent such as polyethylene glycol, or cleavage of a solubilizing tag such as MBP. For multi-condensate reactions each component is likely to show different condensation behaviors, and it is useful to study properties of each separately before assembling reactions. As a general starting point for assembling multi-condensates, we find it useful to start with approximately physiological ratios of each component. The protocol below provides a guidelines for the assembly and imaging of condensates using trace-labeled MEG-3, PGL-3 and RNA.

5.1.1 Materials, equipment and reagents

Labeled protein and RNA
Protein storage buffer
25 mM HEPES (pH 7.5)
1 M NaCl
Glass slides and coverslips
Double sided tape
Thin chambered glass slides (Erie Scientific Company: 30-2066A)
Glass bottom dishes (Fisher Scientific: NC9341562)
Chambered glass coverslips (Sigma-Aldrich: GBL103380)
Confocal microscope

5.1.2 Protocols

1. *Condensation reactions*: The following recipe generates co-assemblies of PGL-3 and MEG-3 condensates containing RNA.

   a. Make a 20 μL reaction containing the following concentrations (25 mM HEPES pH 7.4, 20 ng/μL RNA, 150 mM salt, 1.8 μM PGL-3, 150 nM MEG-3):

      13.65 μL of H_2O.
      1 μL 500 mM HEPES (pH 7.4).

2 μL of 200 ng/μL labeled *nos-2* RNA (Alexa546).
1.75 μL of 1 M NaCl.

b. Initiate the reaction by adding:
1 μL of trace labeled 36 μM PGL-3 (+360 nM PGL-3-Dylight488).
0.6 μL of trace labeled 5 μM MEG-3 (+833 nM MEG-3-Alexa647).
*Addition of MEG-3 and PGL-3 in protein storage buffer will bring the reactions to a final concentration of 150 mM (37.5 mM KCl, 112.5 mM NaCl). The concentration of salt in reactions is critical for condensate assembly, and should always be kept constant. If protein components are removed an equivalent amount of protein storage buffer or salt should be used instead.

c. Incubate reactions at room temperature for 30 min.

2. *Slides for imaging*: Imaging of condensates should be performed on chambered glass slides or in dishes. We have found several slide configurations that work for different assays and imaging setups. For use on both upright or inverted scopes we use thin chambered glass slides or double-sided tape chambered glass slides (Alberti et al., 2018). For inverted microscopes, we use glass bottom dishes or chambered glass cover slips. These options are particularly useful for reactions where a component is added during imaging; however, evaporation may be a problem over time.

3. *Imaging*: Several factors are critical when imaging condensates. (1) Time: condensates continue to grow over time and can interact with slide surfaces. In particular, the number of condensates deposited and the degree of wetting of the condensate will depend on reactions conditions and the time on the surface. (2) Magnification: the appropriate magnification will depend on the purpose. For quantification of condensates, taking a lower resolution image with a large field of view may enable collection of more condensate data; however for analysis of condensate structure, high magnification may be more useful. (3) Image heterogeneity: it is critical to take many fields of view of any condensation reaction. Condensates will show a distribution of sizes, and a visually striking field of view with a large granule may not accurately reflect the entire population.

4. *Surface coating*: The surface of the slide can impact the shape and growth of condensates during imaging, and is known to affect distinct condensates differently (Ceballos et al., 2018; Feric et al., 2016). For many applications, using untreated slides is sufficient to obtain information about the condensates being studied. However, coating of slides may improve image quality and is essential when imaging granules over time. Several protocols for surface coating with PEG-silane, Pluronic F127, and lipid bilayers for condensate assays are available (Alberti et al., 2018; Ceballos et al., 2018; Keenen et al., 2018).

6. Imaging and data analysis

6.1 Imaging and analysis of single or multicomponent images

Condensate images can be used qualitatively; however substantial quantitative information can be acquired by analyzing the data. Individual condensates can be detected, and parameters such as number, size, intensity, localization, or enrichment of components in multicomponent condensates can be measured. The following protocol uses the publicly available FIJI software package (Fig. 4) (Schindelin et al., 2012).

6.1.1 Imaging datasets

For each data set, we recommend analyzing multiple imaging fields for each data point. One field of view may not accurately reflect the distribution of condensates. For every data set, images of buffer only should also be taken to determine a background fluorescence value.

6.1.2 Object identification

1. Threshold image
   a. Measure the intensity histogram of the image (Analyze≫Histogram). Record the mean and standard deviation of the histogram, and close the histogram window.
   b. Open the Brightness/Contrast manager (Image≫Adjust≫Brightness/Contrast). Set the minimum threshold to the mean plus one standard deviation. Set the maximum threshold to the mean plus 3–6 times the standard deviation. Select Apply.
   The minimum and maximum threshold values will depend on the imaging conditions and fluorophores used; however, these values should be set to the same for all images in the same dataset.
2. Convert to binary image
   a. Autothreshold the image (Image≫Adjust≫Auto threshold≫Method: Otsu).

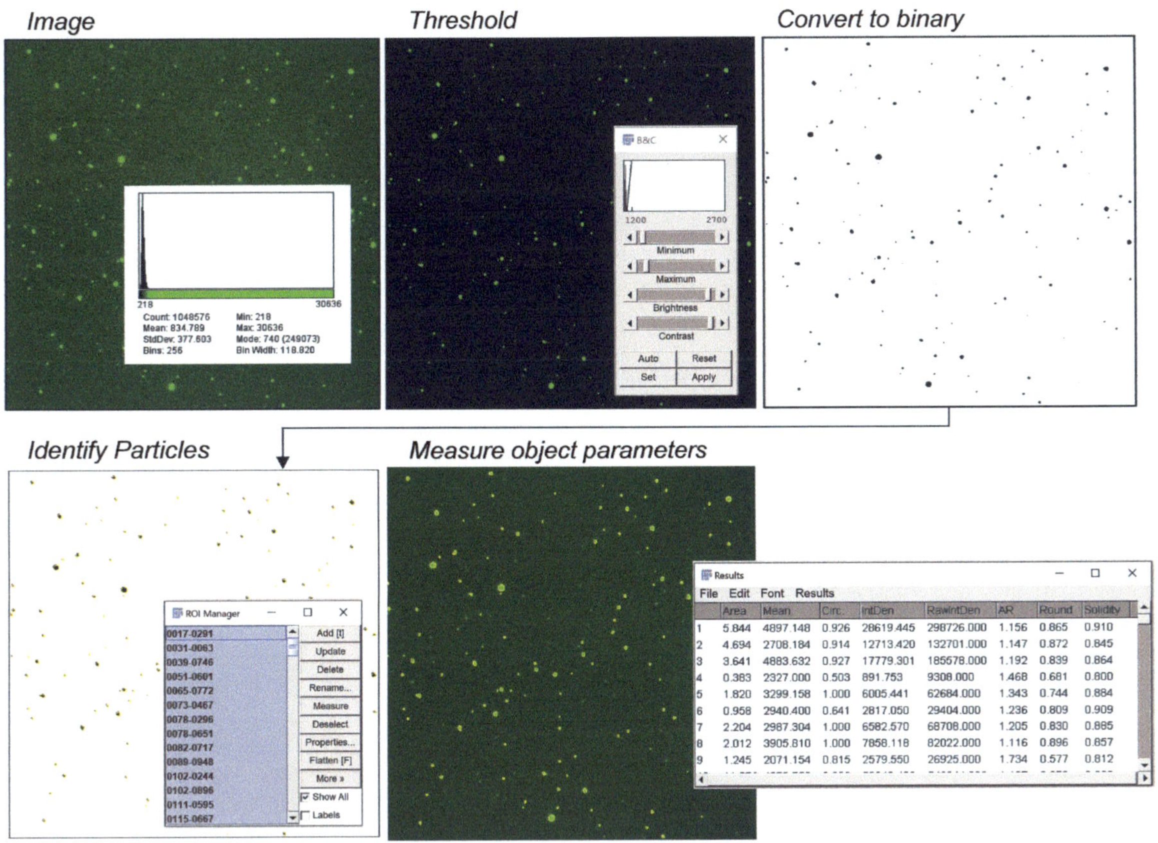

Fig. 4 Schematic of the imaging analysis workflow.

b. Convert to binary (Process≫Binary≫Make Binary). Image should be black and white after this step. If it is not use Image≫Adjust≫ Threshold≫"Apply."

Optional: If many droplets are in close proximity, process the image using a watershed filter (Process≫ Binary≫ Watershed) to improve separation of objects.

Optional: For two or more component analysis: Binary images can be generated from a single components and then merged to a single image. To merge individual binary images, first convert two or more channels to binary images. Then merge the images using the following methods: (1) For two images, the image calculator (OR) function can be used (2) For two or more images, convert the single binary images to a stack. Then make a maximum projection of the stack.

3. Identify particles

   Open the analyze particles menu (Analyze≫Analyze particles). Set the minimum particle size (Check "pixel units," set to 2-infinity). Under the dropdown menu "Show" select "overlay." Check the box to "Add to Manager," no other options should be selected. Select OK.

4. Measure object parameters
   a. Open the unedited image.
   b. Open the ROI manager (Image≫Overlay≫to ROI Manager). To visualize the objects on the original image, click "show all" in the ROI manager (if show all is already selected, reselect).
   c. Set parameters to be analyzed (Analyze≫Set Measurements). We recommend the mean gray value (the average intensity of each object), the integrated density (the total intensity of each object), area, and the Shape descriptors.
   d. Select all objects in the ROI manager. Then select measure in the ROI manager, and a results table containing desired measurements will be displayed.

5. Measure the total image intensity

   Select the unedited image and measure the intensity and area of the entire image (Analyze≫measure).

   Recommended: Measure the intensity and area of a buffer only control image for background correction (see Section 6.1.1).

6. Steps 1–5 can be automated through a macro or using existing plugins such as the Cookbook plugin, nucleus counter.

6.2 Data analysis

6.2.1 Analysis for one or more components

From the generated list of intensities a number of variables can be analyzed including total granule number, intensity of individual condensates or total intensity. Useful parameters to analyze are described below:

1. *Background correction*: We find that frequently the background signal in the absence of fluorophore is significant and intensity measurements should be corrected for background. To subtract background fluorescence, we use the following procedure. Calculate the background fluorescence per unit area (pixel or μm) of a fluorescence background image (i.e., buffer alone) using the following equation:

$$I^{B} = I^{totB} / A^{tot}$$

Where I^{totB} is the total background intensity (measured in step 5 of Section 6.1.2), A^{tot} is the total area of the background image, and I^{B} is the background intensity per unit Area (pixel or μm).

Calculate the background corrected total intensity of each experimental image using the following equation:

$$I^{tot} = \left(I^{totraw} - I^{B}\right)$$

Where I^{tot} is the background corrected total intensity measured for the experimental image and I^{totraw} is the total intensity of the image before background correction. For each condensate identified subtract a background for each object using the following equation:

$$I^{O} = I^{Oraw} - I^{B}A^{O}$$

Where I^{O} is the background corrected intensity detected for each object, I^{Oraw} is the intensity of each object before background correction, and A^{O} is the area of each object.

2. *Total intensity in condensates*: To calculate the total intensity of protein phase use the following equations:

$$I^{totO} = \sum I^{O}$$

Where I^{totO} is the total intensity of detected objects. To calculate the fraction of condensed protein use the following equation:

$$fI^{totO} = I^{totO} / I^{tot}$$

Where fI^{totO} is the fraction of fluorescence intensity in identified objects.

3. *Intensity histograms*: Condensate size and intensity frequently show an exponential size distribution and have a large dynamic range. As a result it is useful to view intensities on a log scale. Calculate the log(I^{totO}) for each object. Bin log(I^{totO}) values into appropriate size bins (i.e., bin size = 0.2 Log(I^{totO}) units) in excel or a graphing program. Histograms can be normalized to the total number of condensates to compare histograms across different conditions.
4. *Condensate shape*: The shape descriptor option in the "Analyze particles" menu provides four parameters for describing condensate shape: circularity, aspect ratio, roundness, and solidity. One of the critical characteristics of a phase separated droplet held together by surface tension is a spherical shape, and these parameters are useful for analyzing *in vitro* droplet shape.

6.2.2 Analysis for two or more components

The calculations in the previous section are useful for single or multicomponent condensates. The analysis presented below are useful specifically for two or more component images.

1. *Enrichment in condensates*: In multicomponent reactions where proteins and/or RNA colocalize, one can measure the degree of enrichment of components within a condensate. Using objects identified in one imaging channel, overlay the generated mask onto a second imaging channel, and measure the intensity. Calculate enrichment of the second component using the method described in Section 6.1.2, step 2.
2. *Colocalization*: In multicomponent reactions where condensates of different components don't fully mix, it may be useful to calculate the degree of colocalization. Using objects identified in the two or more component mask (Section 6.1.2, step 2, optional), measure the intensity of each image channel and correct for background fluorescence. Then compare the variation of the signal between the two channels by Pearson's correlation analysis using software such as Graphpad Prism or Matlab. The Pearson's correlation coefficient value can be determined as a function of time or between different reaction conditions to quantitatively compare the colocalization of multiple components.
3. *Profile plots*: Intensity profile plots can identify and measure regions with condensate substructure (as we observe for MEG-3 and PGL-3 condensates). Choose a representative condensate, and draw a rectangle across the middle of the condensate or over the structural feature to be analyzed. In FIJI select Analyze≫Plot profile, and repeat for each channel

of the image. FIJI will produce an intensity profile plot for each channel. These graphs can be used directly, or alternately select "List" in the profile plot window to generate a list of values which can be used in any graphing program.

6.3 Measuring condensate dynamics by FRAP

Exchange of molecules between condensates and the dilute phase provides critical information for understanding the material properties of a condensate. Using fluorescently trace-labeled solutions, it is possible to measure the dynamics of each component in a multi-condensate assembly individually or simultaneously. Fluorescence recovery after photobleaching (FRAP) is a commonly used and powerful assay to determine molecular dynamics of condensates. Fluorescence in a region of interest is bleached using a laser and monitored over time to estimate the exchange of molecules in and out of the bleached region. Exchange rates vary over several orders of magnitude (McSwiggen, Mir, Darzacq, & Tjian, 2019) and can change as a function of time, environment, and material properties of the condensates. Liquid droplets, where molecules exchange constantly with the dilute phase, recover fluorescence relatively quickly. In contrast gel-like droplets or solid aggregates exhibit slow to non-detectable recovery.

6.3.1 Considerations for FRAP experiments

Bleaching of fluorescent condensates can be performed in two ways: (1) bleaching an entire condensate to monitor exchange of molecules between the condensate and the dilute phase (2) bleaching a small region of a condensate to monitor internal rearrangement of molecules within a condensate (and exchange with the dilute phase). In both methods, care should be taken to maintain consistency between experimental replicates in (1) the extent of bleaching, (2) the size of the bleached region, (3) the position of the bleached region relative to the condensate's center and (4) the size of the condensate, as all of these factors will affect recovery (McSwiggen et al., 2019; Taylor, Wei, Stone, & Brangwynne, 2019).

6.3.2 Imaging protocol

To perform FRAP experiments, condensates are assembled and mounted on slides for imaging as described in Section 5.

1. Image three to five frames to record the intensity before photobleaching.
2. Select regions of condensates to be photobleached. More than one condensate and/or component in a multi-condensate can be photobleached at the same time depending on the capability of the microscope being used.

3. Image condensates at regular intervals after photobleaching. The interval will depend on the exchange rate of molecules within the condensate; however, the total image time should be sufficient to measure the maximum extent of recovery.
4. *Imaging conditions*: Both exposure time and the number of frames acquired should be optimized to minimize non-specific photobleaching during imaging acquisition. Additionally, the extent of photobleaching of the region of interest should be close to 100% and similar between all experiments.
5. *Condensate selection*: Condensates to be photobleached should be carefully selected. In addition to observing droplet size (Section 6.3.1), the time droplets are allowed to adhere to the glass surface is also critical for FRAP measurements. Droplets that are allowed to settle on the surface of the slide can lose dynamics over time depending on the ability of the condensate to wet the surface of the glass. Coating the slide surface may minimize this effect; however, mobile droplets that have not adhered to the glass surface may leave the field of view during imaging making observation of fluorescence recovery difficult.

6.3.3 Data analysis

For data analysis, the mean or total intensity of the bleached spot should be measured over time. This data is provided by most FRAP capable imaging programs; however, the intensity of the photobleached region can also be measured after image collection using FIJI. Also measure the fluorescence of a similarly sized region away from the condensate to determine the background fluorescence of the image. Calculation of the extent and rate of fluorescence recovery can be calculated using a single exponential recovery equation described below. However, for more complicated recovery patterns or to use FRAP recovery rates to calculate additional parameters such as diffusion, additional models for condensate recovery must be considered (McSwiggen et al., 2019; Mueller, Mazza, Stasevich, & McNally, 2010; Sprague & McNally, 2005; Taylor et al., 2019).

1. Calculate the fractional fluorescence recovery for each time point, using the intensity measured from the photobleached region and correct the intensity values for background fluorescence using the following equation:

$$fI(\mathrm{t}) = \left(I(\mathrm{t}) - I^B\right)/\left(I^i - I^B\right)$$

Where fI is the background corrected fractional intensity at a given time (t) in the photobleached region, I^B is the background intensity of an identically sized region, I is the fluorescence intensity measured at a given time (t) after photobleaching, and I^i is the initial fluorescence intensity before photobleaching. It is also useful to measure the intensity of a nearby condensate to detect non-specific photobleaching during image acquisition. Intensity data should be corrected if non-specific photobleaching is an issue (Day, Kraft, Kang, & Kenworthy, 2012).

2. Plot the fractional intensity (fI) vs time to obtain a FRAP time trace. To calculate the extent and rate of recovery fit the data to the following single exponential equation:

$$fI = A \cdot \left(1 - e^{-kt}\right)$$

Where A is the extent or amplitude of fluorescence recovery, and k is the rate of recovery.

3. Frequently, FRAP is reported as the time required for half of the fluorescence to recover ($t^{1/2}$). This can be determined directly from the plot by measuring the time required to recover half of the maximum fluorescence recovery (A). Alternatively, $t^{1/2}$ values can be calculated from the recovery rate using the following equation:

$$t^{1/2} = \ln(2)/(k)$$

7. Concluding remarks

In this chapter, we present techniques to reconstitute and analyze protein/RNA condensates *in vitro*. Our protocols enable maximum flexibility during imaging and minimize the influence of fluorescent tags on condensate formation. We also present a workflow for analyzing condensation experiments. We emphasize that the design of *in vitro* reconstitution assays should be guided by *in vivo* observations to ensure that the reconstituted system is physiologically relevant and based on BC proteins and/or RNAs with demonstrated roles *in vivo*. Once a reconstituted system is created, the protocols presented here can be used to perform structure-function analyses on BC proteins and RNAs to identify key molecular parameters driving assembly. In turn, models derived from *in vitro* reconstitution experiments can be used to guide cell-based experiments to further explore BC assembly and function *in vivo*.

References

Alberti, S., Gladfelter, A., & Mittag, T. (2019). Considerations and challenges in studying liquid-liquid phase separation and biomolecular condensates. *Cell*, *176*(3), 419–434. https://doi.org/10.1016/j.cell.2018.12.035.

Alberti, S., Saha, S., Woodruff, J. B., Franzmann, T. M., Wang, J., & Hyman, A. A. (2018). A user's guide for phase separation assays with purified proteins. *Journal of Molecular Biology*, *430*(23), 4806–4820. https://doi.org/10.1016/j.jmb.2018.06.038.

Banani, S. F., Rice, A. M., Peeples, W. B., Lin, Y., Jain, S., Parker, R., et al. (2016). Compositional control of phase-separated cellular bodies. *Cell*, *166*(3), 651–663. https://doi.org/10.1016/j.cell.2016.06.010.

Banerjee, P. R., Milin, A. N., Moosa, M. M., Onuchic, P. L., & Deniz, A. A. (2017). Reentrant phase transition drives dynamic substructure formation in ribonucleoprotein droplets. *Angewandte Chemie (International Ed. in English)*, *56*(38), 11354–11359. https://doi.org/10.1002/anie.201703191.

Boeynaems, S., Alberti, S., Fawzi, N. L., Mittag, T., Polymenidou, M., Rousseau, F., et al. (2018). Protein phase separation: A new phase in cell biology. *Trends in Cell Biology*, *28*(6), 420–435. https://doi.org/10.1016/j.tcb.2018.02.004.

Boisvert, F.-M., van Koningsbruggen, S., Navascués, J., & Lamond, A. I. (2007). The multifunctional nucleolus. *Nature Reviews Molecular Cell Biology*, *8*(7), 574–585. https://doi.org/10.1038/nrm2184.

Boke, E., Ruer, M., Wühr, M., Coughlin, M., Lemaitre, R., Gygi, S. P., et al. (2016). Amyloid-like self-assembly of a cellular compartment. *Cell*, *166*(3), 637–650. https://doi.org/10.1016/j.cell.2016.06.051.

Brangwynne, C. P., Eckmann, C. R., Courson, D. S., Rybarska, A., Hoege, C., Gharakhani, J., et al. (2009). Germline P granules are liquid droplets that localize by controlled dissolution/condensation. *Science*, *324*(5935), 1729–1732. https://doi.org/10.1126/science.1172046.

Ceballos, A. V., McDonald, C. J., & Elbaum-Garfinkle, S. (2018). *Methods and strategies to quantify phase separation of disordered proteins*. In E. Rhoades (Ed.), *Vol. 611. Methods in enzymology* (pp. 31–50). Elsevier. https://doi.org/10.1016/bs.mie.2018.09.037.

Day, C. A., Kraft, L. J., Kang, M., & Kenworthy, A. K. (2012). Analysis of protein and lipid dynamics using confocal fluorescence recovery after photobleaching (FRAP). *Current Protocols in Cytometry/Editorial Board, J. Paul Robinson, Managing Editor … [et al.]*, *62*(1), 2.19.1–2.19.29. https://doi.org/10.1002/0471142956.cy0219s62.

Ditlev, J. A., Case, L. B., & Rosen, M. K. (2018). Who's in and who's out—Compositional control of biomolecular condensates. *Journal of Molecular Biology*, *430*(23), 4666–4684. https://doi.org/10.1016/j.jmb.2018.08.003.

Elbaum-Garfinkle, S., Kim, Y., Szczepaniak, K., Chen, C. C.-H., Eckmann, C. R., Myong, S., et al. (2015). The disordered P granule protein LAF-1 drives phase separation into droplets with tunable viscosity and dynamics. *Proceedings of the National Academy of Sciences of the United States of America*, *112*(23), 7189–7194. https://doi.org/10.1073/pnas.1504822112.

Feric, M., Vaidya, N., Harmon, T. S., Mitrea, D. M., Zhu, L., Richardson, T. M., et al. (2016). Coexisting liquid phases underlie nucleolar subcompartments. *Cell*, *165*(7), 1686–1697. https://doi.org/10.1016/j.cell.2016.04.047.

He, F. (2011). Total RNA extraction from *C. elegans*. *Bio-Protocol*, *1*(6), e47. https://doi.org/10.21769/BioProtoc.47.

Hubstenberger, A., Courel, M., Bénard, M., Souquere, S., Ernoult-Lange, M., Chouaib, R., et al. (2017). P-body purification reveals the condensation of repressed mRNA regulons. *Molecular Cell*, *68*(1), 144–157.e5. https://doi.org/10.1016/j.molcel.2017.09.003.

Hyman, A. A., Weber, C. A., & Jülicher, F. (2014). Liquid-liquid phase separation in biology. *Annual Review of Cell and Developmental Biology*, *30*(1), 39–58. https://doi.org/10.1146/annurev-cellbio-100913-013325.

Jain, S., Wheeler, J. R., Walters, R. W., Agrawal, A., Barsic, A., & Parker, R. (2016). ATPase-modulated stress granules contain a diverse proteome and substructure. *Cell*, *164*(3), 487–498. https://doi.org/10.1016/j.cell.2015.12.038.

Keenen, M. M., Larson, A. G., & Narlikar, G. J. (2018). *Visualization and quantitation of phase-separated droplet formation by human HP1α*. In E. Rhoades (Ed.), *Vol. 611. Methods in enzymology* (pp. 51–66). Elsevier. https://doi.org/10.1016/bs.mie.2018.09.034.

Khong, A., Matheny, T., Jain, S., Mitchell, S. F., Wheeler, J. R., & Parker, R. (2017). The stress granule transcriptome reveals principles of mRNA accumulation in stress granules. *Molecular Cell*, *68*(4), 808–820.e5. https://doi.org/10.1016/j.molcel.2017.10.015.

Kroschwald, S., Maharana, S., & Simon, A. (2017). Hexanediol: A chemical probe to investigate the material properties of membrane-less compartments. *Matters*, *3*(5), e201702000010. https://doi.org/10.19185/matters.201702000010.

Kryndushkin, D., Pripuzova, N., Burnett, B. G., & Shewmaker, F. (2013). Non-targeted identification of prions and amyloid-forming proteins from yeast and mammalian cells. *Journal of Biological Chemistry*, *288*(38), 27100–27111. https://doi.org/10.1074/jbc.M113.485359.

Kuznetsova, I. M., Turoverov, K. K., & Uversky, V. N. (2014). What macromolecular crowding can do to a protein. *International Journal of Molecular Sciences*, *15*(12), 23090–23140. https://doi.org/10.3390/ijms151223090.

Lee, C. Y. S., Putnam, A., Lu, T., He, S. X., Ouyang, J. P. T., & Seydoux, G. (2020). Recruitment of mRNAs to P granules by condensation with intrinsically-disordered proteins. *eLife*, *9*e52896 https://doi.org/10.7554/eLife.52896.

Maharana, S., Wang, J., Papadopoulos, D. K., Richter, D., Pozniakovsky, A., Poser, I., et al. (2018). RNA buffers the phase separation behavior of prion-like RNA binding proteins. *Science*, *360*(6391), 918–921. https://doi.org/10.1126/science.aar7366.

Marnik, E. A., & Updike, D. L. (2019). Membraneless organelles: P granules in *Caenorhabditis elegans*. *Traffic*, *20*(6), 373–379. https://doi.org/10.1111/tra.12644.

McSwiggen, D. T., Mir, M., Darzacq, X., & Tjian, R. (2019). Evaluating phase separation in live cells: Diagnosis, caveats, and functional consequences. *Genes & Development*, *33*(23–24), 1619–1634. https://doi.org/10.1101/gad.331520.119.

Mueller, F., Mazza, D., Stasevich, T. J., & McNally, J. G. (2010). FRAP and kinetic modeling in the analysis of nuclear protein dynamics: What do we really know? *Current Opinion in Cell Biology*, *22*(3), 403–411. https://doi.org/10.1016/j.ceb.2010.03.002.

Nanda, J. S., & Lorsch, J. R. (2014). *Chapter seven—Labeling of a protein with fluorophores using maleimide derivitization*. In J. Lorsch (Ed.), *Vol. 536. Methods in enzymology* (pp. 79–86). Academic Press. https://doi.org/10.1016/B978-0-12-420070-8.00007-6.

Nott, T. J., Craggs, T. D., & Baldwin, A. J. (2016). Membraneless organelles can melt nucleic acid duplexes and act as biomolecular filters. *Nature Chemistry*, *8*(6), 569–575. https://doi.org/10.1038/nchem.2519.

Ouyang, J. P. T., Folkmann, A., Bernard, L., Lee, C.-Y., Seroussi, U., Charlesworth, A. G., et al. (2019). P granules protect RNA interference genes from silencing by piRNAs. *Developmental Cell*, *50*(6), 716–728.e6. https://doi.org/10.1016/j.devcel.2019.07.026.

Palmer, I., & Wingfield, P. T. (2004). Preparation and extraction of insoluble (inclusion-body) proteins from *Escherichia coli*. *Current Protocols in Protein Science/Editorial Board, John E Coligan [et al]*, *38*(1), 6.3.1–6.3.18. https://doi.org/10.1002/0471140864.ps0603s38.

Patel, A., Malinovska, L., Saha, S., Wang, J., Alberti, S., Krishnan, Y., et al. (2017). ATP as a biological hydrotrope. *Science*, *356*(6339), 753–756. https://doi.org/10.1126/science.aaf6846.

Protter, D. S. W., Rao, B. S., Van Treeck, B., Lin, Y., Mizoue, L., Rosen, M. K., et al. (2018). Intrinsically disordered regions can contribute promiscuous interactions to RNP granule assembly. *Cell Reports*, *22*(6), 1401–1412. https://doi.org/10.1016/j.celrep.2018.01.036.

Putnam, A., Cassani, M., Smith, J., & Seydoux, G. (2019). A gel phase promotes condensation of liquid P granules in *Caenorhabditis elegans* embryos. *Nature Structural & Molecular Biology*, *26*(3), 220–226. https://doi.org/10.1038/s41594-019-0193-2.

Rai, A. K., Chen, J.-X., Selbach, M., & Pelkmans, L. (2018). Kinase-controlled phase transition of membraneless organelles in mitosis. *Nature*, *559*(7713), 211–216. https://doi.org/10.1038/s41586-018-0279-8.

Riback, J. A., Katanski, C. D., Kear-Scott, J. L., Pilipenko, E. V., Rojek, A. E., Sosnick, T. R., et al. (2017). Stress-triggered phase separation is an adaptive, evolutionarily tuned response. *Cell*, *168*(6), 1028–1040.e19. https://doi.org/10.1016/j.cell.2017.02.027.

Saha, S., Weber, C. A., Nousch, M., Adame-Arana, O., Hoege, C., Hein, M. Y., et al. (2016). Polar positioning of phase-separated liquid compartments in cells regulated by an mRNA competition mechanism. *Cell*, *166*(6), 1572–1584.e16. https://doi.org/10.1016/j.cell.2016.08.006.

Schindelin, J., Arganda-Carreras, I., Frise, E., Kaynig, V., Longair, M., Pietzsch, T., et al. (2012). Fiji: An open-source platform for biological-image analysis. *Nature Methods*, *9*(7), 676–682. https://doi.org/10.1038/nmeth.2019.

Seydoux, G. (2018). The P granules of *C. elegans*: A genetic model for the study of RNA–protein condensates. *Journal of Molecular Biology*, *430*(23), 4702–4710. https://doi.org/10.1016/j.jmb.2018.08.007.

Sprague, B. L., & McNally, J. G. (2005). FRAP analysis of binding: Proper and fitting. *Trends in Cell Biology*, *15*(2), 84–91. https://doi.org/10.1016/j.tcb.2004.12.001.

Tauber, D., Tauber, G., Khong, A., Treeck, B. V., Pelletier, J., & Parker, R. (2020). Modulation of RNA condensation by the DEAD-box protein eIF4A. *Cell*, *180*(3), 411–426.e16. https://doi.org/10.1016/j.cell.2019.12.031.

Taylor, N. O., Wei, M.-T., Stone, H. A., & Brangwynne, C. P. (2019). Quantifying dynamics in phase-separated condensates using fluorescence recovery after photobleaching. *Biophysical Journal*, *117*(7), 1285–1300. https://doi.org/10.1016/j.bpj.2019.08.030.

Tropea, J. E., Cherry, S., & Waugh, D. S. (2009). Expression and purification of soluble His(6)-tagged TEV protease. *Methods in Molecular Biology (Clifton, N.J.)*, *498*, 297–307. https://doi.org/10.1007/978-1-59745-196-3_19.

Uebel, C. J., & Phillips, C. M. (2019). Phase-separated protein dynamics are affected by fluorescent tag choice. *MicroPublication Biology*, *2019*(08), 143. https://www.micropublication.org/journals/biology/micropub-biology-000143/.

Wang, J., Choi, J.-M., Holehouse, A. S., Lee, H. O., Zhang, X., Jahnel, M., et al. (2018). A molecular grammar governing the driving forces for phase separation of prion-like RNA binding proteins. *Cell*, *174*(3), 688–699.e16. https://doi.org/10.1016/j.cell.2018.06.006.

Wang, J. T., Smith, J., Chen, B.-C., Schmidt, H., Rasoloson, D., Paix, A., et al. (2014). Regulation of RNA granule dynamics by phosphorylation of serine-rich, intrinsically disordered proteins in *C. elegans*. *eLife*, *3*, e04591. https://doi.org/10.7554/eLife.04591.

Wang, Z., Zhang, G., & Zhang, H. (2019). Protocol for analyzing protein liquid–liquid phase separation. *Biophysics Reports*, *5*(1), 1–9. https://doi.org/10.1007/s41048-018-0078-7.

West, J. A., Mito, M., Kurosaka, S., Takumi, T., Tanegashima, C., Chujo, T., et al. (2016). Structural, super-resolution microscopy analysis of paraspeckle nuclear body organization. *The Journal of Cell Biology*, *214*(7), 817–830. https://doi.org/10.1083/jcb.201601071.

CHAPTER FIVE

Practical considerations for generation of multi-compartment complex coacervates

Gregory A. Mountain and Christine D. Keating*
Department of Chemistry, Pennsylvania State University, University Park, PA, United States
*Corresponding author: e-mail address: keating@chem.psu.edu

Contents

Abstract

We discuss preparation of experimental models for multi-compartment membraneless organelles in which distinct compositions are maintained indefinitely for macromolecule-rich phases in contact with each other. These model systems are based on the physical chemistry phenomenon of complex coacervation. In complex coacervation, liquid-liquid phase separation occurs due to ion pairing interactions between oppositely charged polyelectrolytes. This mechanism can drive the associative phase separation of proteins and nucleic acids, the major macromolecular components of membraneless organelles. Here we provide examples, advice and practical considerations for the design, generation, and analysis of multi-compartment complex coacervates. These structures are of interest to compartmentalize the interior of artificial cells and as models for the intracellular membraneless organelles of biological cells.

1. Introduction

A number of non-membranous intracellular organelles form via liquid-liquid phase separation (LLPS) of their biomacromolecular

Methods in Enzymology, Volume 646
ISSN 0076-6879
https://doi.org/10.1016/bs.mie.2020.09.001

components (typically nucleic acid and/or protein). Multiple types of intermolecular interactions including charge-charge, cation-π, dipole-dipole, and π-π stacking can contribute to the associative phase separation process (Brangwynne, Tompa, & Pappu, 2015). The resulting intracellular condensates can have varying degrees of fluidity, with some exhibiting more liquidlike and others more gel- or solid-like apparent viscosities (Hyman, Weber, & Julicher, 2014; Mitrea & Kriwacki, 2016). Abnormalities are associated with a range of diseases including Alzheimer's disease, ALS, and type 2 diabetes (Brangwynne et al., 2015; Shin & Brangwynne, 2017). Multiple types of membraneless organelles, with different biomolecular compositions and functions, exist simultaneously in both the nucleoplasm and cytoplasm of eukaryotic cells. Some intracellular condensates occur as directly contacting subcompartments with characteristic compositions. For example, the multiple subcompartments of the nucleolus have been identified as immiscible liquid phases (Feric et al., 2016). Intracellular compartmentalization by numerous distinct membraneless organelles is increasingly appreciated as an important mechanism by which the rich biochemistry of living cells is organized. The complexity of these intracellular condensates *in vivo* makes model systems appealing as a means of studying the underlying principles that influence their properties. Relatively simpler compartments formed by LLPS are also appealing as a means to organize and increase functionality of artificial cells (Crowe & Keating, 2018) and bioreactors (Dewey, Strulson, Cacace, Bevilacqua, & Keating, 2014; Nakashima, Vibhute, & Spruijt, 2019; Rowland, Cacace, Pulati, Gulley, & Keating, 2019).

Experimental model systems for biomacromolecular phase separation generally focus on associative phase separation, also called coacervation, which leads to one phase rich in macromolecules (termed the coacervate phase) and another, usually much larger volume phase that is comparatively dilute (often called the dilute phase or equilibrium liquid) (Aumiller, Davis, & Keating, 2014). In general, the liquidlike properties of coacervates are due to the existence of numerous weak, dynamic interactions between components; when intermolecular associations become too strong, gel-like or solid condensates result (Wang & Schlenoff, 2014). Coacervation can be driven by different types of intermolecular interactions, and it is important to understand the intermolecular interactions that lead to phase separation for a particular molecular system in order to design approaches for multiphase coexistence. For example, elastin-like polypeptides (ELPs) are a well-known phase separating system in which temperature-driven phase separation results from changes in solvation (MacEwan & Chilkoti, 2010;

Wang, Ozsvar, Yeo, & Weiss, 2019). By tuning the hydrophobicity of pentapeptide repeat sequences, Chilkoti, Lopez and coworkers demonstrated control over the critical temperatures for phase separation and were able to generate multiphase, nested droplets with multiple phases formed from different sequences of ELPs (Quiroz & Chilkoti, 2015; Simon, Carroll, Rubinstein, Chilkoti, & Lopez, 2017). For many intracellular condensates that form due to associations of intrinsically disordered proteins and RNAs, charge-charge interactions between the molecular components are important (Brangwynne et al., 2015); Dignon, Best, & Mittal, 2020. These membraneless compartments can be modeled by a type of phase coexistence called complex coacervation (Fig. 1) (Bungenberg de Jong, 1949; Bungenberg de Jong & Kruyt, 1929; Gucht, Spruijt, Lemmers, & Cohen Stuart, 2011; Perry, 2019; Wang & Schlenoff, 2014).

Complex coacervate systems refer to those driven by ion pairing interactions between oppositely charged polyions ("complexation"), and the concomitant entropy gain associated with release of counterions (Bungenberg de Jong, 1949). These polyelectrolyte complexes can exist in solution, or phase-separate into a dense, polyelectrolyte-rich liquid coacervate phase or a solid precipitate depending on the relative strength of the intermolecular interactions (Wang & Schlenoff, 2014). Multiphase complex coacervate coexistence with two or more distinct dense, polymer-rich phases, was recently demonstrated by several groups and provides new models for these coexisting liquid organelles in biology (Boeynaems et al., 2019; Lu & Spruijt, 2020; Mountain & Keating, 2020).

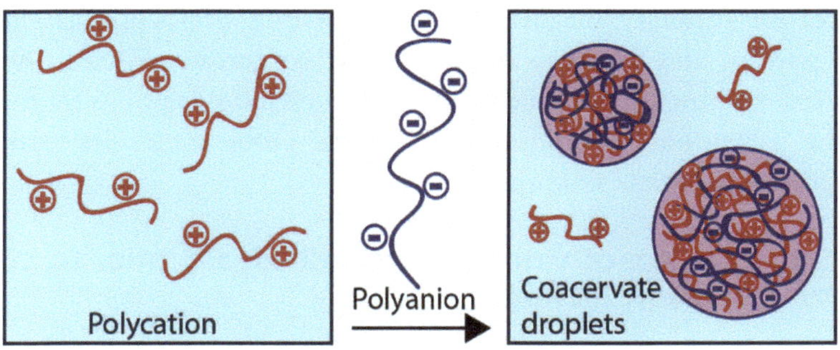

Fig. 1 Oppositely charged polyelectrolytes in solution can undergo an associative phase separation event known as complex coacervation resulting in aqueous polymer-rich droplets and a polymer-poor dilute phase. Counterions and solvent water molecules, which are not shown here, are also important to the physical chemistry of the phase separation process.

The polyelectrolytes used in these model systems for membraneless organelles have fewer interaction types than are seen in biological systems, as this allows for more straightforward interpretation of physical phenomena. Even so, as biomacromolecules such as polypeptides and nucleic acids are employed, the fundamental interaction types go beyond ion pairing. For example, poly(arginine) contains positively charged guanidinium groups that also have the capacity to participate in cation-π and π-π stacking interactions with the nucleic acid bases, or with the amino acids phenylalanine, tyrosine, and tryptophan (Beaver & Waters, 2016; Boeynaems et al., 2019; Hughes & Waters, 2006; Tatko & Waters, 2003). Our lab has recently employed complex coacervate systems to generate coexisting compartments using synthetic and biologically inspired polyelectrolytes which show tunable partitioning behavior based on composition (Mountain & Keating, 2020). Here we detail protocols and practical considerations for creating multi-compartment liquid phase separated polymer systems based on complex coacervation.

2. Procedure

Sections 2.1–2.5 will provide context and specific examples to use as a guide in designing multiphase coacervate systems. They are intended as a starting point for readers to adapt to their own experimental requirements based on their scientific goals. We begin with a discussion of why and how the polyions and solution condition should be pre-tested pairwise to understand likely outcomes upon mixing the full set, and then describe procedures for turbidity measurements, preparation protocols that include three-polyion and four-polyion multiphase coacervate systems, methods for avoiding formation of solid condensates, and considerations for measuring the partitioning of fluorescently labeled probe molecules across the multiphase coacervate systems.

2.1 Screening coacervate systems individually prior to multi-coacervate formation

Successful design of a multiphase coacervate system requires knowledge of how the component molecules interact in all of their possible combinations. Characteristics of the polyelectrolytes such as multivalency and charge density are important in determining their interaction strength and phase separation behavior (Chang et al., 2017; Choi, Holehouse, & Pappu, 2020; Spruijt, Westphal, Borst, Cohen Stuart, & van der Gucht, 2010).

The phenomenon of complex coacervation is also dependent on a number of factors including ionic strength, pH, and polyelectrolyte concentration (Bungenberg de Jong, 1949; Priftis et al., 2014). These factors can be used to tune the interaction strength between polyelectrolytes to optimize the behavior of the overall multiphase system. A charge matching condition of 1:1 cationic:anionic moieties is usually a good starting point for determining relative amounts of the oppositely charged polyions, particularly for polyelectrolytes with similar multivalency. When multivalencies are mismatched, excess of the smaller polyion may be needed to drive coacervation; an example is the spermidine/polyU RNA system (Aumiller, Cakmak, Davis, & Keating, 2016).

We will discuss methods using the set of example polyelectrolytes shown in Fig. 2. Abbreviations and chemical properties for these polyelectrolytes

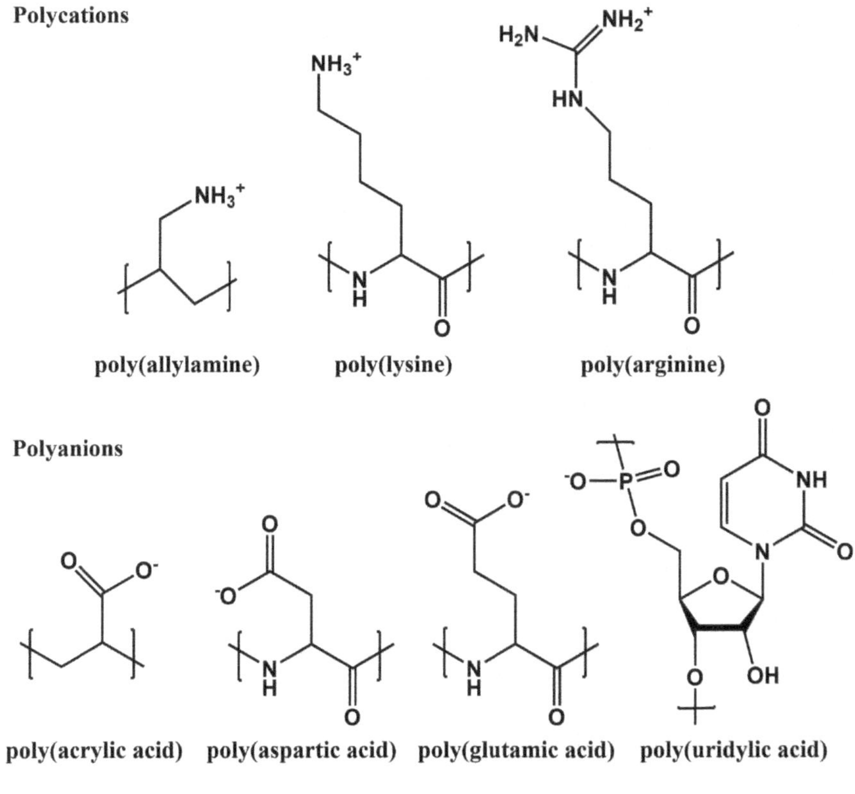

Fig. 2 Monomer chemical structure of several polyelectrolytes for complex coacervate experiments, including both synthetic and biologically inspired monomers.

Table 1 Properties of several polyelectrolytes used to prepare coexisting complex coacervate systems.

Polyelectrolyte	Abbrev.	Molecular weight (g/mol)	Charge/ molecule	Mass per charge (g/mol)
RRASLRRASL	2xRRASL	1185	(+) 4	296
Protamine sulfate[a]	Prot	4236[b]	(+) 21	202
Poly(L-lysine)	Lys20	3300[b]	(+) 20	160
	Lys100	16,000[b]	(+) 100	160
Poly(allylamine hydrochloride)	PAH	17,500[b]	(+) 300	58
Poly(acrylic acid)	PAA	1800[b]	(−) 25	72
Poly(L-glutamic acid)	Glu100	15,000[b]	(−) 100	150
Poly(L-aspartic acid)[c]	Asp100	14,000[b]	(−) 100	140
Poly(uridylic acid)	polyU	600–1000k[b]	(−) 1850–3085	324

[a]Protamine is a mixture of naturally occurring arginine rich peptides of similar sequence (Hoffmann, Chance, & Johnson, 1990).
[b]Average molecular weight.
[c]Poly(L-aspartic acid) synthesis results in random α–β isomerization of the polymer (Aswad, Paranandi, & Schurter, 2000; Clarke, 1987).
Reprinted from Mountain, G. A., & Keating, C. D. (2020). Formation of multiphase complex coacervates and partitioning of biomolecules within them. *Biomacromolecules*, *21*(2), 630–640 with permission from the American Chemical Society.

are displayed in Table 1. We have found for several different collections of polyelectrolytes that if a given solution environment (i.e., pH, buffer, ionic strength, presence and identity of any divalent metal cations) supports LLPS of two separate pairs of oppositely charged polyelectrolytes, this solution environment will also support coexistence of coacervate phases when all polyelectrolytes are added together (Mountain & Keating, 2020). An exception occurs when interactions between any two polyelectrolytes are strong enough to result in gel or solid formation (further discussed in Section 2.4). There are two suggested approaches to screening polymers for use in multi-coacervate systems: (1) with a specific buffer system in mind, for example, if physiological pH and ionic strength, or some other particular solution conditions are important or (2) with specific polyelectrolytes in mind. Flexibility in either the solution conditions or the chemical identity of the polyions will generally enable multiphase coacervate systems to be achieved.

Polyelectrolyte pairs are screened for presence of liquid phase separated droplets based on visual presence of turbidity and optical microscopy.

Turbidity can be performed rapidly and indicates phase separation but does not differentiate between liquid droplets and solid aggregates. Optical microscopy is used to evaluate whether the resulting complexes are solid, gel-like, or liquid in composition. Generally smooth, spherical microstructures indicate that the material has sufficient fluidity to adopt a surface energy minimizing morphology, although it is possible for solidification to occur after such a morphology has been adopted. Observing droplet coalescence is an excellent way to confirm that structures are in fact liquid, and can be used to monitor changes in phase fluidity (Kaur et al., 2019; Wang et al., 2018). Fluorescence recovery after photobleaching (FRAP) of fluorescently labeled polyelectrolytes or a fluorescently labeled probe molecule can be used to determine the mobility of labeled molecules if desired (Boeynaems et al., 2019; Feric et al., 2016).

Solid or gel-like morphologies are most likely to be observed between the two polyelectrolytes of a given collection that have the strongest and/or most numerous associative interactions. Therefore, if screening of these "best" oppositely charged polyelectrolytes (i.e., those having the highest charge density and/or number of charges/molecule) results in liquid coacervate formation, non-liquid assemblies are unlikely to be observed in systems where additional polyelectrolytes are also present. In the event that solid formation is observed between the polyelectrolytes that interact most strongly, the solution ionic strength can be increased. However, it is important to consider that coacervates formed from weakly interacting polyelectrolytes may be dissolved by increasing ionic strength. Therefore, knowing the ionic strength range that supports phase separation between the various possible molecular pairings is helpful before manipulating the solution ionic strength. The temperature of the system can also be adjusted in order to tune interaction strengths. Depending on the system, coacervation may display lower critical solution temperature (LCST), upper critical solution temperature (UCST), or both, and adjusting the temperature in a given direction may not have the same effect for all polyelectrolytes (Nott et al., 2015; Quiroz & Chilkoti, 2015; Simon et al., 2017). We often find it convenient to poise a coacervate system close to a phase transition via tuning one parameter (e.g., ionic strength) so that a small change in another parameter (e.g., temperature) is sufficient to drive the phase transition. Polyelectrolyte concentration can also be increased to encourage phase separation under a given set of solution conditions (ionic strength, pH, temperature); this approach is most useful for systems where one or both polyions has low multivalency.

2.1.1 Screening polyelectrolytes for phase separation in a desired buffer system

In order to select polymers for multiphase coacervate systems with a particular buffer system in mind, one should screen individual polyelectrolyte pairs for phase separation and construct a small library of possible polymer pairs resulting in coacervation in the desired buffer. Aside from the functional groups of a given polymer, the overall length of the polymer also influences the strength of ion pairing. Thus, the polymer length can be tuned to influence the buffer parameters under which complex coacervation will occur. For instance, Spruijt et al. show that the polymer length of polyacrylic acid (PAA)/poly(N,N,-dimethylaminoethyl methacrylate) (PDMAEMA) coacervate has a strong relationship with the critical salt concentration under which coacervation will occur where increasing length of both polymers results in a higher critical salt concentration (higher salt stability) (Spruijt et al., 2010). In general increasing the length (multivalency) of polyelectrolytes will result in a higher ionic strength tolerance but may decrease partitioning of small charged molecules as they compete for charged sited on the oppositely charged polyelectrolyte (Mountain & Keating, 2020). If a particular peptide or nucleic acid sequence is desired such as the "RRASL" pentapeptide sequence, increasing the number of repeats (i.e., RRASLRRASL to RRASLRRASLRRASL) will provide higher ionic strength tolerance (Aumiller & Keating, 2016). In the case of $RRASL_n$ with polyanion poly(uridylic acid), (polyU), addition of one extra RRASL unit from 2xRRASL to 3xRRASL increases the salt tolerance at which the coacervates dissolve by nearly fourfold, from 100 mM NaCl with 2xRRASL/polyU to 350 mM NaCl for 3xRRASL/polyU (at 50 mM HEPES pH 7.4, 4 mM $MgCl_2$) (Aumiller & Keating, 2016). Tables 2 and 3 summarize the phase separation screening data for several polyelectrolytes in two different solution media, one a relatively low ionic strength system (Table 2) and one at physiological ionic strength (Table 3). Screening was performed at three temperatures; we observed different screening results for only one of the systems (2xRRASL/PAA), which formed solids at 37 °C but had no visible phase separation at the lower temperatures tested. Despite the lack of obvious thermal sensitivity in these datasets, many coacervating systems are known to have temperature dependent phase behavior near room temperature, including, for example, the spermine/polyU system (Aumiller et al., 2016). Indeed, this system can exhibit phase separation due to LCST in response to the heat from a researcher's fingers touching the Eppendorf tube that contains them (Aumiller et al., 2016).

Table 2 Example of polyelectrolyte screening table for several polyelectrolyte pairings.[a,b]

		Charge concentration (mM)		Temperature		
Polycation	**Polyanion**	**Polycation**	**Polyanion**	**5 °C**	**20 °C**	**37 °C**
Lys20	PAA	5.5	5	L	L	L
Lys20	PolyU	6.88	1.4	L	L	L
Lys100	Asp100	2.5	5	L	L	L
Lys100	PolyU	5	1.4	L	L	L
Lys100	PAA	5	5	S	S	S
Protamine	PAA	4.8	5	L	L	L
2xRRASL	Asp100	4	5	–	–	–
2xRRASL	Glu100	4	5	–	–	–
2xRRASL	PolyU	4	1.4	L	L	L
2xRRASL	PAA	4	5	–	–	S
PAH	PAA	3.9	5	L	L	L
PAH	PAA	5.35	8.33	L	L	L
PAH	Asp100	3.9	5	L	L	L
PAH	Glu100	3.9	5	L	L	L
PAH	PolyU	3.9	1.4	S	S	S

[a]All samples in 50 mM HEPES pH 7.4, 25 mM NaCl, 2 mM $MgCl_2$.
[b]L, liquid; S, solid; dash, no phase separation observed.

2.1.2 Screening solution parameters for phase separation of desired polyelectrolytes

If specific polyelectrolytes are desired, parameters of the buffer system can be screened in order to find conditions that will allow for liquid phase separation of all of the intended polyelectrolyte pairs. Polyelectrolyte complexation can result in liquid coacervate phase, but sufficiently strong interaction strength between polyelectrolytes can also result in gel-like and/or solid complexation (Wang & Schlenoff, 2014). If non-liquid complexation is observed, increasing the ionic strength of the buffer can result in more liquid complexes through additional charge screening of polyelectrolytes. Temperature can also be adjusted in order to reduce interaction strength between polyelectrolytes, bearing in mind that while some LLPS systems

Table 3 Polyelectrolyte screening table for higher (physiological) ionic strength.[a,b]

		Charge concentration (mM)		Temperature		
Polycation	**Polyanion**	**Polycation**	**Polyanion**	**5 °C**	**20 °C**	**37 °C**
Lys20	Asp100	5	5	L	L	L
Lys20	PolyU	2	1.38	L	L	L
Lys20	PAA	4	3.33	L	L	L
Lys100	Glu10	10	10	L	L	L
Lys100	Glu20	2	2	–	–	–
Lys100	Glu20	10	10	L	L	L
Lys100	Glu20	20	20	S	S	S
Lys100	Glu100	5	5	S	S	S
Lys100	Asp100	5	5	L	L	L
Lys100	PolyU	2	1.38	L	L	L
Lys100	PAA	4	3.33	S	S	S
3xRRASL	Asp100	5	5	–	–	–
3xRRASL	Asp100	10	10	L	L	L
3xRRASL	PolyU	6	1.38	L	L	L

[a]All samples in 50 mM HEPES pH 7.4, 150 mM NaCl, 2 mM $MgCl_2$.
[b]L, liquid; S, solid; dash, no phase separation observed.

exhibit the more intuitive UCSTs, where heating above a critical point results in coacervate dissolution, many LCSTs, where cooling below a critical point results in coacervate dissolution. If your system of interest has an LCST, heating will favor phase separation. In either case, moving closer to the critical temperature point of coacervate dissolution will result in softening of polyelectrolyte interactions and decrease the probability of solid or gel-like complexes. If no coacervation is observed with the desired polyelectrolytes, ensuring that the solution pH is sufficiently far away from the pKa of ionizable groups of the polymers, such that the polyelectrolytes are highly charged and decreasing the ionic strength of the solution, will increase ion pairing strength of the polyelectrolytes. Note that while we have emphasized relatively rapid screens for whether or not LLPS occurs between particular polymer pairs under particular conditions, it is often useful to more carefully map out the salt, temperature, pH, concentration, and/or cation:anion concentration ratio to more fully understand a given coacervation system.

2.2 Turbidity measurements

High turbidity is observed in response to scattering of light from generation of polymer rich-droplets in complex coacervate samples. Measurements of sample turbidity can be made quickly and are helpful in characterizing coacervate systems as the turbidity of a coacervate sample at a set wavelength (here 500 nm) depends on the presence and amount of coacervation in the sample. An alternative wavelength can be used if, for example, molecules within the system have natural absorbance at 500 nm; it is important to report the wavelength at which turbidity is measured. The amount of light scattering from a turbid sample depends on many factors from the wavelength of light to the size, shape, and refractive index of the suspended droplets or particles. We caution against overinterpreting turbidity data, since changes in turbidity can arise for multiple reasons. Turbidity is nonetheless a valuable tool in monitoring for the presence of coacervate droplets in a sample. Some precautions should be taken to ensure consistency due to the possibility of changes in coacervate suspensions over time, for example, as droplets coalesce and sediment. In our experience, some coacervate samples show rapid decreases in turbidity over time, due to sedimentation, while others form slowly after initial mixing and therefore show increases in turbidity for the first few minutes.

Turbidity is calculated using the % transmittance of the sample at a set wavelength as shown in Eq. (1).

$$Turbidity = 100 - \%transmittance \tag{1}$$

In practice, a low transmittance means that most of the light entering the sample has been scattered due to coacervate droplets or other particulates suspended in the sample. Such samples will appear visibly cloudy (opaque). Because settling of droplets can have a large effect on the measured turbidity of a coacervate sample over time, it is imperative to be consistent with the lifetime and handling of the coacervate samples prior to measurement in order to decrease error in replicate measurements. In practice, we typically generate complex coacervate samples, and immediately transfer to a cuvette for turbidity measurement. If the sample is held in the cuvette for an extended period of time, to observe effects of changing temperature for instance, then adequate pipette mixing should be performed prior to each measurement to ensure that settling does not affect the turbidity data. If the sample is held in a cuvette for an extended period of time as measurements are taken (i.e., varying sample temperature), it is recommended to repeat the measurement in both directions (hot to cold, and cold to hot) in order to check for artifacts due to droplet coalescence and settling. Alternatively, several points can be repeated in the middle of the measurement range to ensure that the sample lifetime is not affecting measured results.

Note: Turbidity measurements can also be taken using a microwell plates and a plate reader, which can be useful if you have many samples to run.

2.3 Preparation of coexisting systems

Multiphase complex coacervates can be made in a number of ways, and the order of polymer addition may in some cases allow additional control over the resulting coacervate system. Double coacervate systems can be made using pairs of oppositely charged polymers (i.e., two polycations + two polyanions) or through addition of a single polyelectrolyte to multiple oppositely charged polyelectrolytes. Examples of each are discussed below.

2.3.1 Shared polymer complex coacervates

It is possible to generate coexisting coacervate phases that share a single type of polyion of one sign (e.g., the polycation) and have multiple oppositely charged polyions (e.g., >1 polyanion) as depicted in Fig. 3. For instance, addition of 6.4 mM charge Lys100, and 5 mM charge Asp100 polypeptides to 1.4 mM charge polyU RNA ($n = \sim 2500$) results in two distinct coacervate phases that share Lys100 as a common polycation (Fig. 4) (Mountain & Keating, 2020). When generating these kind of shared polymer coacervate systems the order of addition does not appear to affect the resulting coacervates as much as for systems having multiple polycations and multiple polyanions, where metastable complexes may need to be disrupted for equilibrium to be reached. In practice, we premix like-charged polyelectrolytes prior to addition of oppositely charged polyelectrolytes (i.e., polyanions Asp100, and polyU are premixed prior to addition of polycation Lys100). It seems likely that more than two coexisting coacervate phases could be generated using this type of shared polymer system by including additional polyions (in this example, a third or fourth polyanion). Practical considerations for such efforts include: (1) The total charge concentrations for all polycations should approximately match the total charge concentration for all polyanions. This is to provide sufficient excess polycation to interact with even the least-favorable of the multiple polyanions. (2) The resulting

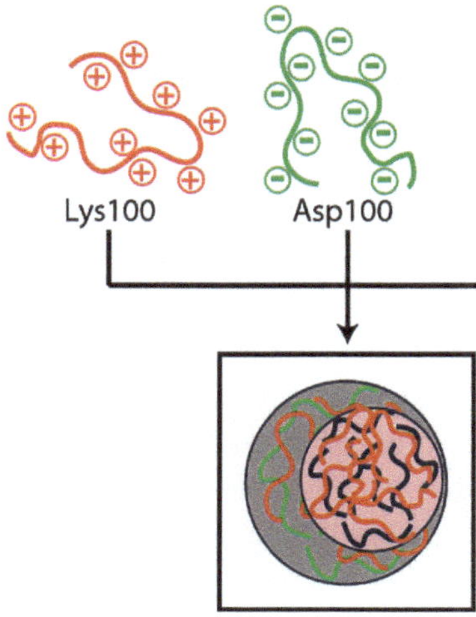

Fig. 3 Combining a single polycation (Lys100) with a mixture of polyanions (Asp100, and polyU) at concentrations such that the total charge concentration of polycation to polyanion is approximately 1:1, results in multiphase coacervates that share the polycation. *Reprinted from Mountain, G. A., & Keating, C. D. (2020). Formation of multiphase complex coacervates and partitioning of biomolecules within them.* Biomacromolecules, *21(2), 630–640 with permission from the American Chemical Society.*

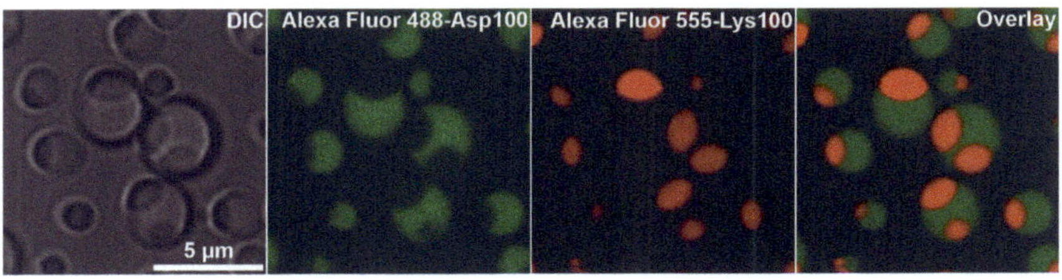

Fig. 4 Fluorescence confocal microscopy images of double coacervate system of Lys100, Asp100, and polyU. Individual channels false colored to indicate which labeled molecule was present and an overlay of fluorescence channels are displayed. Both polyanions (Asp100, polyU) were added simultaneously, followed by the polycation (Lys100). *Reprinted from Mountain, G. A., & Keating, C. D. (2020). Formation of multiphase complex coacervates and partitioning of biomolecules within them.* Biomacromolecules, *21(2), 630–640 with permission from the American Chemical Society.*

coacervate phases must be sufficiently different to form multiple phases. Different lengths of the same polyelectrolytes may not be sufficient to induce multiple coacervate phases to form (i.e., Lys20 + Lys100 + polyU formed uniform rather than multiphase coacervates in our experiments)

(Mountain & Keating, 2020). (3) The distribution of the shared polyelectrolyte across coacervate phases will reflect the relative strength of its ion pairing interactions with other polyelectrolytes of the system. Within the Lys100, Asp100, polyU system shown in Fig. 4 we observe a much higher magnitude of Lys100 in the polyU phase (red, due to high labeled Lys100 signal) than in the Asp100 phase (green, due to labeled Asp100 signal). Although there is some Lys100 present even in the Asp100-rich phase, the PolyU used in these experiments is comparatively a much longer, more multivalent, polyelectrolyte than Asp100. Polyelectrolyte characteristics such as multivalency and charge density will tune the resulting polymer distributions observed in these shared coacervate systems (Mountain & Keating, 2020). Below we detail the procedure for preparing the multiphase coacervates shown in Figs. 3 and 4. Both systems correspond to the buffer condition of Table 2.

2.3.1.1 Materials

1. 500 mM HEPES, pH 7.4
2. 1 M NaCl
3. 50 mM $MgCl_2$
4. 500 mM charge aspartic acid sodium salt ($n = 100$, Alamanda Polymers), (Asp 100) in dI water (spiked with ~1% fluorescently labeled Asp 100)
5. 500 mM charge lysine hydrochloride salt ($n = 100$, Alamanda Polymers), (Lys100) in dI water (spiked with ~1% fluorescently labeled Lys100)
6. 145 mM charge polyuridylic acid potassium salt, (polyU, 600–1000 kDa, Sigma-Aldrich) in dI water
7. #1.5 glass coverslips
8. 0.12 mm × 9 mm adhesive spacers (secure-seal, Life Technologies)

Note: Concentrations are given in mM charge. This is done because for complex coacervation, ion pairing drives the intermolecular interactions and hence is more readily interpreted than molarity for polyions of varying size and charge density. Charge concentration is equal to molar concentration multiplied by the net charge of the molecule at the systems pH.

Note: Fluorescently labeled peptides are prepared as described in Section 2.5.1.

Note: The coacervate phase may wet the glass coverslip such that droplets are greatly deformed by spreading, or even form a uniform film over the entire surface. Coverslips can be silanized to reduce wetting, for example, using an organosilane with a "nonstick" terminal moiety. To silanize, first clean the slides by base treatment (e.g., saturated KOH in isopropanol for 30 min). React clean, dry slides with a toluene solution of 3 mg/mL N-(triethoxysilylpropyl)-O-polyethylene oxide urethane for 4 h, rinse and dry before use (Mountain & Keating, 2020).

2.3.1.2 Procedure

For a final sample volume of 250 μL with final concentrations of 50 mM HEPES pH 7.4, 25 mM NaCl, 2 mM $MgCl_2$, 6.4 mM charge Lys100, 5 mM charge Asp100, and 1.4 mM charge polyU.

1. Add 200.6 μL of dI water, 25 μL of 500 mM HEPES buffer, 6.25 μL of 1 M NaCl, and 10 μL of 50 mM $MgCl_2$ solutions to the centrifuge tube and vortex to mix
2. Add 2.41 μL poly(uridylic acid) and 2.5 μL 500 mM charge poly(aspartic acid) ($n = 100$, Alamanda Polymers) then vortex to mix
3. Add 3.2 μL of 500 mM charge poly(lysine) ($n = 100$, Alamanda Polymers) and mix the solution by gentle pipetting
4. Immediately seal ~8 μL between two #1.5 glass coverslips using a 0.12 mm × 9 mm adhesive spacer to prevent evaporation
5. Wait approximately 15–30 min for droplets to settle to the bottom coverslip and then image using inverted fluorescence or optical microscopy

2.3.2 Pairwise multiphase complex coacervates

Multiphase complex coacervate systems can also be generated from mixtures of multiple polycations and multiple polyanions (Lu & Spruijt, 2020; Mountain & Keating, 2020). Based on screening of individual polyelectrolyte pairs as discussed earlier in Section 2.1.1, if liquid phase separation is intended, polymer pairs should be chosen such that no combination of selected polyelectrolytes results in solid formation. Resulting multiphase coacervates will not retain binary pairings of polymer but will instead reach some equilibrium with all present polyelectrolytes in the system determined by the relative interaction strength of polyelectrolytes present (discussed further in Section 2.5) (Mountain & Keating, 2020). As a starting point, a charge ratio for each intended polymer pairing should be close to 1:1 (positive to negative charge), with potential optimization necessary especially for polyelectrolyte pairs of non-equal length. Like-charged polyelectrolytes can be premixed prior to simultaneous addition of all polycations and all polyanions to the sample (Fig. 5A), or added in a stepwise fashion (Fig. 5B) with no observable effect on the morphology or composition of the resulting system (Mountain & Keating, 2020). Sequential order of addition (Fig. 5B) can be used in order to mitigate solid/gel formation in non-compatible polyelectrolyte systems where one of the potential polyelectrolyte pairings leads to formation of a solid (see Section 2.4).

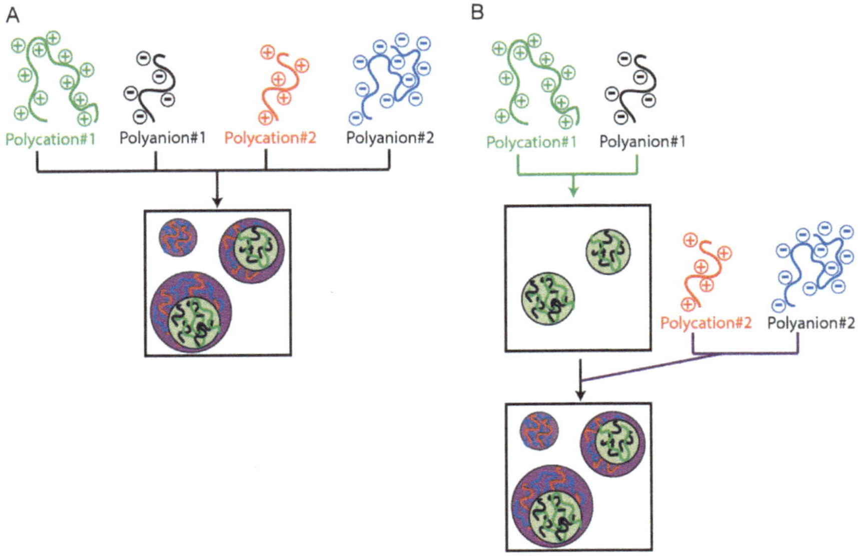

Fig. 5 Mixing scheme for coexisting complex coacervate systems: (A) Addition of all polyelectrolytes simultaneously by premixing of like-charged polyelectrolytes prior to addition to the sample. (B) Polymers are added in pairs, stepwise resulting in generation of one coacervate at a time. *Reprinted from Mountain, G. A., & Keating, C. D. (2020). Formation of multiphase complex coacervates and partitioning of biomolecules within them.* Biomacromolecules, *21(2), 630–640 with permission from the American Chemical Society.*

2.4 Methods of overcoming solid complexation

In cases where solid formation results from sufficiently strong ion pairing interactions between polyelectrolytes, it is still possible to achieve multi-compartment systems by generating phases in a stepwise fashion (Fig. 5B), or by generating the individual coacervates in separate containers prior to mixing the two coacervate systems. In either case it is important to allow some time for equilibration of the initial coacervate phase before introduction of the second coacervate phase components. In practice we find that 3–5 min is an appropriate amount of time to allow coacervates to form and equilibrate before addition to a multiphase system, allowing otherwise incompatible polymers to coexist in multi-compartment coacervate systems.

This is demonstrated in systems containing PAH, PAA, RRASLRRASL peptide, and polyU, where PAH and polyU will undergo solid complexation if no control is placed over order of addition for the system due to strong ion pairing interactions (Fig. 6A). Generation of each coacervate phase in a stepwise manner results in more spherical coacervate droplets and greatly reduced presence of large aggregated solids (Fig. 6B). However, it is important to note that although these systems are metastable for more than 3 days, introducing sheer force through vigorous pipette or vortex mixing will result in solid aggregates. Here, this is likely due to the PAH and polyU gaining access to each other, but may also arise in part from the effect of shear force itself on the interactions between polyelectrolytes. Shear-induced change in physical properties has been observed in comparisons of droplets produced in microfluidic channels vs free solution (van Swaay, Tang, Mann, & de Mello, 2015), and has now been well characterized across a range of biomolecular condensate compositions in a recent report from Knowles and coworkers (Shen et al., 2020).

2.4.1 Materials

1. 500 mM HEPES, pH 7.4
2. 1 M NaCl
3. 50 mM $MgCl_2$
4. (278 mM charge) poly(acrylic acid), 1.8 kDa in dI water
5. (214 mM charge) poly(allylamine hydrochloride), 17.5 kDa, (PAH) in dI water (spiked with ~1% Alexa Fluor-488 fluorescently labeled PAH, 17.5 kDa)
6. 40 mM charge RRASLRRASL peptide (Genscript, spiked with ~1% TAMRA labeled RRASLRRASL)
7. 145 mM charge poly(uridylic acid potassium salt), (polyU, 600–1000 kDa, Sigma-Aldrich) in dI water
8. #1.5 glass coverslips
9. 0.12 mm × 9 mm adhesive spacers (secure-seal, Life Technologies)

*Note: The coacervate phase may wet the glass coverslip such that droplets are greatly deformed by spreading, or even form a uniform film over the entire surface. Coverslips can be silanized to reduce wetting, for example, using an organosilane with a "nonstick" terminal moiety. To silanize, first clean the slides by base treatment (e.g., KOH in isopropanol for 30 min). React clean, dry slides with a toluene solution of 3 mg/mL N-(triethoxysilylpropyl)-O-polyethylene oxide urethane for 4 h, rinse and dry before use (*Mountain & Keating, 2020).

2.4.2 Procedure

Two separate samples will be made for (1) 4.03 mM charge PAH and 5.0 mM charge PAA (charge ratio of 0.81, PAH/PAA) and (2) 4 mM charge

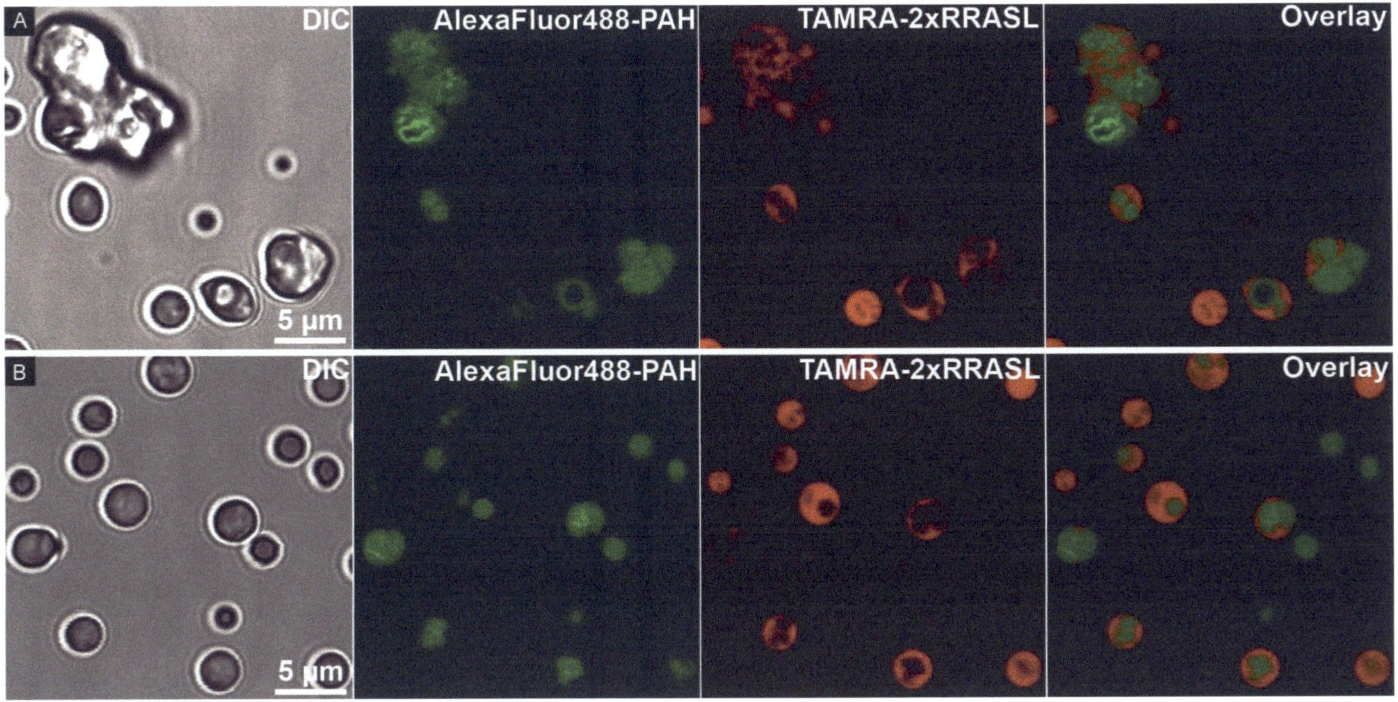

Fig. 6 Fluorescence confocal microscopy image of double coacervate system of PAH/PAA and 2xRRASL/polyU. Fluorescently labeled Alexa Fluor 488-PAH (green) and TAMRA-2xRRASL (red) were doped in to aid in visualization of individual phases. Brightness has been enhanced to aid in visualization. (A) Gel-like aggregates of polyelectrolytes are obtained rather than liquid droplets if the order of addition of polyelectrolytes is not controlled for this set of polyelectrolytes. (B) Making each coacervate system in separate tubes and allowing a 3 min equilibration period before mixing reduces the occurrence of aggregate like assemblies.

2xRRASL and 1.4 mM charge polyU each to a final sample volume of 250 μL each, and final concentrations of 50 mM HEPES pH 7.4, 25 mM NaCl, 2 mM $MgCl_2$. This the buffer condition corresponds to Table 2 screening data.

PAH/PAA coacervate sample

1. Add 199.54 μL of dI water, 25 μL of 500 mM HEPES buffer, 6.25 μL of 1 M NaCl, and 10 μL of 50 mM $MgCl_2$ solutions to the centrifuge tube and vortex to mix
2. Add 4.71 μL of 214 mM charge poly(allylamine hydrochloride), (PAH) and vortex to mix
3. 4.5 μL of 278 mM charge poly(acrylic acid), (PAA) and gently pipette mix 2–3 times

2xRRASL/polyU coacervate sample

1. Add 181.34 μL of dI water, 25 μL of 500 mM HEPES buffer, 6.25 μL of 1 M NaCl, and 10 μL of 50 mM $MgCl_2$ solutions to the centrifuge tube and vortex to mix
2. Add 25 μL of 40 mM charge 2xRRASL and vortex to mix
3. Add 2.41 μL of 145 mM poly(uridylic acid), (polyU) and gently pipette mix 2–3 times

Generating double-coacervate sample

1. Place 100 μL of each coacervate system into a third tube and gently pipette mix 2–3 times
2. Immediately seal ~8 μL of mixed sample between two #1.5 glass coverslips using a 0.12 mm × 9 mm adhesive spacer to prevent evaporation
3. Wait approximately 15–30 min for droplets to settle to the bottom coverslip and then image using inverted fluorescence or optical microscopy

We found that tuning the polyelectrolyte ratios of individual polyelectrolyte pairs can further reduce polyelectrolyte incompatibilities resulting in solid formation. PAH/PAA coacervates at a charge ratio near 1:1 (0.95—PAH/PAA) exhibit a positive surface charge, measured as zeta potential $(\zeta) = +17 \pm 17$ mV. 2xRRASL/polyU coacervates generated at a charge ration of 2.9 (2xRRASL/polyU) exhibit a ζ potential of -15 ± 8 mV. PAH/PAA ($\zeta = +17 \pm 17$ mV) and 2xRRASL/polyU coacervates ($\zeta = -15 \pm 8$ mV) prepared in separate tubes and mixed following a 3 min equilibration window (Fig. 5A) show some degree of aggregate formation reminiscent of Fig. 6A. Decreasing the charge ratio of the PAH/PAA coacervate system to 0.81 (less cation) shown in Fig. 7B shifts the droplet surface charge to near neutral ($\zeta = +2 \pm 9$ mV) although aggregates are still observed when mixing with 2xRRASL/polyU coacervates. Further decreasing the

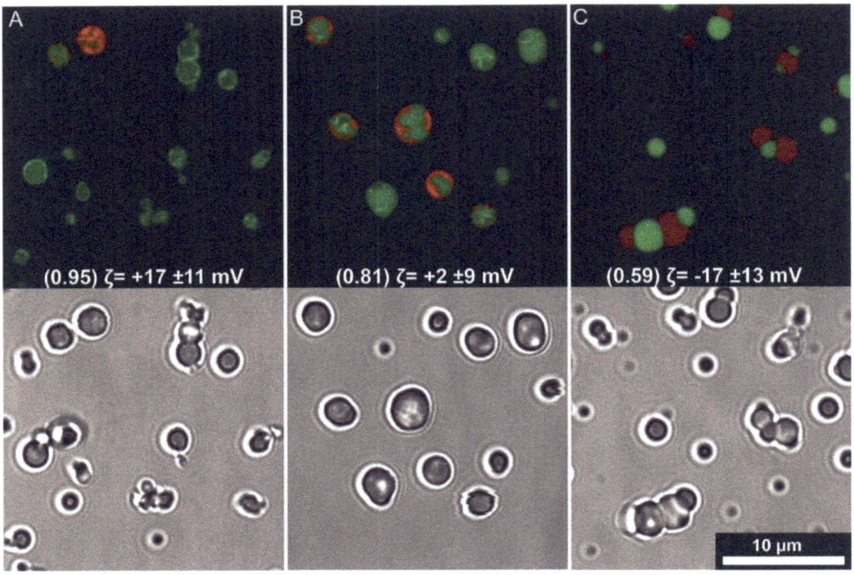

Fig. 7 The characteristics of interactions between two unique coacervate systems can be influenced by altering the charge ratio of the coacervate phase. The charge ratio of 2xRRASL/polyU coacervate system (with TAMRA-2xRRASL, false colored red) is held at 2.9 yielding a surface charge, zeta potential $(\zeta) = -15 \pm 8$ mV. The charge ratio of PAH/PAA (with Alexa Fluor 488-PAH, false colored green) is altered which in turn affects the observed zeta potential of the PAH/PAA phase (charge ratio and ζ of the PAH/PAA phase in each case is shown). Fluorescence and DIC channels are shown for each formulation. Error represents the standard deviation in ζ potential of triplicate measurements of separately made samples.

charge ratio of PAH/PAA to 0.59 results in a ζ potential of -17 ± 13 mV which when mixed with 2xRRASL/polyU coacervates (Fig. 7C) no longer generates observable aggregate material but instead produces alternating chains of spherical droplets that are observed to be stable for up to 3 days. This droplet chaining may indicate electrostatic repulsion between droplets of the same composition.

2.5 Probe partitioning

The partitioning of solutes in complex coacervate systems depends on the composition of the coacervate phase, and can be at least partially understood in terms of ion pairing interactions between the solute of interest and the coacervate polyelectrolyte components (Aumiller et al., 2016; Aumiller & Keating, 2016; Mountain & Keating, 2020). For example, we found that the partitioning of small oligonucleotide and peptide solute probes into complex coacervate systems depends on: (1) the magnitude of ion pairing interactions between the probe and the oppositely charged polyelectrolyte of the coacervate, which drives accumulation, (2) the ion pairing capability of the coacervates like-charged polyelectrolyte (with respect to the probe), which opposes accumulation, and (3) the availability of specific binding interactions such as Watson-Crick base pairing, which can greatly increase solute partitioning into coacervates (Aumiller et al., 2016; Mountain & Keating, 2020).

When performing partitioning experiments within a singular coacervate phase, the most important issue is ensuring that adequate time is allowed for equilibrium to be reached within the sample. In practice, this can be difficult when there is a very strong interaction between the probe molecule and the polyelectrolyte polymers of the coacervate system. One way to alleviate this concern is to introduce the fluorescent probe into solution prior to the addition of either polyelectrolyte ensuring an even dispersion of the fluorescent probe prior to induction of coacervation by adding polyelectrolytes to the sample. Additionally, allowing an equilibration time of 30 min or more after generation of the coacervate allows for settling of most coacervate droplets for imaging with inverted microscopy. Probe fluorescence intensity is measured in raw fluorescence via confocal microscopy and compared to a standard curve of fluorescent probe in buffer (in the absence of polyelectrolytes) which is serially diluted to create a calibration curve. Because partitioning of some fluorescent probes is so strong ($\geq$1000 ×), it is important to make sure that the amount of probe added to the sample (typically to final concentration of 0.01–0.1 μM for our samples) is sufficiently small that the fluorescence intensity of the coacervate phase does not surpass the fluorescence intensity of the stock solution of fluorescent probe, such that it is outside the range of the calibration curve. Additionally, be sure that the maximum pixel intensity for the coacervate droplets is below saturation. If too many pixels are saturated the data will not accurately be translated to a concentration value through the calibration curve of fluorescent probe. If necessary, repeat the experiment with a reduced concentration of added probe to bring the fluorescence intensity in the coacervate phase into the range for which a calibration curve can be prepared. Once labeled probe concentrations and microscope settings are appropriate to allow meaningful fluorescence measurements, images of the coacervate sample can be taken. To achieve an accurate representation of the fluorescence intensity of a coacervate sample, a total of 10+ random droplets from the microscopy image are measured for fluorescence intensity and the region measured is slightly smaller than the full area of the coacervate droplet (to eliminate inconsistencies in fluorescence resulting from curvature at the edge of the droplets). When using a confocal microscope, the slice thickness of the image can be compared to the diameter of the coacervate droplet to yield insight on the curvature of the droplet and the appropriate size of measurement area. All coacervate samples analyzed are prepared in triplicate to obtain a standard deviation for probe concentration in the coacervate phase.

2.5.1 Fluorescently labeling polyelectrolytes

When working with more than a single pair of polyelectrolytes the composition of the coacervate phase becomes complicated as polyelectrolytes distribute across the multiple coacervate phases based on the comparative properties of the polyelectrolytes involved. One way to probe this behavior is by introducing a small amount of fluorescently tagged polyelectrolyte and observing its relative distribution throughout the sample. Quantification of partitioning coefficients by fluorescence microscopy is often limited by the very low intensity from the dilute phase; this can be addressed by performing bulk fluorescence measurements (i.e., in a small-volume cuvette), which is generally possible due to the relatively large volume of the dilute supernatant phase compared to the coacervate phase. The commercial availability of fluorescently labeled polyelectrolytes is limited. It therefore may be necessary, and will likely be more cost efficient, to fluorescently label polyelectrolytes in-house. We generally choose to use small molecule fluorophores (like rhodamine, fluorescein, or Alexa Fluor® dyes) rather than large macromolecule-based fluorophores like green fluorescent protein (GFP) because their smaller size (in terms of molecular weight) minimizes impact on the net molecular properties of the molecule, and hence minimizes any difference in its partitioning compared to that of the unlabeled polyelectrolyte. For the same reason, labels on higher-molecular weight polyelectrolytes are considered less perturbative than those on smaller molecular weight polyions, where the label consequently makes up a larger fraction of the overall labeled molecule and can have a larger impact on its behavior. For low molecular weight solutes, radioisotope labeling can be a good option that avoids changing the solute chemistry and allows quantification over a very large dynamic range; (Dewey et al., 2014; Frankel, Bevilacqua, & Keating, 2016) drawbacks include the inability to perform while imaging and the need for careful attention safety considerations.

Longer wavelength dyes (i.e., Alexa Fluor 647 vs Alexa Fluor 488) typically have a larger extinction coefficient allowing for greater signal at lower degrees of labeling. A variety of reactive fluorescent dyes can be purchased with specific reactivity for chemical functional groups such as primary amines, and carboxylic acid groups which are common in polyelectrolyte systems. These reactive dyes typically come with detailed manufacturer suggested protocols for labeling. For instance, labeling of primary amine containing polyelectrolytes with N-succinimidyl (NHS) ester functionalized reactive dyes selectively reacts with primary amine groups of polyelectrolytes and forms a stable amide bond between the amine containing molecule and the dye as shown in Fig. 8.

2.5.2 Purification of fluorescently labeled polyelectrolytes

Small charged molecules, such as unreacted dyes remaining after polyelectrolyte labeling reactions, can be expected to partition differently than the labeled polyelectrolytes (Aumiller et al., 2016; Aumiller & Keating, 2016; Marianelli, Miller, & Keating, 2018; Mountain & Keating, 2020). Therefore, it is very important to ensure that all unreacted dye remaining from polyelectrolyte labeling has been removed. Several premanufactured options exist for purification of fluorescently labeled macromolecules such as Zebra™ desalting columns (Thermo Fischer Scientific) at various molecular weight cutoff points. In our experience, superior purification was achieved at the cost of convenience by using conventional glass columns packed with size exclusion purification resin, such as GE Healthcare Life Sciences Sephadex® resins, or Bio-Rad Laboratories Bio-Gel® resins. Various size exclusion ranges are available for these options and can be selected is such a way that the labeled polyelectrolyte is quickly and easily separated from the unreacted dye so long as the polyelectrolyte molecular weight is sufficiently larger than the dye molecular weight. For example,

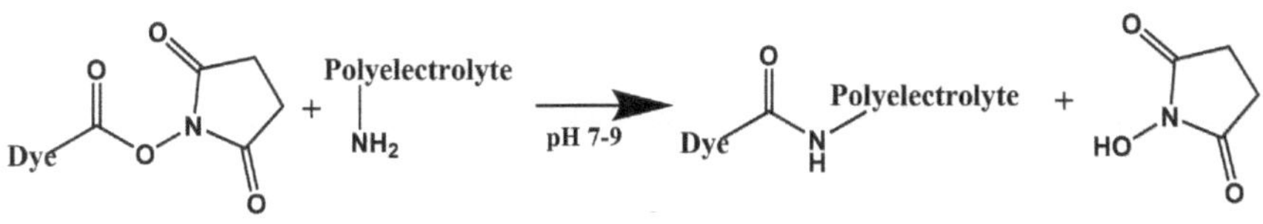

Fig. 8 Reaction scheme of N-succinimidyl ester reactive dyes with primary amine containing molecules results in stable amide bond formation between the dye and polyelectrolyte. The reaction relies on the primary amine containing molecule to act as a nucleophile, therefore an elevated pH (pH: 7–9) is required for efficient conjugation.

Fig. 9 Purification of Alexa Fluor® 546-NHS ester labeling of Lys100 using Bio-Gel® P4 size exclusion resin. The faster moving pink band (bottom) corresponds to functionalized Lys100 which is separated from unfunctionalized dye (slower moving pink band, top), and can be collected as it elutes from the column.

Fig. 9 shows a purification of Alexa Fluor 546 labeled poly(L-lysine) reaction mixture using Bio-Rad Bio-Gel® P-4 resin. Alexa Fluor 546-NHS ester has a molecular weight of 1159.6 g/mol, and poly(L-lysine) $n = 100$ has an average molecular weight of 16,000 g/mol. Employing Bio-Rad Bio-Gel® P-4 resin with a size exclusion range of 800–4000 g/mol allows the functionalized Lys100 flow quickly though the column separating it from the unreacted dye which falls within the size exclusion range of the resin and travels much more slowly. The column diameter can be modified to suit different scale labeling reactions (i.e., larger labeling volumes would require a larger diameter column for optimum separation and reduced sample dilution), and the column bed volume can be modified to achieve better separation and minimize sample dilution (i.e., a larger bed volume would increase separation at the cost of increased sample dilution as the sample band broadens over time).

2.5.3 Importance of avoiding over-labeling

Finally, one of the most important practical consideration when fluorescently labeling molecules is that the change in net charge and molecular weight of the resulting labeled macromolecule or polyelectrolyte should be sufficiently small as to preserve the physical and chemical characteristics of the polymer. Over-labeling of polyelectrolytes with dye moieties can alter their partitioning and phase behavior. For example, introducing

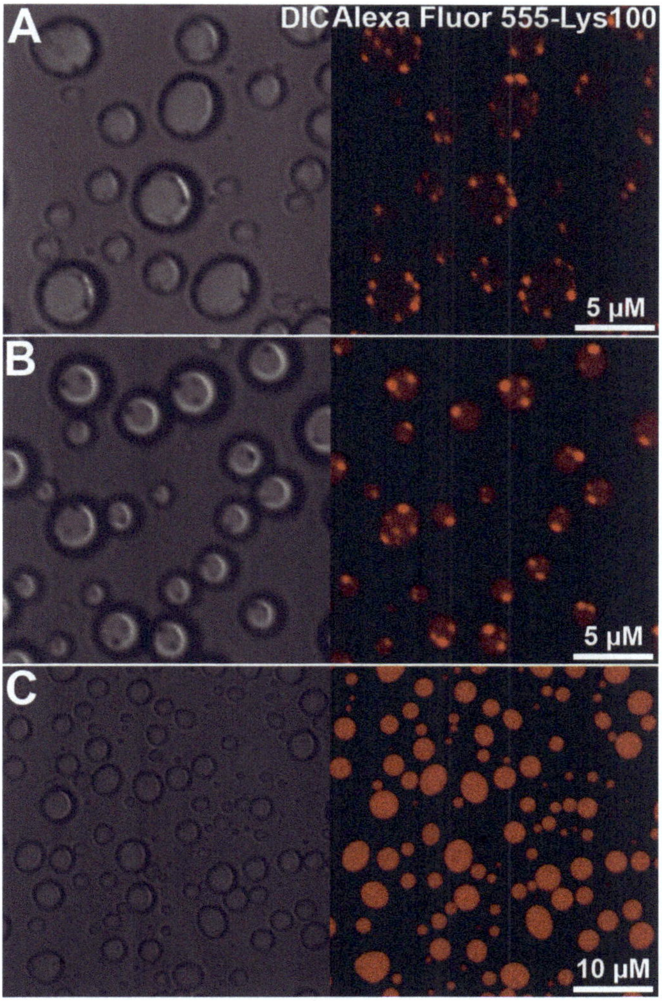

Fig. 10 Coacervates composed of 5 mM charge Lys100 and 5 mM charge Asp100 with added Alexa Fluor 555 labeled Lys100 accounting for approximately 1% of the total Lys100 concentration at the following theoretical degrees of labeling: (A) 20 labels per molecule, (B) 10 labels per molecule, and (C) 3 labels per molecule.

fluorescently labeled Lys100 with a theoretical degree of labeling (DOL) of ~20 dyes per polymer using Alexa Fluor 555-NHS ester into a coacervate system of unlabeled 5 mM charge Lys100 and 5 mM charge Asp100 results in a low level of diffuse fluorescence throughout the coacervate droplets and formation of highly fluorescent puncta within coacervate droplets (Fig. 10A). Labeled Lys100 accounts for approximately 1% of the total

Lys100 in the sample. It turns out these puncta are the result of the over-labeled Lys100, which behaves as a chemically distinct species and drives formation of a new phase. Labeled and unlabeled Lys100 molecules may interact differently with the polyanion and/or intramolecular interactions between the negatively charged Alexa Fluor dye and the remaining unreacted Lys sidechains could be effectively generating an amphiphilic molecule as a result of over-labeling. Reducing the theoretical DOL from 20 to 10 does not eliminate puncta but does result in increased diffuse fluorescence throughout the coacervate droplets (Fig. 10B). Reducing the theoretical DOL further to three dyes per polymer of Lys100 eliminates puncta formation and results in homogenous fluorescence throughout the coacervate droplet (Fig. 10C). In practice, the DOL should be minimized to avoid any change in polymer characteristics, typically a DOL of 1–3% of monomers yields sufficient fluorescence signal. Using fluorophores with higher extinction coefficients like Alexa Fluor 647 ($\varepsilon = 270{,}000\,\mathrm{cm}^{-1}\,\mathrm{M}^{-1}$) allows for lower DOL while retaining high fluorescence signal.

Acknowledgments

This work was supported by the National Science Foundation, grant MCB-1715984. The authors thank Jacob Shaffer for helpful comments on the manuscript.

References

Aswad, D. W., Paranandi, M. V., & Schurter, B. T. (2000). Isoaspartate in peptides and proteins: Formation, significance, and analysis. *Journal of Pharmaceutical and Biomedical Analysis*, *21*(6), 1129–1136.

Aumiller, W. M., Cakmak, F. P., Davis, B. W., & Keating, C. D. (2016). RNA-based coacervates as a model for membraneless organelles: Formation, properties, and interfacial liposome assembly. *Langmuir*, *32*(39), 10042–10053.

Aumiller, W. M., Jr., Davis, B. W., & Keating, C. D. (2014). Phase separation as a possible means of nuclear compartmentalization. *International Review of Cell and Molecular Biology*, *307*, 109–149.

Aumiller, W. M., & Keating, C. D. (2016). Phosphorylation-mediated RNA/peptide complex coacervation as a model for intracellular liquid organelles. *Nature Chemistry*, *8*(2), 129–137.

Beaver, J. E., & Waters, M. L. (2016). Molecular recognition of Lys and Arg methylation. *ACS Chemical Biology*, *11*(3), 643–653.

Boeynaems, S., Holehouse, A. S., Weinhardt, V., Kovacs, D., Van Lindt, J., Larabell, C., et al. (2019). Spontaneous driving forces give rise to protein−RNA condensates with coexisting phases and complex material properties. *Proceedings of the National Academy of Sciences of the United States of America*, *116*(16), 7889.

Brangwynne, C. P., Tompa, P., & Pappu, R. V. (2015). Polymer physics of intracellular phase transitions. *Nature Physics*, *11*, 899.

Bungenberg de Jong, H. G. (1949). Complex colloid systems. In H. R. Kruyt (Ed.), *Vol. II. Colloid Science* (pp. 335–432). Amsterdam: Elsevier.

Bungenberg de Jong, H. G., & Kruyt, H. R. (1929). Coacervation (partial miscibility on colloid systems) (preliminary communication). *Proceedings of the Royal Academy of Sciences at Amsterdam, 32*, 849–856.

Chang, L.-W., Lytle, T. K., Radhakrishna, M., Madinya, J. J., Vélez, J., Sing, C. E., et al. (2017). Sequence and entropy-based control of complex coacervates. *Nature Communications, 8*(1), 1273.

Choi, J.-M., Holehouse, A. S., & Pappu, R. V. (2020). Physical principles underlying the complex biology of intracellular phase transitions. *Annual Review of Biophysics, 49*(1), 107–133.

Clarke, S. (1987). Propensity for spontaneous succinimide formation from aspartyl and asparaginyl residues in cellular proteins. *International Journal of Peptide and Protein Research, 30*(6), 808–821.

Crowe, C. D., & Keating, C. D. (2018). Liquid-liquid phase separation in artificial cells. *Interface Focus, 8*(5), 20180032.

Dewey, D. C., Strulson, C. A., Cacace, D. N., Bevilacqua, P. C., & Keating, C. D. (2014). Bioreactor droplets from liposome-stabilized all-aqueous emulsions. *Nature Communications, 5*, 4670.

Dignon, G. L., Best, R. B., & Mittal, J. (2020). Biomolecular phase separation: From molecular driving forces to macroscopic properties. *Annual Review of Physical Chemistry, 71*(1), 53–75.

Feric, M., Vaidya, N., Harmon, T. S., Mitrea, D. M., Zhu, L., Richardson, T. M., et al. (2016). Coexisting liquid phases underlie nucleolar subcompartments. *Cell, 165*, 1686–1697.

Frankel, E. A., Bevilacqua, P. C., & Keating, C. D. (2016). Polyamine/nucleotide coacervates provide strong compartmentalization of Mg^{2+}, nucleotides, and RNA. *Langmuir, 32*(8), 2041–2049.

Gucht, J.v.d., Spruijt, E., Lemmers, M., & Cohen Stuart, M. A. (2011). Polyelectrolyte complexes: Bulk phases and colloidal systems. *Journal of Colloid and Interface Science, 361*(2), 407–422.

Hoffmann, J. A., Chance, R. E., & Johnson, M. G. (1990). Purification and analysis of the major components of chum salmon protamine contained in insulin formulations using high-performance liquid chromatography. *Protein Expression and Purification, 1*(2), 127–133.

Hughes, R. M., & Waters, M. L. (2006). Arginine methylation in a β-hairpin peptide: Implications for Arg−π interactions, ΔCp°, and the cold denatured state. *Journal of the American Chemical Society, 128*(39), 12735–12742.

Hyman, A. A., Weber, C. A., & Julicher, F. (2014). Liquid-liquid phase separation in biology. *Annual Review of Cell and Developmental Biology, 30*, 39–58.

Kaur, T., Alshareedah, I., Wang, W., Ngo, J., Moosa, M. M., & Banerjee, P. R. (2019). Molecular crowding tunes material states of ribonucleoprotein condensates. *Biomolecules, 9*(2), 71.

Lu, T., & Spruijt, E. (2020). Multiphase complex coacervate droplets. *Journal of the American Chemical Society, 142*, 2905–2914.

MacEwan, S. R., & Chilkoti, A. (2010). Elastin-like polypeptides: Biomedical applications of tunable biopolymers. *Peptide Science, 94*(1), 60–77.

Marianelli, A. M., Miller, B. M., & Keating, C. D. (2018). Impact of macromolecular crowding on RNA/spermine complex coacervation and oligonucleotide compartmentalization. *Soft Matter, 14*(3), 368–378.

Mitrea, D. M., & Kriwacki, R. W. (2016). Phase separation in biology; functional organization of a higher order. *Cell Communication and Signaling: CCS, 14*(1), 1–20.

Mountain, G. A., & Keating, C. D. (2020). Formation of multiphase complex coacervates and partitioning of biomolecules within them. *Biomacromolecules, 21*(2), 630–640.

Nakashima, K. K., Vibhute, M. A., & Spruijt, E. (2019). Biomolecular chemistry in liquid phase separated compartments. *Frontiers in Molecular Biosciences*, *6*, 21.

Nott, T. J., Petsalaki, E., Farber, P., Jervis, D., Fussner, E., Plochowietz, A., et al. (2015). Phase transition of a disordered nuage protein generates environmentally responsive membraneless organelles. *Molecular Cell*, *57*(5), 936–947.

Perry, S. L. (2019). Phase separation: Bridging polymer physics and biology. *Current Opinion in Colloid & Interface Science*, *39*, 86–97.

Priftis, D., Xia, X., Margossian, K. O., Perry, S. L., Leon, L., Qin, J., et al. (2014). Ternary, tunable polyelectrolyte complex fluids driven by complex coacervation. *Macromolecules*, *47*(9), 3076–3085.

Quiroz, F. G., & Chilkoti, A. (2015). Sequence heuristics to encode phase behaviour in intrinsically disordered protein polymers. *Nature Materials*, *14*(11), 1164–1171.

Rowland, A. T., Cacace, D. N., Pulati, N., Gulley, M. L., & Keating, C. D. (2019). Bioinspired mineralizing microenvironments generated by liquid–liquid phase coexistence. *Chemistry of Materials*, *31*(24), 10243–10255.

Shen, Y., Ruggeri, F. S., Vigolo, D., Kamada, A., Qamar, S., Levin, A., et al. (2020). Biomolecular condensates undergo a generic shear-mediated liquid-to-solid transition. *Nature Nanotechnology*. https://doi.org/10.1038/s41565-020-0731-4.

Shin, Y., & Brangwynne, C. P. (2017). Liquid phase condensation in cell physiology and disease. *Science*, *357*(6357), eaaf4382.

Simon, J. R., Carroll, N. J., Rubinstein, M., Chilkoti, A., & Lopez, G. P. (2017). Programming molecular self-assembly of intrinsically disordered proteins containing sequences of low complexity. *Nature Chemistry*, *9*(6), 509–515.

Spruijt, E., Westphal, A. H., Borst, J. W., Cohen Stuart, M. A., & van der Gucht, J. (2010). Binodal compositions of polyelectrolyte complexes. *Macromolecules*, *43*(15), 6476–6484.

Tatko, C. D., & Waters, M. L. (2003). The geometry and efficacy of cation–π interactions in a diagonal position of a designed β-hairpin. *Protein Science*, *12*(11), 2443–2452.

van Swaay, D., Tang, T.-Y. D., Mann, S., & de Mello, A. (2015). Microfluidic formation of membrane-free aqueous coacervate droplets in water. *Angewandte Chemie International Edition*, *54*(29), 8398–8401.

Wang, J., Choi, J.-M., Holehouse, A. S., Lee, H. O., Zhang, X., Jahnel, M., et al. (2018). A molecular grammar governing the driving forces for phase separation of prion-like RNA binding proteins. *Cell*, *174*(3), 688–699.e16.

Wang, R., Ozsvar, J., Yeo, G. C., & Weiss, A. S. (2019). Hierarchical assembly of elastin materials. *Current Opinion in Chemical Engineering*, *24*, 54–60.

Wang, Q., & Schlenoff, J. B. (2014). The polyelectrolyte complex/coacervate continuum. *Macromolecules*, *47*(9), 3108–3116.

CHAPTER SIX

Methods for characterizing the material properties of biomolecular condensates

Ibraheem Alshareedah, Taranpreet Kaur, and Priya R. Banerjee*
Department of Physics, University at Buffalo, Buffalo, NY, United States
*Corresponding author: e-mail address: prbanerj@buffalo.edu

Contents

Methods in Enzymology, Volume 646
ISSN 0076-6879
https://doi.org/10.1016/bs.mie.2020.06.009

Abstract

Biomolecular condensates are membrane-less sub-cellular compartments that perform a plethora of important functions in signaling and storage. The material properties of biomolecular condensates such as viscosity, surface tension, viscoelasticity, and macromolecular diffusion play important roles in regulating their biological functions. Aberrations in these properties have been implicated in various neurodegenerative disorders and certain types of cancer. Unraveling the molecular driving forces that control the fluid structure and dynamics of biomolecular condensates across different length- and time-scales necessitates the application of innovative biophysical methodologies. In this chapter, we discuss major experimental techniques that are widely used to study the material states and dynamics of biomolecular condensates as well as their practical and conceptual limitations. We end this chapter with a discussion on more advanced tools that are currently emerging to address the complex fluid dynamics of these condensates.

1. Introduction

Biomolecular condensates are liquid-like bodies harboring multiple protein and RNA molecules and are believed to form *via* liquid-liquid phase separation (Alberti, 2017a; Banani, Lee, Hyman, & Rosen, 2017; Forman-Kay, Kriwacki, & Seydoux, 2018; Hyman, Weber, & Jülicher, 2014; Mitrea & Kriwacki, 2016). The molecular components of a biomolecular condensate can be categorized as either scaffolds or clients (Banani et al., 2016; Ditlev, Case, & Rosen, 2018). Scaffolds are multivalent proteins/RNA/DNA that are necessary to form a biomolecular condensate, their absence directly results in the disappearance of such condensates in the cell. Clients are molecules that localize within biomolecular condensates due to preferable interactions with the scaffolds but may not directly contribute to their formation/stabilization. The fluid properties of a condensate *in vivo* are determined by inspecting whether these condensates undergo coalescence, display jetting or dripping behavior, and exchange molecules with the surrounding environment. Furthermore, the liquid state of a condensate can be assessed by characterizing the diffusion of constituent components within the condensates' internal microenvironment (Brangwynne et al., 2009; Feric et al., 2016; Nott et al., 2015).

It is becoming increasingly clear that the fluid properties of biomolecular condensates are intricately connected to the sequence composition and structure of the scaffolding biopolymers (Alshareedah et al., 2019; Boeynaems et al., 2019; Franzmann et al., 2018; Onuchic, Milin, Alshareedah, Deniz, & Banerjee, 2019; Wang et al., 2018; Zhou et al., 2019). Furthermore, environmental factors such as macromolecular crowding, salt concentration, and pH have been shown to substantially influence the material properties of these condensates (Alberti, 2017b; Kaur et al., 2019). These molecular and environmental cues are essential to the physiological regulation of a condensate's biological function. Alterations in the fluidity and hence the internal dynamics of a condensate by means of gelation and/or aggregation have been implicated in many neurodegenerative disorders (Aguzzi & Altmeyer, 2016; Alberti & Hyman, 2016; Forman-Kay et al., 2018; Gitler, Dhillon, & Shorter, 2017; King, Gitler, & Shorter, 2012; Ramaswami, Taylor, & Parker, 2013; Shin & Brangwynne, 2017). Therefore, understanding the molecular rules governing the material properties of biomolecular condensates is indispensable for mitigating their pathological transformation at the onset of disease.

The majority of the current techniques used to assess the dynamics of biomolecular condensates involve the application of fluorescence microscopy. Recently, optical tweezers have been utilized to manipulate biomolecular condensates and infer some key mechanical properties (Alshareedah et al., 2019; Kaur et al., 2019; Patel et al., 2015; Wang et al., 2018). Here, we present a detailed account of the current methods to probe the physical properties of biomolecular condensates (see Table 1). Our description will

Table 1 Summary of the experimental techniques discussed in this chapter.

Technique	Measured quantity
Active droplet coalescence	• Viscosity:Surface tension
Fluorescence recovery after photobleaching (FRAP)	• Macromolecular diffusion
Single particle tracking (SPT)	• Condensate viscosity
Active microrheology	• Surface tension • Viscosity • Viscous and elastic moduli

focus on conceptual and procedural details of applying these techniques and not on the instrumental aspect, since almost all of these experiments can be done utilizing commercially available instruments without any special requirements.

2. Measuring mesoscale physical properties of liquid droplets using optical trap-induced droplet coalescence: Concepts

One of the hallmarks of liquid droplets is the ability to coalesce and relax to different shapes in response to the governing forces. This is due to the flexibility of the liquid surface and the relatively weak and reversible binding between the molecules constituting the liquid. The formation of protein/RNA condensates by liquid-liquid phase separation is usually manifested in the form of spherical liquid droplets that are enriched in these biomolecules (Brangwynne, Tompa, & Pappu, 2015). These condensates assume spherical geometry (in the absence of an external field) to minimize their surface energy (Berthier & Brakke, 2012; Rowlinson & Widom, 2013). Any deformation from the spherical shape of a droplet is thermodynamically unfavorable due to the creation of pressure gradients on either side of the surface (Leal, 2007). However, such deformities may be stable in the presence of external forces such as shear flow or adhesion to a solid surface (Berthier & Brakke, 2012). The relaxation of a deformed droplet into the equilibrium spherical shape occurs on a time scale that is dictated by the interfacial tension and viscosity of the droplet (Leal, 2007). Specifically, the interfacial tension is the driving force for the relaxation process while the viscosity opposes such a process *via* a drag/friction force. Simply put, a droplet with high surface tension and low viscosity will have a short relaxation time (fast relaxation) while a droplet with low surface tension and high viscosity will have a long relaxation time (slow relaxation) (Leal, 2007). The method presented here utilizes the event of droplet coalescence to extract the characteristic time-scale of relaxation. When two droplets come in contact with each other, a capillary bridge forms between them, joining the droplets together into a single body of liquid that has an ellipsoidal shape (Paulsen, Carmigniani, Kannan, Burton, & Nagel, 2014). The relaxation of this deformed liquid drop back to a spherical shape can be monitored in real time. Conventionally, the parameter used to quantify the relaxation process is the aspect ratio of the droplet (Brangwynne, Mitchison, & Hyman, 2011; Eggers, Lister, & Stone, 1999; Elbaum-Garfinkle et al., 2015; Jeon et al., 2018).

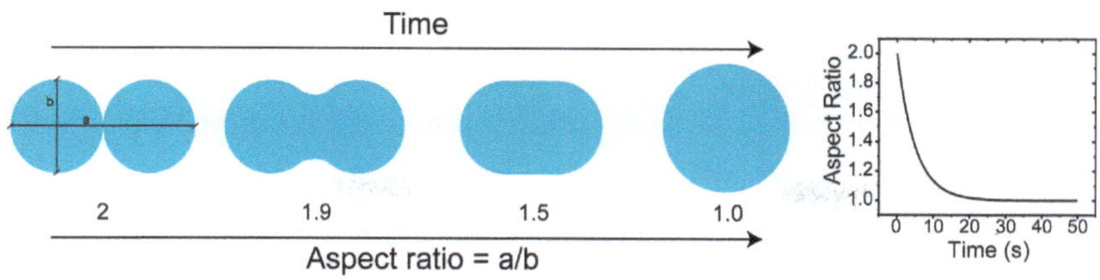

Fig. 1 Probing droplet coalescence using aspect ratio analysis. a and b represent the lengths of the orthogonal axes.

The aspect ratio is defined as the ratio between the lengths of the orthogonal axes of the deformed droplet. During coalescence, the aspect ratio relaxes from an initial value of 2 to a final value of 1 given that the two droplets are identical in size (Fig. 1). In a mathematical form, the aspect ratio (A) can be written as

$$A = 1 + (A_0 - 1)e^{-\frac{t}{\tau}} \tag{1}$$

where τ is the characteristic time of relaxation. In the case of a Newtonian liquid, where there is a linear relationship between stress and strain, we can relate the relaxation time τ to the viscosity η and surface tension γ as

$$\tau = l\left(\frac{\eta}{\gamma}\right) \tag{2}$$

where l is the average diameter of the two droplets (Brangwynne et al., 2011). It is clear from Eq. 2 that the relaxation time also depends on the size of the coalescing droplets. This calls for normalization with respect to the droplet size in order to extract the mesoscale fluid properties of the condensate. For Newtonian liquids, plotting the relaxation time against the average diameter of the coalescing droplets should yield a linear relation with a slope of η/γ. The quantity η/γ is also called the *inverse capillary velocity* (Brangwynne et al., 2011). Droplets that have viscoelastic properties may not display a simple relaxation process. This is due to the relaxation process being accompanied by an increase in the elastic energy of the fluid due to compression, which hinders the relaxation process leading to arrested coalescence (Pawar, Caggioni, Hartel, & Spicer, 2012). In other words, for predominantly elastic condensates, the relaxation process is expected to take an indefinitely long time with no visible change (Dahiya, Caggioni, & Spicer, 2016). Accordingly, this assay is advantageous for judging the condensate material property since it displays a clearly distinct behavior for viscous and viscoelastic condensates.

Conventionally, analyzing droplet coalescence events is done using video microscopy (Brangwynne et al., 2009, 2011; Elbaum-Garfinkle et al., 2015; Jeon et al., 2018). A sample containing biomolecular condensates is deposited on a non-adhesive glass surface and imaged for a long period. Passive fusion events occur when droplets within proximity of each other touch and undergo coalescence. Image analysis is then used to extract and monitor the aspect ratio of the deformed droplet as a function of time (Fig. 1). Although widely used, this method suffers from some technical and practical issues:

a. Droplet coalescence is opposed by the friction between the condensate and the glass surface, which may alter the characteristic relaxation time.

b. The sample requires a high number density of droplets to efficiently detect a statistically significant number of fusion events since these events are stochastically driven.

c. For droplets with low viscosity and/or high surface tension, the fusion processes are fast (typically in millisecond time scale) which poses a limitation on the detection capabilities of the instrument used and may require the use of expensive imaging devices (for higher time resolution).

d. The initial shape of the two coalescing droplets is not usually a perfect ellipsoid and fitting errors may reduce the quality of the data.

To overcome such limitations, a series of recent studies have used optical tweezers to perform controlled coalescence of suspended droplets (Alshareedah et al., 2019; Kaur et al., 2019; Patel et al., 2015; Wang et al., 2018). In these experiments, two droplets are optically trapped in two different highly focused laser beams. The optical trapping is possible due to the refractive index mismatch between the interior of the droplet and the external dilute phase. Trapped droplets are then brought into contact to initiate coalescence. The laser signal from the optical trap is able to detect the relaxation process and is used to extract the characteristic relaxation time. In what follows, we will describe the concept of this method as well as its application on biomolecular condensates *in vitro*.

The ability of an optical trap to detect the relaxation of a deformed droplet stems from the time-dependent deflection of the trapping laser through the droplet. When two optically trapped droplets undergo fusion, the intermediate state of the fusion process constitutes a deformed droplet and the final state is a spherical droplet shifted from the center of the optical trap (Fig. 2A). To illustrate the concept of detection, we consider the deflection of a single ray of light passing through the droplet at the deformed and relaxed states.

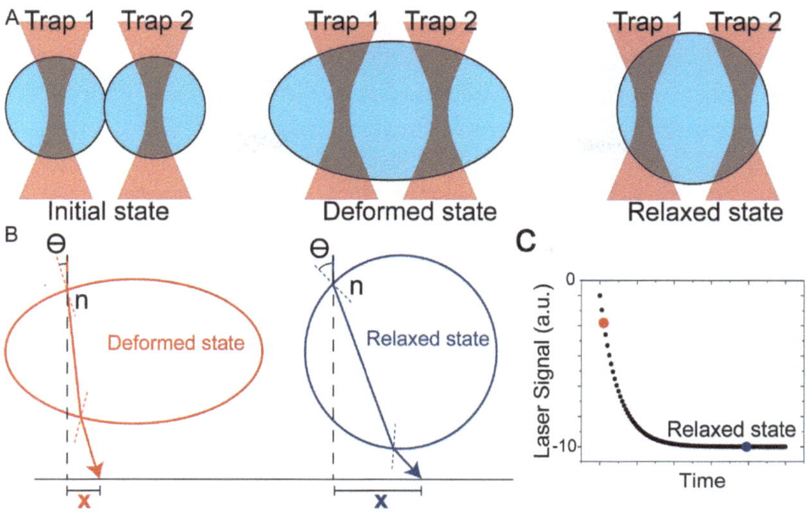

Fig. 2 Optical trap-induced droplet coalescence as a method to measure the relaxation time. (A) A schematic diagram showing that the fusion process is induced by two optical traps. (B) A schematic ray diagram showing how the laser deflection (x) of a single ray of light varies with droplet shape due to the change in the angle of incidence θ. (C) A representative laser signal from Trap-1 as a function of time during the relaxation process.

Due to the change in the droplet shape from deformed to relaxed, the curvature of the droplet surface increases, which leads to variation in the angle of incidence of the ray from small to large, respectively (Fig. 2B). Thus, the deflection of the ray by the condensate is enhanced at the relaxed state. A similar principle can be applied to the optical trap (constituting millions of rays) to understand how the deflection of the trap varies during the condensate shape relaxation. As the shape of the droplet relaxes to a sphere, the deflection of the laser relaxes to a steady-state value (Fig. 2C). The advantages of measuring the relaxation time *via* optical tweezer-controlled fusion are listed below:

a. Droplets are suspended in solution and hence there are no spurious friction effects as those caused by surface adhesion in the case of passive fusion (Jeon et al., 2018).

b. Optical traps (with sampling rate > 70 kHz) usually offer higher time resolution than video cameras [$\sim 10^2$ fps for CCD-based cameras and 10^3 fps for SPAD detectors (Shen et al., 2017)], thereby allowing the accurate detection of very fast fusion events with sub-millisecond relaxation time.

c. The sample does not require a high density of droplets since one can actively trap droplets and induce fusion events. This offers a key practical advantage over passive fusion assays as it allows a feasible collection of a large number of fusion events for statistical analysis.

3. Measuring mesoscale physical properties of liquid droplets using optical trap-induced droplet coalescence: Methods

3.1 Definition

Measuring the characteristic relaxation time of protein/RNA droplets using optical tweezer-induced coalescence.

3.2 Rationale

Fusion-relaxation time is a measure of the inverse capillary velocity (or the ratio of viscosity/surface tension; Eq. 2). Conventionally, the passive fusion of condensates that occurs stochastically is imaged with a video camera and analyzed in terms of the condensate aspect ratio. Here, we describe a protocol to measure the fusion-relaxation time using optical tweezers to provide higher time resolution, eliminate surface adhesion effects, and ensure practicality in collecting statistically large data sets using 1 μL sample volume.

3.3 Materials, equipment, and reagents

Dual-trap optical tweezers (1064 nm laser/1–10 mW power), coupled with a bright-field or confocal microscope, a sample containing phase-separated protein/RNA droplets, 1 mm thick glass slide, microscope coverslips, double-sided tape, Tween20 or Pluronic F-127, Ethanol (190–200 proof).

Note: The experimental data discussed here were collected using a dual-trap optical tweezer operating at a 78 kHz sampling frequency. However, a similar analysis could also be done with any sampling frequency as long as the fusion rate is considerably slower than the sampling rate (Jahnel, Behrndt, Jannasch, Schäffer, & Grill, 2011; Patel et al., 2015).

3.4 Protocols

1. Put two strips of double-sided tape on a microscope glass slide as shown in Fig. 3.
2. Add 1–1.5 μL of the droplet-containing sample between the two sheets of double-sided tape.

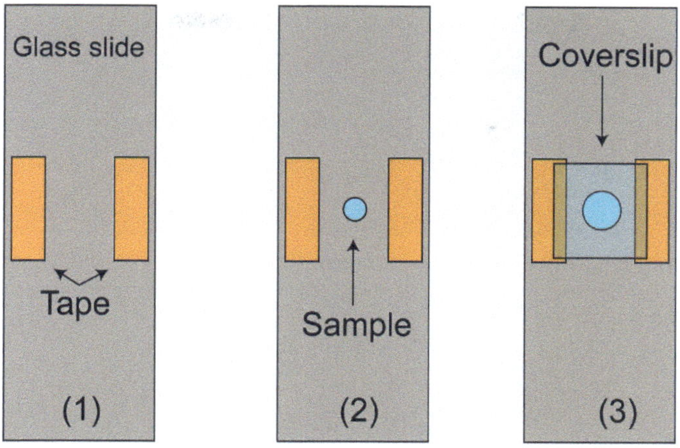

Fig. 3 Sample holder preparation for optical trap-induced coalescence experiment.

3. Place the coverslip on top of the sample and press firmly to "sandwich" the sample between the glass slide and coverslip. If the sample touches the sides of the double-sided tape, reduce the volume of the sample and ensure that it is placed at the center.
4. Load the sample onto the optical trap-microscope system.
5. Turn on the traps at minimum power. If the droplets have a higher refractive index than the surrounding medium, they will be attracted to the trap. Make sure to keep the trapped droplets away from the surface at the lower and upper edges of the sample, to avoid surface adhesion.
6. Slowly increase the power of the trapping laser until droplets are not visibly fluctuating around the center of each trap (around 5–15% of the total trap power of ~10 mW split equally between two traps, excluding the power lost in the optics). However, avoid trapping droplets with too much power (20–100% of the total trap power of ~10 mW, excluding the power lost in the optics), which may heat the droplets and change their material properties or cause a visible droplet "boiling."
7. Once you have two trapped condensates, take an image of the two droplet system (confocal or bright-field). This image will be used to measure the diameter of each of the fusing droplets (see Fig. 4).
8. Set Trap-1 to move at a constant velocity toward Trap-2. Once the fusion starts, you will see a clear change in the laser signal followed by an exponential relaxation phase (see Fig. 5A for an example). Allow sufficient time for the fusion-relaxation process to be completed.

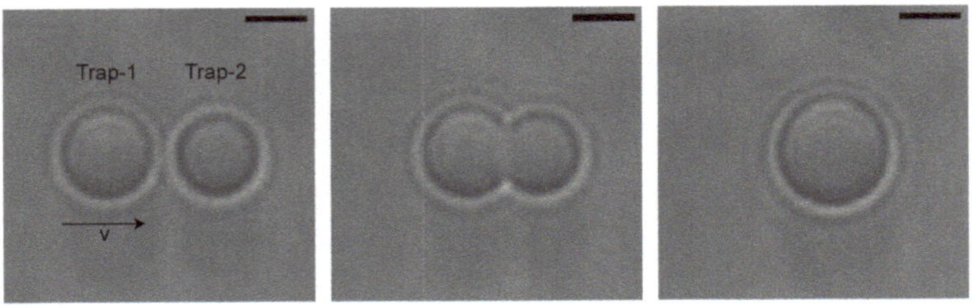

Fig. 4 Bright-field images of two optically trapped protein condensates as they are driven to undergo coalescence. Trap-1 velocity was set at 40 nm/s. Scale bar represents 5 μm.

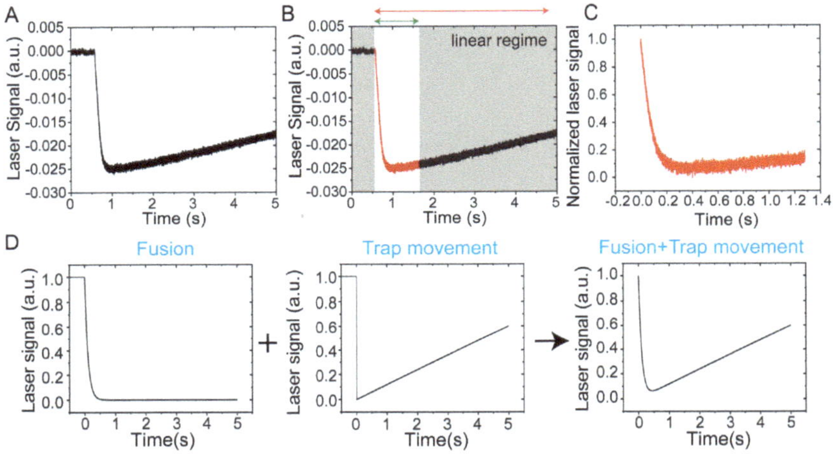

Fig. 5 (A) Representative raw laser signal recorded during the fusion of two optically trapped droplets. (B) Trimming the laser signal to selectively extract the fusion-relaxation signal (the exponential phase). The green double-sided arrow shows a reasonable trim. The red double-sided arrow shows an unrecommended trim since most of the points lie in the linear part of the signal and may result in a biased fit. (C) The trimmed signal in (B) is re-plotted. (D) The final laser signal functional form is a combination of an exponential relaxation process and a linear process.

9. Once the fusion is completed, save the detected laser signal from Trap-1 for further analysis. Release the relaxed droplet from the traps by using the shutter or simply switching the traps off.
10. Repeat steps 1–9 at different locations within the sample to collect the desired number of fusion events. Fusion events should be collected with droplets of varying sizes. Varying the size of the trapped droplets can be easily achieved by allowing the traps to collect more droplets before initiating fusion. It is also advisable to keep the two fusing droplets similar in size for each fusion. This allows for more accurate analysis (see Section 3.7).

3.5 Precursor techniques

3.5.1 Coating glass slides and coverslips

1. Prepare a 20% (vol/vol) solution of Tween20 or 1% (wt/vol) solution of Pluronic F-127.
2. (Optional) Clean the microscope glass slides and coverslips with 70% ethanol. You may sonicate glass slides and coverslips for 15 min for efficient cleaning.
3. Dry the residual ethanol under nitrogen gas.
4. Immerse the glass slides in the coating solution and leave them as such for 30 min.
5. Rinse the glass slides and coverslips with milliQ water for 8–10 times to remove excess coating solution.
6. Dry the glass slides in a vacuum chamber or a heated oven at 40–50 °C overnight.
7. Dry coated glass slides and coverslips can be wrapped in lens paper and stored at room temperature for later use.

3.6 Safety considerations and standards

Follow the standard safety guidelines when dealing with optical traps. This includes wearing appropriate PPE as required by the manufacturer's safety guidelines.

3.7 Analysis and statistics

The analysis for this technique involves three steps: (i) extracting the relaxation time of each fusion event by fitting the laser signal with an appropriate relaxation model, (ii) plotting the relaxation time as a function of droplet size, and (iii) fitting the relaxation time *vs* droplet size data to extract the inverse capillary velocity. These steps are described in detail below.

A typical fusion-relaxation laser signal is shown in Fig. 5A. The data can be trimmed to effectively analyze only the relaxation process (Fig. 5B and C). Care should be taken not to include an excessive amount of data points at the linear regime where the relaxation process is completed, since this will bias the fit towards the linear part. Typically, the trimmed data should have a time span that is approximately twice the time taken for the signal to reach the lowest value (see Fig. 5B). The trimmed data is then normalized to facilitate the fitting process. The normalized laser signal (S) is fitted with the following function

$$S(t) = ae^{-\frac{t}{\tau}} + bt + d \tag{3}$$

where a, τ, b and d are fitting parameters. The fusion process is expected to exhibit an exponential relaxation trend which is accounted for by the first term in Eq. 3. If we consider the case of an infinitely slow trap, then the signal will predominantly show a relaxation trend (Fig. 5D). The second term is linear in time and is added to account for the signal coming from the movement of the trap across the droplet. The signal for an infinitely fast fusion will only display the signal coming from the trap movement (Fig. 5D). Combining these two terms gives a functional form that effectively describes the signal coming from a fusion-relaxation experiment (Fig. 5A and D). The linearity in time in the signal coming from the trap movement requires that throughout the fusion process, the trap must remain in the linear regime (i.e., the displacement of the particle from the center of the trap is linear with the laser signal). This implies an additional consideration for conducting these experiments, *i.e.*, the velocity of the moving trap should be close to the velocity of fusion to ensure that the trap remains in the linear regime throughout the fusion-relaxation process. Failure to do that may result in erroneous fusion curves and/or the loss of the signal coming from the first trap due to the droplet being completely drawn out of the trap.

Once the fitting is done (Fig. 6A), the size of the fusing droplets (before fusion) should be determined using image analysis. Although confocal images are preferred for more accurate determination, bright-field images can also be used. The size determination of the fusing droplets can be done manually using Fiji-ImageJ software (Rueden et al., 2017) or through the use of programming scripts such as Python or Matlab. For the manual case, Fiji-ImageJ software is used to measure the diameters of the fusing droplets as follows: using the "measure tool," a line is drawn across the center of the droplet which has its endpoints on the periphery of the droplet and the droplet diameter is recorded. This measurement is repeated for several lines on the same droplet and their average is taken as the droplet diameter. Similarly, the diameter of the second droplet is measured. Finally, the average diameter of the two droplets is recorded and used for subsequent analysis. A more accurate and perhaps a practical way is the use of coding scripts to measure the diameter of the droplets. This can be done using the function *regionprops()* in Matlab or the function *skimage.measure.regionprops()* in python (skimage python library is required). These scripts fit the droplets in the image with ellipsoids (Fig. 6B). An example of a Matlab script to detect the droplet diameter is given below:

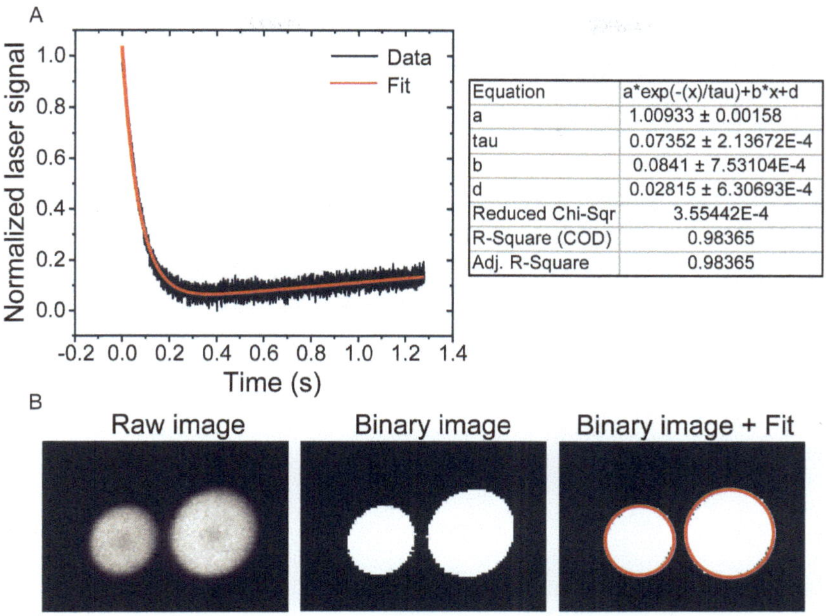

Fig. 6 (A) Fitting (red curve) of the laser signal (black curve) to extract the characteristic time of the relaxation process. (B) Image processing of the two droplets under investigation to extract their average diameter. The fluorescence micrograph is converted to a binary image by applying a threshold and the droplets are fitted with ellipses to extract the diameters (see the provided Matlab script below).

```
%Read Image
a = imread('Fluorescence Image.png');
%Convert to a binary image by applying a threshold
bw = a > 100;
%show binary image
imshow(bw);
saveas(gcf,'Binary Image.png');
% find regions and save the center and major axes length
props = regionprops('Table',bw,'Centroid','MajorAxisLength',
'MinorAxisLength');
%calculate diameters of each circle by averaging the major axis length
%(this should give you an array of two with diameters in pixels)
diameters = mean([props.MajorAxisLength props.MinorAxisLength],2);
pixelsize= 0.1 %micron
display(diameters*pixelsize)
%for visualization
centers = props.Centroid;
```

```
radii = diameters/2;
hold on
viscircles(centers,radii);
hold off
saveas(gcf,'Droplet Fit.png')
```

The presented analysis is repeated for multiple fusion events with varying droplet diameters. Plotting the relaxation time against the average diameter of the fusing droplets for each fusion event gives a linear relation (Fig. 7) for purely viscous droplets. The slope of the linear fit gives the scaled relaxation time which is equal to the inverse capillary velocity η/γ (viscosity to surface tension ratio) for the condensates under investigation (Ceballos, McDonald, & Elbaum-Garfinkle, 2018; Elbaum-Garfinkle et al., 2015; Feric et al., 2016).

$$slope = \frac{\tau}{l} = \frac{\eta}{\gamma} \tag{4}$$

It is important to only consider fusion events for which the fitting procedure yields a good fit. Any irregularities in the fusion-relaxation signal should be excluded unless they are consistently appearing for all trapping powers and all velocities. One important consideration to note is that the linear relation between the fusion-relaxation time and the droplet diameter

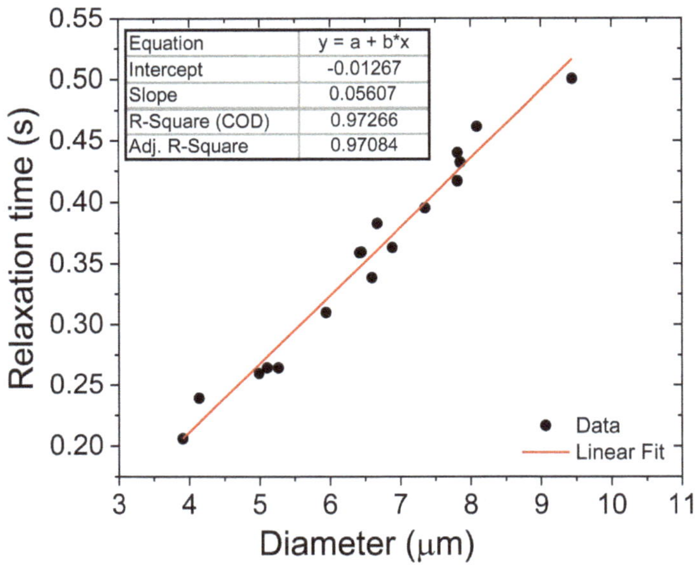

Fig. 7 The relaxation time scales linearly with droplet size, with a slope equal to inverse capillary velocity η/γ (I. Alshareedah and W. Wang, unpublished).

is a property of purely viscous droplets. This method is useful for relative comparison between droplets in different conditions (as in protein/nucleic acid sequence variations, mixture composition variation, varied buffer conditions including pH, molecular crowding, *etc.*). However, it should be understood that standing alone, this method only gives the ratio between viscosity and surface tension and does not provide any direct information on either quantity.

3.8 Related techniques

Passive fusion analysis using video microscopy is an alternative method for determining the inverse capillary velocity. This method consists of imaging fusion events that occur on the surface of the glass slide and conducting droplet aspect ratio analysis to extract the fusion-relaxation time. A description of this method can be found elsewhere (Brangwynne et al., 2011; Ceballos et al., 2018).

3.9 Pros and cons

Pros	Cons
1. Eliminates surface adhesion effects that are present when analyzing passive fusion events on a glass surface using video microscopy 2. Allows for rapid and efficient data collection of multiple fusion events for statistical analysis 3. Does not require fluorescence labeling of molecules	1. Does not provide information on either viscosity or surface tension but only the ratio between the two quantities 2. The method is quantitative for liquid droplets but binary for solid-like droplets (can only reveal if a droplet is solid-like or not)

3.10 Troubleshooting and optimization

Problem	Solution
Droplet is lost from the trap during fusion	This might indicate that the fusion process is much faster than the velocity of the moving trap. Increasing the velocity of the moving trap should resolve the issue. Alternatively, repeating this assay with larger or smaller droplets may resolve this issue

Continued

—cont'd

Problem	Solution
Droplet leaps from Trap-1 to Trap-2 during fusion	Increase the power of the trapping laser. Increase the velocity of the moving trap
Laser signal from the optical trap is noisy	This is an indication that the trapping power is too low. Increasing the power of the trapping laser should eliminate this problem. Also, make sure that the two traps have equal power
Droplets are falling to the glass surface rapidly, impeding the chance to collect a large number of fusion events	This can be overcome in two ways: (a) Increasing the thickness of the sample chamber by using two layers of double-sided tape instead of one. This creates a deeper channel with a lower surface area on the glass slide-sample interface (b) The use of hydrophobic surface coatings such as SigmaCote or Silane may resolve this issue. These types of coatings prevent certain droplets from sticking to the surface and hence can be easily picked up by the optical traps
Sample is drying too fast which impedes the collection of large numbers of fusion events	Seal the sample chamber open ends with a coverslip sealant. Avoid using excessive amounts of the sealant as it may touch the microscope objective
Data is not reliably reproducible	Make sure the conditions such as trapping laser power, the temperature of the sample, and the velocity of the moving trap are as consistent as possible across samples. Also, make sure the measurements are taken within a consistent time span as sample drying/aging may alter the material properties of the condensates

3.11 Summary

The use of optical tweezers to measure the fusion-relaxation time and inverse capillary velocity of biomolecular condensates *in vitro* has been demonstrated as a powerful tool to assess their material properties.

Optical traps provide higher time resolution of the process and allow for practical and efficient collection of large numbers of fusion events, a feature that is absent in methods that use video microscopy to extract these physical quantities.

4. Determining nanoscale biomolecular diffusion within liquid droplets using fluorescence recovery after photobleaching (FRAP): Concepts

Fluorescence recovery after photobleaching (FRAP) is a widely used technique to probe the diffusivity of fluorescently tagged protein/RNA molecules within biomolecular condensates *in vitro* and *in vivo* (Banerjee, Milin, Moosa, Onuchic, & Deniz, 2017; Elbaum-Garfinkle et al., 2015; Feric et al., 2016; Jain et al., 2016; Kroschwald et al., 2015; Li et al., 2012; Molliex et al., 2015). The concept of FRAP is based on monitoring the exchange rate of labeled molecules between a region of interest and the surrounding environment, which is subsequently used to quantify their diffusivity dynamics. To achieve this, a region of interest (ROI) within the droplet is irreversibly photo-bleached through a high power laser excitation. Bleaching is the result of photochemical damage of the fluorescence probes such that it permanently loses its ability to fluoresce (Kubitscheck & Peters, 2013). This results in the ROI having a significantly lower fluorescence (or zero) intensity as compared to the surrounding regions (Fig. 8A). The recovery of fluorescence intensity in the bleached region is then monitored as a function of time. If the labeled species undergoes fast diffusion, the rate of exchange between the bleached region and the surrounding area would be high and the resultant recovery time would be short. Conversely, the recovery time for a molecular species with slow diffusivity dynamics would be long.

The information that can be extracted from FRAP depends on two parameters, the recovery half time and the fraction of mobile phase. The recovery half time is defined as the time taken for the bleached region to recover half of its final intensity, which provides insights into how fast the labeled molecules are diffusing (Fig. 8B). The fraction of mobile phase is estimated by the ratio of the recovered fluorescence intensity of the bleached spot to the bleaching depth (Fig. 8C). If the labeled species is completely mobile (as in the case of a purely viscous liquid), the fraction of mobile phase is expected to be ~100% (Fig. 8B). If the labeled species has two populations, mobile and immobile, the fraction of mobile phase is typically much less than 100%, which is classically interpreted as apparent

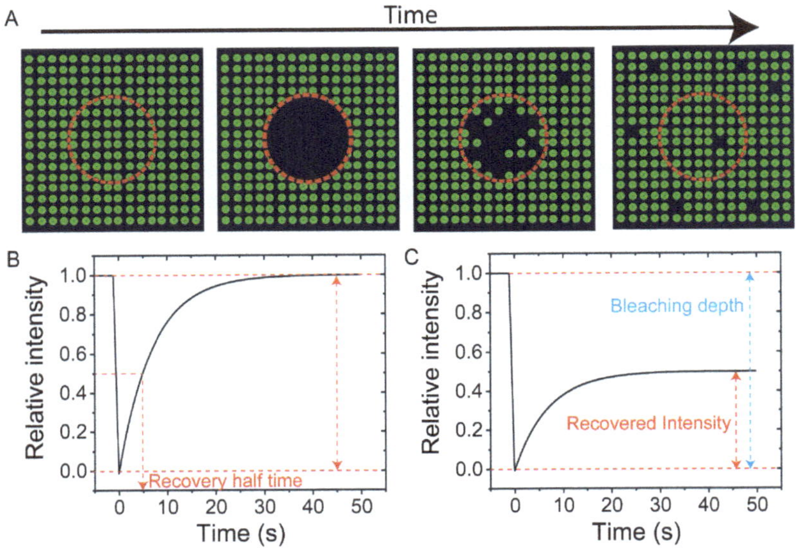

Fig. 8 (A) Diagram illustrating the concept of a FRAP experiment. Each green dot represents a single fluorescently-labeled molecule. (B) FRAP recovery curve showing full recovery. The single-headed red arrow indicates the recovery half time. The double-headed red arrow indicates the recovered intensity. (C) FRAP recovery curve showing partial recovery. The red dashed double-sided arrow indicates the recovered intensity [fraction of mobile phase is 50% in (C) and 100% in (B)]. The blue dashed double-sided arrow indicates the bleaching depth.

viscoelasticity (Fig. 8C). Ideally, apparent diffusion coefficients can be extracted from FRAP curves by fitting the recovery time trace with an appropriate diffusion model, however, FRAP is more readily used only to qualitatively assess the mobility of the molecules under investigation.

Many studies have utilized FRAP to measure the diffusivity of molecules within biomolecular condensates using purified *in vitro* systems and in cells (Alshareedah et al., 2019; Banerjee et al., 2017; Elbaum-Garfinkle et al., 2015; Feric et al., 2016; Jain et al., 2016; Kaur et al., 2019; Kroschwald et al., 2015; Li et al., 2012; Molliex et al., 2015; Onuchic et al., 2019; Patel et al., 2015; Taylor, Wei, Stone, & Brangwynne, 2019; Wang et al., 2018). While the implementation of the experiment and analysis is fairly straight forward, the interpretation of FRAP results has proven tricky and often needs careful considerations on a case-by-case basis. For example, the FRAP recovery rate is sensitively dependent on the molecular identity of the labeled species in the case of heterotypic condensates harboring multiple molecular species (Boeynaems et al., 2019; Chalupníková et al., 2008; Weidtkamp-Peters et al., 2008). In the same biomolecular condensate, FRAP measurements on one component might reveal liquid-like behavior (complete recovery), while the same analysis on another component might reveal viscoelastic behavior (partial recovery) (Chalupníková et al., 2008; Weidtkamp-Peters et al., 2008). This variation of FRAP results with the probe molecules for multi-component condensates is interpreted as a result of differential intermolecular interactions and spatial heterogeneities within the droplet. This differential FRAP behavior can be rationalized in light of a scaffold-client model. As noted earlier, scaffolds are biopolymers that feature multivalent binding domains. Clients, on the other hand, do not require multivalency and can partition within a condensate through monovalent scaffold-binding modules (Banani et al., 2017). Therefore, scaffold-client interaction can be more transient than scaffold-scaffold interactions. Consequently, scaffolds are expected to have intrinsically slower mobility than the clients since a scaffold makes substantially more and possibly stronger intermolecular contacts compared to a client (Alberti, 2017b; Banani et al., 2016). Apart from such component-specific diffusivity dynamics, a recent study has shown that the extracted diffusion coefficients are highly dependent on the dimensionality of the diffusion model used to fit the data as well as the boundary conditions, with large variance occurring between different models (Taylor et al., 2019). Therefore, it is important to take additional considerations into account when interpreting FRAP results for biomolecular condensates. Nevertheless, FRAP remains widely used since it is a simple, accessible, and convenient technique that can be done to get a qualitative idea on the diffusivity of molecules within a condensate. More importantly, FRAP is one of the few techniques for probing condensate dynamics that can be performed *in vivo* (Brangwynne et al., 2009; Feric et al., 2016; Jain et al., 2016). Moreover, FRAP experiments are conveniently done in a confocal fluorescence microscope, which is widely accessible.

5. Determining diffusion dynamics within liquid droplets using fluorescence recovery after photobleaching: Methods

5.1 Definition

Measuring the molecular diffusivity of a fluorescently tagged molecule by monitoring the fluorescence recovery after photobleaching (FRAP).

5.2 Rationale

FRAP experiments provide a simple way to make a preliminary judgment on whether a condensate is a liquid or solid. It is also one of the few experiments on molecular diffusivity that can be done *in vivo* for biomolecular condensates.

5.3 Materials, equipment, and reagents

Confocal fluorescence microscope, a sample containing phase-separated protein/RNA droplets with desired fluorescently labeled species, 1.0 mm thick glass slide or microscope coverslips, Tween20 or Pluronic F-127, Ethanol (190–200 proof).

5.4 Protocols

1. Place the condensate sample containing fluorescently labeled species on a tween-coated microscope glass slide/coverslip. A 5 μL sample is sufficient for ≥30 min of imaging provided the sample is kept covered to prevent evaporation/drying.
2. Load the sample onto the microscope stage.
3. Take a fluorescence image of the sample at the glass surface to make sure that the labeled molecules are positively partitioning into the droplets and that the fluorescent intensity is appropriate (i.e., no detector saturation is observed). Adjust the laser power and/or detection settings to ensure that the sample brightness is well within the detectors' dynamic range.
4. Focus the confocal volume on a region where you can see at least two droplets stationary on the glass slide surface.
5. Set the bleaching spot at the center of one of the droplets, make sure that the size of the bleaching spot is significantly smaller than the size of the droplet [at least three times smaller (Taylor et al., 2019)]. This is to improve the accuracy of the analysis and minimize interfacial resistance effects (see the analysis section for more details).
6. Start the bleaching experiment by shining the excitation laser with maximum intensity for 1–2 s. You can set the parameters of the bleaching experiment using the software of your microscope or a customized automation script, if available.
7. Image the two droplets after bleaching for a fixed time interval. If the recovery is slow, it is advisable to increase the time interval between consecutive frames to minimize photofading effects (see Section 5.7

for details). Keep imaging until the intensity of the bleached region reaches a plateau.

8. Repeat steps 4–7 for multiple droplets to ensure the statistical significance of the results. Notice that the recovery time depends on the size of the bleaching region (see the analysis section for details), hence it is advisable to keep the bleaching region to droplet diameter ratio similar throughout the experiment (if possible). This will allow for online monitoring of the consistency of the data.

5.5 Precursor techniques

5.5.1 Coating glass slides and coverslips

1. Prepare a 20% (vol/vol) solution of Tween20 or 1% (wt/vol) solution of Pluronic F-127.
2. (Optional) Clean the microscope glass slides and coverslips with 70% ethanol. You may sonicate glass slides and coverslips for 15 min for efficient cleaning.
3. Dry the residual ethanol under nitrogen.
4. Immerse the glass slides in the coating solution and leave them as such for 30 min.
5. Rinse the glass slides and coverslips with milliQ water for 8–10 times to remove excess coating solution.
6. Dry the glass slides in a vacuum chamber or a heated oven at 40–50 °C overnight.
7. Dry coated glass slides and coverslips can be wrapped in lens paper and stored at room temperature for later use.

5.5.2 Notes

1. Make sure that the labeled species is added to the sample before forming the condensates. This is important to eliminate interfacial resistance effects and ensure good partitioning of the labeled molecule within the droplet.
2. Depending on the detector sensitivity of the microscope, the optimal probe concentration may vary. A probe is defined as a site-specifically labeled protein/RNA (or DNA) molecule. Such site-specific labeling protocols are thoroughly reviewed in the literature and will not be discussed here (Banerjee & Deniz, 2014). Usually, probe concentrations that we use in our laboratory range between 300 nM and ~1 μM depending on the imaging instrument. The brightness of the fluorescent dye may also be a factor in optimizing the concentration of the

fluorescent probe. It is not advisable to utilize ~100% of the labeled protein in the sample since labeling may alter the biomolecular properties and hence may change the behavior of condensates. A ratio of the labeled species to unlabeled species of ~1% or less is typically found to be a good practice in our experiments. However, this consideration generally excludes proteins containing an exogenous fluorescence protein tag, such as GFP-tag.

5.6 Safety considerations and standards

Follow the standard safety guidelines when dealing with biological samples including wearing appropriate PPE.

5.7 Analysis and statistics

To analyze FRAP experiments, one needs to carefully consider the experimental conditions in order to choose a suitable model. Here, we will demonstrate the analysis using a two-dimensional diffusion model with infinite boundary conditions (Kaur et al., 2019; Taylor et al., 2019). The prerequisites for this analysis is that the bleached spot radius is smaller than that of the droplet by a factor of 3 or more. Furthermore, the laser used for bleaching has a Gaussian profile. These are necessary to ensure that the interface has a reduced effect on the recovery (since it is far from the bleached spot). Another issue that needs attention when analyzing FRAP data is photofading. Photofading can be described as progressive decay in the fluorescence intensity of the whole imaging area during the FRAP experiment due to non-specific bleaching. This usually occurs due to repeated excitations of fluorescent molecules over a long duration of time. To correct for photofading, a reference droplet that is not bleached or altered but included within the imaging area is recorded and analyzed (Fig. 9A). The decrease in intensity of the reference droplet is assumed to be due to photofading, and hence, can be used to correct for the photofading effect for the bleached droplet. Typically, the intensity time trace of the reference droplet shows a linear decrease with time (Fig. 9B and D) due to photofading. Below, we will show two FRAP curves to illustrate important details in this analysis.

To correct for photofading effect, we calculate a correction factor C from the intensity time trace of the reference droplet. For each time point t, the correction factor is calculated as

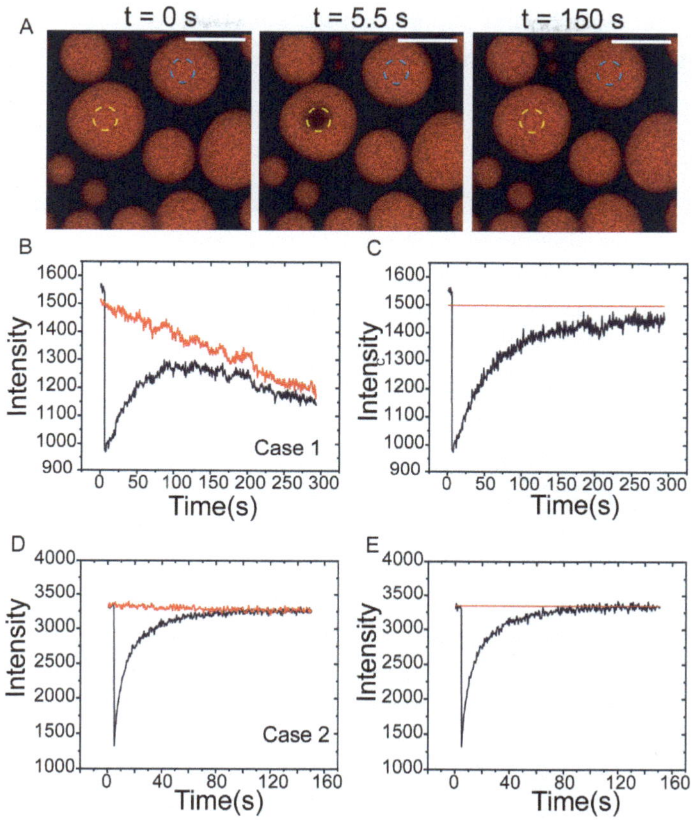

Fig. 9 (A) Confocal images of a protein-RNA droplet before bleaching, after bleaching, and at a full recovery. The yellow circle indicates the bleaching spot; the cyan circle indicates a reference region in an unbleached droplet. (B and C) Intensity time traces of the reference droplet (red) and the bleached droplet (black) before correction (B) and after correction (C) for the case of significant photofading effects. (D and E) Same as (B and C) but for a sample with weak photofading effects.

$$C(t) = \frac{R(0)}{R(t)} \tag{5}$$

where $R(t)$ is the fluorescence intensity of the reference droplet at time t. The corrected FRAP curves are then calculated using

$$I_{correct}(t) = C(t)^* I(t) \tag{6}$$

where $I(t)$ is the raw intensity signal coming from the bleached region at time t. The corrected FRAP curves are shown in Fig. 9C and E. Once FRAP curves are corrected, they can be normalized with respect to the bleaching depth using the following equation:

$$I_{noramlized}(t) = \frac{I_{correct}(t) - \min(I_{correct})}{I_{correct}(0) - \min(I_{correct})} \tag{7}$$

This normalization makes it easier to extract the relevant quantities from the FRAP curve (see Fig. 10). Notice that $I_{correct}(0)$ is the value of intensity before bleaching starts (usually we take few frames before bleaching in order to quantify the bleaching depth and the normalization). Fitting FRAP curves to extract the recovery half time requires the use of the appropriate fitting function. Several functions have been suggested in the literature (Axelrod, Koppel, Schlessinger, Elson, & Webb, 1976; Braeckmans, Peeters, Sanders, De Smedt, & Demeester, 2003; Day, Kraft, Kang, & Kenworthy, 2012; Kang, Day, Kenworthy, & DiBenedetto, 2012; Mazza et al., 2008; Soumpasis, 1983; Sprague, Pego, Stavreva, & McNally, 2004), with each corresponds a specific set of conditions that depend on the experimental conditions such as laser profile, the relative size of the bleached droplet, the size and shape of the bleached region, the dimensionality of the diffusion process, the existence of reaction processes (*via* binding and unbinding events) that couple with free diffusion (Kang & Kenworthy, 2008; Taylor et al., 2019), *etc.* We will solely focus here on fitting the FRAP recovery time trace and extract the recovery half time and the fraction of the mobile phase (Fig. 8). We will use two independent cases (see Fig. 9B and D) to illustrate how one fitting function may work for some systems but not others. For the FRAP curve in case 1, we find that a single exponential function is sufficient for fitting the data (Fig. 10A)

$$I(t) = a\left(1 - \exp\left(-\frac{t}{\tau}\right)\right) \tag{8}$$

where a, τ are fitting parameters and $\tau_{1/2} = \tau \ln 2$. In case 2, the exponential function does not give a good fit while the function (Day et al., 2012)

$$I(t) = \frac{a + b\left(\frac{t}{\tau_{1/2}}\right)}{1 + \frac{t}{\tau_{1/2}}} \tag{9}$$

fits the data significantly better (Fig. 10B). Here, a, b, $\tau_{1/2}$ are fitting parameters.

These examples serve to illustrate that the choice of fitting function is important for the final values of the apparent diffusion coefficient (to be

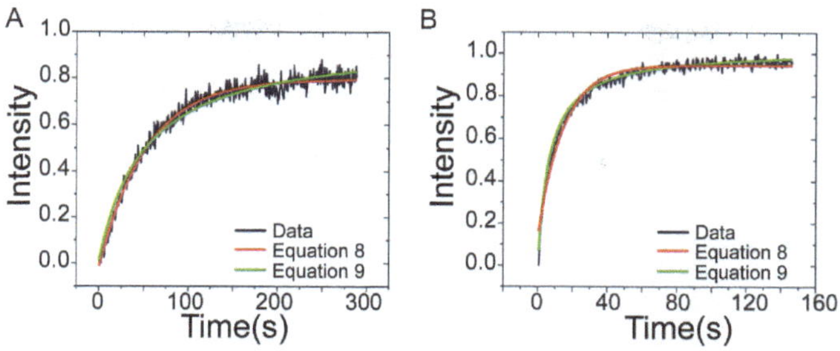

Fig. 10 (A) FRAP data fitting for case 1, Fig. 9C using Eqs. 8 (red) and 9 (green). (B) FRAP data fitting for case 2, Fig. 9E using Eqs. 8 (red) and 9 (green).

estimated using the $\tau_{1/2}$ obtained from fitting). The values of the fitting parameters for each of these cases are shown in Table 2.

To extract the diffusion coefficient, one needs to estimate the area of the bleached spot. Qualitatively speaking, the diffusion coefficient is proportional to the area of the bleaching spot divided by the recovery half-time (Kang et al., 2012)

$$D \sim \frac{R^2}{\tau_{1/2}} \tag{10}$$

where R is the radius of the bleaching spot. Due to the Gaussian shape of the laser as well as the diffusion process within the bleached region, the post-bleaching intensity profile across the bleached region assumes a Gaussian shape. The active diffusion processes into the bleached region during the bleaching period results in differences between the predetermined bleaching region (set by the software of the microscope) and the effective bleaching area. To account for this discrepancy between predetermined and effective bleaching areas, we define two radii, R_n and R_e, identifying the predetermined and effective bleaching spots, respectively. R_n is taken from the experimental parameter used or can be determined using Fiji-ImageJ software (provided the software exports the predetermined regions to Fiji-ImageJ). R_e can be measured by fitting the intensity profile across the bleached area immediately after bleaching with a Gaussian function (Fig. 11A and B). An example of a function that has been used to fit the post-bleaching profile is (Kang et al., 2012)

$$F(x) = a - a_1\, exp\left(-\frac{(x-b)^2}{R_e^2}\right) \tag{11}$$

Table 2 Extracted recovery half time and the R^2 (Coefficient of determination) for the two cases (Fig. 10A and B) as fitted by models 1 & 2 (Eqs. 8 and 9, respectively).

Model 1				Model 2			
Case 1		Case 2		Case 1		Case 2	
$\tau_{1/2}$	R^2 (COD)	$\tau_{1/2}$	R^2 (COD)	$\tau_{1/2}$	R^2 (COD)	$\tau_{1/2}$	R^2 (COD)
35.9 ± 0.5 s	0.98072	10.1 ± 0.2 s	0.97102	49 ± 2 s	0.96991	6.8 ± 0.2 s	0.9829

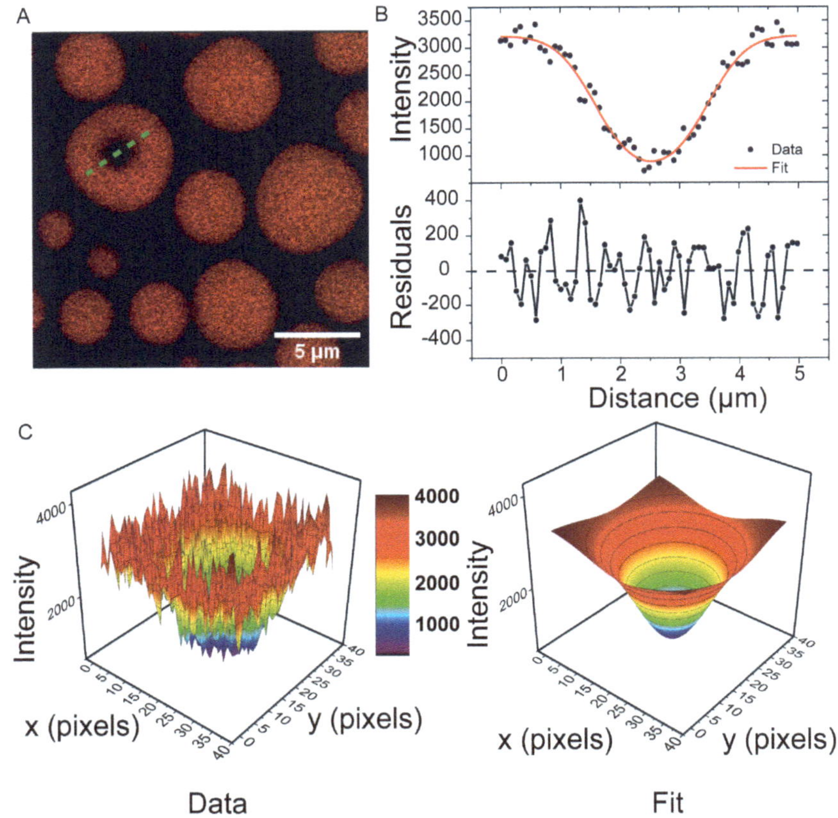

Fig. 11 (A) Fluorescence image of a protein-RNA droplet immediately after bleaching. (B) Intensity profile along the green dashed line in (A) fitted to a Gaussian function (Eq. 11). Residuals are shown. (C) 2D surface plot for the raw intensity values as a function of position (left). The surface fit of the data (right) is generated using Eq. 12. 1 pixel represents 100 nm.

where a, a_1, b, R_e are fitting parameters. The parameter that gives the effective width of the Gaussian function and hence the effective diameter of the bleaching spot is R_e. One can also perform two-dimensional fitting by considering the surface plot of intensity in the XY plane (Fig. 11C). And then fitting it using the function (Day et al., 2012)

$$F(x,y) = a\exp\left(-Kexp\left(-2\,\frac{(x-b)^2+(y-c)^2}{R_e^2}\right)\right) \tag{12}$$

where a, b, c, K, R_e are fitting parameters.

Finally, the diffusion coefficient can now be calculated using the following modified formula (Kang et al., 2012)

$$D_{apparent} = \frac{R_n^2 + R_e^2}{8\tau_{1/2}} \tag{13}$$

The presented analysis to extract diffusion coefficients from FRAP is one of the most widely used but not unique. In fact, the literature on FRAP analysis is plentiful with different models that can be used to fit the data and extract diffusion coefficients. A recent study has summarized some of the models that have been used for liquid-liquid phase separation in biology (Taylor et al., 2019). One of the important conclusions was that the FRAP model has to be chosen according to the shape and relative size of the effective bleaching spot with respect to the droplet size. One of the common problems in FRAP is that many systems form condensates that are small in size (~1 μm). This makes it inevitable for the bleaching region to have a comparable size to the droplet, which results in both finite boundary conditions and the interfacial resistance to contribute to the resulting FRAP recovery profile. For such cases, a theoretical model has been developed which requires complementing FRAP with fluorescence correlation spectroscopy (FCS) to extract the diffusion coefficient (Taylor et al., 2019). Furthermore, the choice of the labeled molecule has been shown to affect FRAP results drastically, to the point where one molecular species displays liquid-like behavior and other molecular species display solid-like behavior (Boeynaems et al., 2019; Weidtkamp-Peters et al., 2008). These drastic differences between molecules within the same condensate challenges the simplistic views of classical material characterization ideas such as viscosity and diffusion. Overall, for FRAP analysis to be reliable, we recommend the following:

1. The FRAP model should be chosen based on the experimental conditions regarding the relative size and shape of the effective bleaching region.
2. The chosen model should be used consistently throughout the study.
3. The extracted values of diffusion coefficients should only be interpreted relatively and not to be claimed as absolute values.
4. A statistically large data set (*i.e.*, collecting multiple FRAP events for the same sample) can improve the reliability of the results.

5. FRAP should not be the only method used to determine the physical property of the condensates. Complementary techniques are discussed throughout the chapter.
6. Performing FRAP with each component is warranted for heterotypic multi-component condensates.

5.8 Related techniques

Fluorescence correlation spectroscopy (FCS) and single and multiple particle tracking.

5.9 Pros and cons

Pros	Cons
1. Ease of implementation *in vitro* and in cell culture models 2. Does not require specialized equipment 3. Can be a useful tool for initial and preliminary characterization of biomolecular condensates	1. Does not report on the material state of the condensate but only the diffusion dynamics of the labeled molecules 2. Highly sensitive to the experimental conditions, instrumentation, and parameters such as the area of the bleached region 3. Highly sensitive to the diffusion model being used to fit the data 4. Highly sensitive to the identity of the probe molecule

5.10 Troubleshooting and optimization

Problem	Solution
Bleaching of the droplet does not occur	This problem indicates that the recovery is too fast such that the droplet is already fully recovered by the time the first post-bleaching frame is imaged. In that case, reducing the bleaching time may resolve the issue. Also, make sure the detectors are not saturated due to a high concentration of the fluorescent probe

—cont'd

Problem	Solution
Bleaching profile is not Gaussian	This can sometimes be a result of excessive bleaching that leads to a post-bleach profile that is closer to a square well rather than a Gaussian. The following suggestions may resolve this issue: (a) Decreasing the intensity of the bleaching laser (b) Decreasing the pixel time (dwell time) of the bleaching laser (c) Decreasing the size of the bleaching spot
Sample is photofading at a fast rate	(a) Increase the time duration between consecutive post-bleaching frames (b) Decrease the intensity of the excitation laser (c) Increase the fluorescent probe concentration in the sample

5.11 Summary

FRAP is a quick and simple method to get an idea of the diffusivity dynamics of fluorescently tagged molecules in a biomolecular condensate. However, care should be taken when conducting the experiment and choosing the fitting model. It is recommended that FRAP is complemented with other techniques to provide an orthogonal verification of the material state of the biomolecular condensates.

6. Single particle tracking (SPT)

Particle tracking is a classical technique for measuring the viscous properties of liquids. Several studies have used particle tracking techniques to measure the viscosity of biomolecular condensates (Elbaum-Garfinkle et al., 2015; Feric et al., 2016; Jawerth et al., 2018; Taylor et al., 2016; Zhang et al., 2015). The experiment is done by tracking an exogenous spherical particle diffusing inside a biomolecular condensate (Fig. 12A). The tracking is usually done through video microscopy, by imaging the motion of the particle over a set duration of time. The imaging series is then analyzed using particle tracking software or custom-built programs that

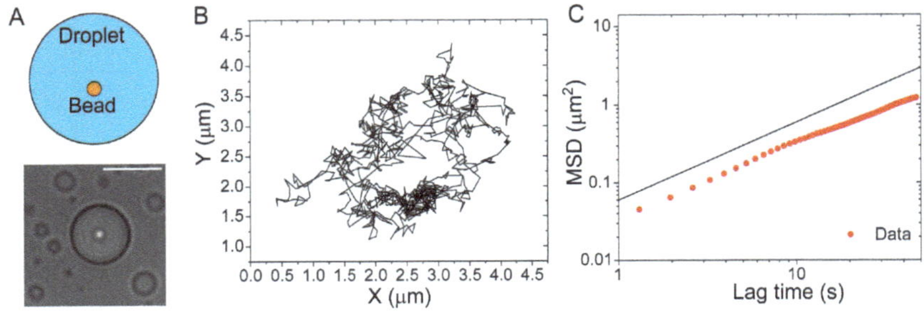

Fig. 12 (A) A schematic diagram and a bright-field image showing a microsphere is localized inside a protein droplet. Scale bar represents 5 μm. (B) A representative trajectory of a microsphere in a particle tracking experiment. (C) Mean square displacement (MSD) calculated for the trajectory in (B). Red: MSD data; Black: a line with a slope equal to 1.

implement a tracking algorithm (Feric & Brangwynne, 2013; Zhang et al., 2015). The output of the particle tracking experiment gives the location of the particle as a function of time (the trajectory, Fig. 12B), from which the mean square displacement (MSD) can be calculated and plotted for a set of lag times τ (Fig. 12C).

The relation between the mean square displacement MSD and the lag time τ is given by

$$MSD(\tau) = < r^2 > (\tau) = 4D\tau^{\alpha} \tag{14}$$

where D is the diffusion coefficient of the bead and α is the diffusive exponent. For classical Brownian motion in a purely viscous fluid, the value of α is equal to unity. The viscosity η of the condensate can be determined using the Stokes-Einstein equation ($\eta = \frac{k_B T}{6\pi DR}$), where R is the radius of the bead. Furthermore, the MSD functional relation with the lag time τ provides useful information on the type of motion the bead is exhibiting inside the condensate. This can be done by inspecting the value of the diffusivity exponent α, which is different from unity for anomalous diffusion. The value of α is also used to distinguish between sub-diffusion ($0 < \alpha < 1$) and superdiffusion ($\alpha > 1$) (Metzler, Jeon, Cherstvy, & Barkai, 2014).

Particle tracking is a useful and reliable technique for measuring the viscosity of the condensate. One of the possible issues regarding the implementation of particle tracking in protein condensates is non-specific interactions between the tracer particle and the molecules within the condensate. To resolve this, the beads used for particle tracking are coated with passive/inert polymers such as poly(ethylene glycol) (PEG), which can prevent such non-specific interactions. The coating procedure utilizes well established techniques such as carboxylate-amine or biotin-streptavidin coupling chemistry (Elbaum-Garfinkle et al., 2015; Feric & Brangwynne, 2013; Feric et al., 2016). One of the advantages of the SPT method is the fact that it is a *label-free* approach for probing condensate dynamics. While the tracer particle may be fluorescently tagged, the biopolymers constituting the condensates need not be fluorescently labeled. Furthermore, SPT can be used to trace nanoparticles and individual molecules in living cells using advanced labeling techniques and super-resolution microscopy. This makes SPT an important tool for studying single-molecule dynamics *in vivo* (Hansen et al., 2018; Shen et al., 2017). Finally, combining particle tracking methods to obtain the condensate viscosity with active coalescence assays to measure the inverse capillary velocity allows for the inference of the surface tension of the condensate (Eq. 2).

6.1 Single particle tracking: Methods

6.1.1 Definition

Measuring the mean square displacement of a tracer microsphere within a biomolecular condensate and detecting the condensate's apparent viscosity.

6.1.2 Rationale

Particle tracking is a robust method of estimating the viscosity of a condensate. It does not require special instrumental setup or fluorescent labeling.

6.1.3 Materials, equipment, and reagents

Confocal/epifluorescence/bright-field microscope, a sample containing phase-separated protein/RNA mixture, carboxylate functionalized polystyrene microspheres (typically 20–500 nm in diameter, passivated with PEG), Tween20/Pluronic F-127 coated microscope cover glass.

6.1.4 Protocol

1. Prepare the sample by adding the passivated polystyrene beads to the buffer followed by the addition of the protein/RNA to form the condensates. The condensates must be formed in a buffer that already contains passivated polystyrene beads (in order to encapsulate the beads within droplets). Bead concentration should be adjusted to have few beads within each droplet (ideally, 1–2 beads per droplet).

2. Load the sample onto the microscope and make sure that the beads are exhibiting Brownian motion. If the beads are sticking to the surface of the glass slide, consider other surface coatings such as SigmaCote (Sigma-Aldrich) or surface PEGylation.
3. Focus the microscope on a droplet that is stationary on the surface of the glass slide and contains one or two beads and collect a movie of the bead motion inside the droplet. For single particle tracking, one bead inside a droplet is ideal so that the motion of the bead can be traced over a larger area without interference from other particles in the field of view, even though some algorithms can detect multiple particles simultaneously (Crocker & Grier, 1996). Also, make sure that the bead being imaged is not close to the droplet interface as the latter will constrain the bead motion and give inaccurate results.
4. Repeat step 3 for a statistically significant number of trials. It is advisable to track different beads in different droplets in each of the trials.

6.1.5 Precursor techniques

The coating of microscope glass chambers is described in Section 2. Bead passivation is a standard procedure and has been described elsewhere (Elbaum-Garfinkle et al., 2015; Feric & Brangwynne, 2013; Feric et al., 2016).

6.1.6 Analysis and statistics

The outcome of the particle tracking experiment is a movie capturing the bead movement inside a droplet. The analysis constitutes tracking this particle and measuring its position as a function of time. This can be done using custom-made tracking programs or commercially available tracking software. Here, we used Fiji-ImageJ software along with the *TrackMate plugin* to track the bead's movement inside a droplet. Once tracking is completed, the output constitutes the XY coordinates of the particle as a function of time. The particle trajectory can then be plotted as shown in Fig. 12B. The calculation of mean square displacement $<r^2>$ for a lag time τ from a trajectory of a total time T can be done by considering the mathematical definition of the MSD

$$\langle r^2 \rangle(\tau) = \left\langle |r(t_0 + \tau) - r(t_0)|^2 \right\rangle_{t_0} \tag{15}$$

where the brackets denote averaging over all initial time points t_0 [for multiple particle tracking, the averaging is done over time and particles (Feric & Brangwynne, 2013)]. The values of τ are multiples of the time resolution of the trajectory. A plot of the MSD as a function of the lag time τ is shown in Fig. 12C. For a Newtonian fluid droplet in equilibrium without internal directional flow, the MSD functional dependence on τ follows Brownian dynamics and can be expressed as

$$MSD = \langle r^2 \rangle(\tau) = 4D\tau^{\alpha} \tag{16}$$

where D is the diffusion coefficient and α is the diffusive exponent. Fitting the data in Fig. 12C with this equation yields a diffusion coefficient of $D = 0.012 \pm 0.009\,\mu m^2/s$ and a diffusivity exponent $\alpha = 0.85$ (Fig. 13). This means that the motion is possibly sub-diffusive ($\alpha < 1$). We note that this analysis should be repeated for multiple trajectories to get accurate information on the condensates under study. Furthermore, to estimate the errors in MSD calculation, the particles can be fixed on the sample surface and imaged for a duration of time. The same analysis of MSD will give the MSD noise signal, any value of MSD larger than that noise is expected to occur due to the actual motion of the particle (Feric & Brangwynne, 2013). The viscosity can be calculated by assuming Brownian dynamics and using the Stokes-Einstein equation

$$\eta = \frac{k_B T}{6\pi a D} \tag{17}$$

where a is the size of the particle and T is temperature. Using the obtained value of the diffusion coefficient, the viscosity is calculated by Eq. 17 to be $\eta = 0.18$ *Pa. s.* The particle tracking experiment could be repeated with beads of different sizes to verify the scaling of the diffusion coefficient D with the particle diameter a (Elbaum-Garfinkle et al., 2015; Feric & Brangwynne, 2013).

6.1.7 Pros and cons

Pros	Cons
1. Provides unbiased information on the viscosity of the condensate 2. Does not require fluorescence labeling of biomolecules	1. Sensitive to the quality of passivation, especially when using surface-modified beads since they may interact with the proteins/RNAs in the condensate 2. Impractical to apply if the system forms condensates of relatively small size (droplet size comparable to bead size)

6.2 Troubleshooting and optimization

Problem	Solution
Beads do not stay in the focal plane of the microscope for a long time and diffuse in the z-direction. This limits the ability to follow the motion of the bead for a sufficient time	Increase the hydrophilicity of the surface by modifying the coating procedure to enhance droplet wetting on the glass slide surface. This would provide lesser volume in the z-direction and enable longer tracking
Beads are constrained in the droplet and are not showing diffusive motion	This problem indicates two possibilities: (a) biomolecules are forming a network within the droplet with mesh size less than the bead diameter, and/or (b) the bead is interacting with biomolecules and changing the droplet properties. Improving the passivation and decreasing the bead diameter may help to resolve the issue. Furthermore, consider different coatings for the glass slide such as coating with PEG or SigmaCote, since the stationary beads maybe sticking to the glass surface.

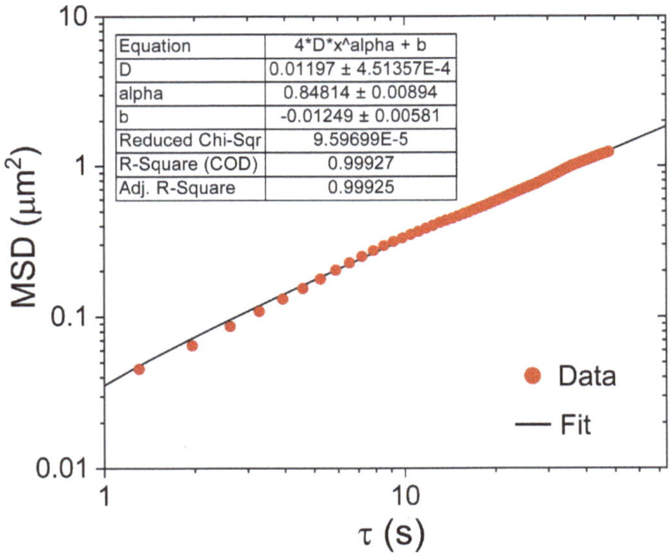

Fig. 13 MSD data extracted from the trajectory in Fig. 12. Red: MSD data. Black: a fitting model using Eq. 16.

6.3 Summary

Particle tracking is a classical experimental technique to measure the apparent viscosity of biomolecular condensates. It constitutes the imaging of a bead diffusing inside a condensate for a finite duration of time. The bead is tracked and quantities such as the MSD, diffusion coefficient, and apparent viscosity are extracted from the particle trajectory. When complementing particle tracking with droplet fusion assays, one can estimate both viscosity and surface tension of a biomolecular condensate.

7. Advanced methods

7.1 Measuring frequency-dependent viscoelasticity and surface tension using dual optical traps

Many cellular condensates exhibit properties of a complex viscoelastic fluid rather than purely viscous liquid. Jawerth et al. recently reported a new method of using optical traps to measure the viscoelastic properties of protein droplets (Jawerth et al., 2018). This method involves trapping a droplet between two tracer particles and applying controlled oscillatory stresses to the condensate (Fig. 14A and B). By monitoring the response of the system, the authors demonstrated that the droplet viscoelastic properties, as well as interfacial tension, could be measured. In brief, droplets are formed in a buffer containing microspheres (similar to the SPT assay described in Section 6.1). The resulting droplets will contain microspheres immersed within. Two optical traps are then used to trap two beads and then place them on a radial distance from the center of a suspended droplet, thereby slightly stretching the droplet. In this configuration, the beads can now act as handles to apply stress on the droplet. The traps are then set to oscillate at fixed frequencies and the corresponding forces from the optical traps are recorded (see Fig. 14C).

From the forces on the optical traps, one can calculate the frequency-dependent spring constant of the system using the relation

$$\chi^*_{sys}(\omega) = \left(\widetilde{F}_1 - \widetilde{F}_2\right)/2\widetilde{x}_{sys} \tag{18}$$

where $\widetilde{F}_1$ is the force measured by Trap-1,$\widetilde{F}_2$ is the force measured by Trap-2, $\widetilde{x}_{sys}$is the distance between the two traps (all as a function of frequency). One can derive the spring constant of the droplet from the relation

$$\chi^*_{drop}(\omega) = \frac{\chi^*_{sys}(4k_1k_2 + i\zeta\omega[k_1 + k_2])}{2k_1(2k_2 + i\zeta\omega) - 4\chi^*_{sys}(k_1 + k_2 + i\zeta\omega)} \tag{19}$$

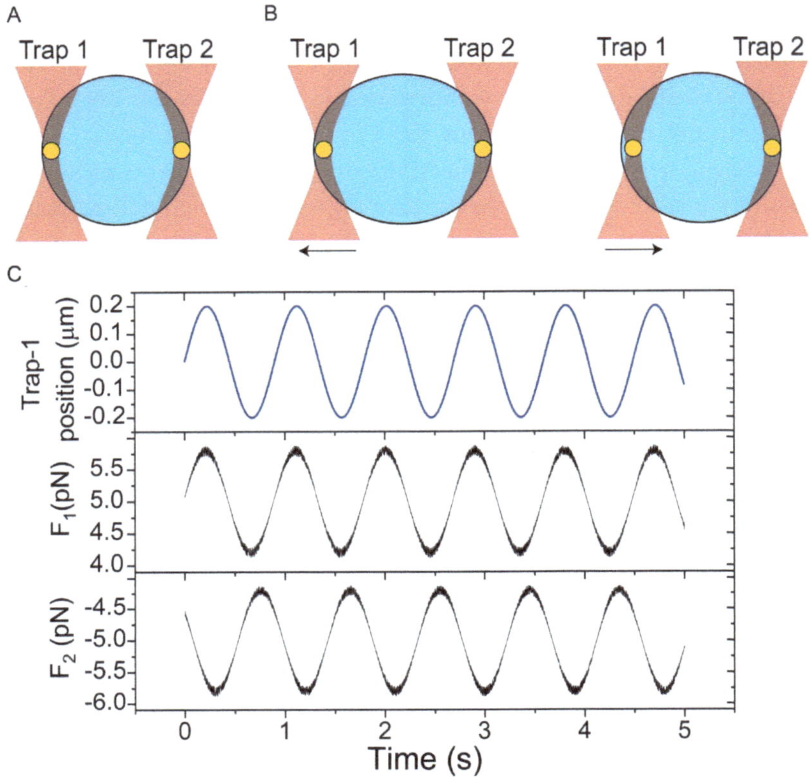

Fig. 14 (A) A diagram showing a trapped droplet between two optically trapped microspheres. (B) Driving Trap-1 to undergo oscillation applies stress on the condensate due to repetitive stretching. (C) Simulated data of Trap position, the force from Trap-1 (F_1), the force from Trap-2 (F_2) as a function of time during the oscillatory stress.

where $k_{1,\ 2}$ is the trap stiffness of trap 1, 2. ζ is the drag coefficient of the surrounding medium (can be calculated using $\zeta = 6\pi\eta_{medium}R_{droplet}$). At the low-frequency regime, corresponding to the slow oscillations, the spring constant of the droplet is dominated by surface tension. The surface tension can be calculated from

$$\gamma \approx \frac{\chi'(\omega)}{1.25 + 4.36\theta_0^2} \tag{20}$$

where $\theta_0 = r_{bead}/R_{droplet}$. This relation is valid for $\theta_0 \ll 1$. The complex modulus is then calculated by subtracting the surface tension effects from the droplet spring constant (Jawerth et al., 2018)

$$G^*(\omega) = \frac{\left[\chi^*_{drop}(\omega) - \left(1.75 + 6.31\theta_0^2\right)\gamma\right]}{R_{droplet}\left(0.58 + 3.42\theta_0^2\right)} = G'(\omega) + i\,G''(\omega) \tag{21}$$

The complex modulus G^* provides a rich description of the mechanical properties of materials. First, it allows the quantification of the elastic (G′) and viscous (G″) response of the material, thereby allowing for a precise elucidation of viscoelasticity (Rubinstein & Colby, 2003). For a purely viscous condensate, G″ dominates, whereas for an elastic condensate, G′ dominates. Furthermore, quantities such as viscosity and terminal relaxation time can be extracted from the material's complex modulus. For example, a Newtonian liquid has a complex modulus of the form $G^*(\omega) = i\eta\omega$, where η is the liquid viscosity (Rubinstein & Colby, 2003). In such a case, the droplet spring constant will have both real and imaginary parts ($\chi_{droplet} = \chi' + i\chi''$). The real part is independent of frequency and quantifies the contribution of the surface tension (see Fig. 15). Notwithstanding the limitation of implementing such a procedure that requires the formation of large droplets (~10–20 μm), and ensuring that the microsphere's refractive index is higher than that of the droplets (to ensure the feasibility of trapping a microsphere within a condensate), the method provides unique and novel insights into the viscoelastic properties as well as the surface tension of biomolecular condensates.

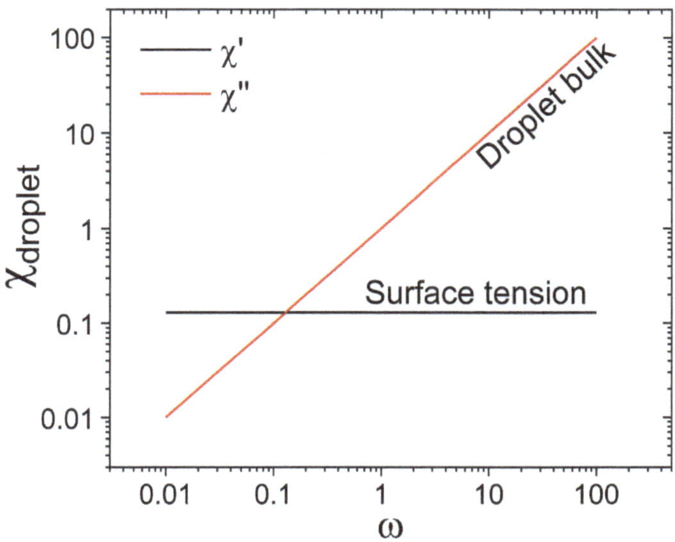

Fig. 15 Representative diagram for the frequency-dependent spring constant of a Newtonian fluid droplet. The real part of the spring constant (black) contains information on the surface tension of the droplet and the imaginary part of the spring constant (red) gives information on the viscosity.

8. Conclusion

Here, we discussed the common experimental methods to determine the fluid properties of a biomolecular condensate. Optical trap-induced coalescence provides information on the droplet dynamics at the mesoscale level. FRAP provides information on the molecular diffusivity in the nanoscale regime. Single particle tracking experiments can measure the apparent viscosity of a condensate. Combining MSD and fusion assays on the same sample enables the measurement of the interfacial tension by inference (Elbaum-Garfinkle et al., 2015; Feric et al., 2016). Combining FRAP and coalescence assays provides a multiscale picture of how condensate material properties may vary with length-scales (Alshareedah et al., 2019; Kaur et al., 2019). One major advantage of FRAP is that it could be utilized to study cellular condensates *in vivo*. However, it should be stressed that these techniques do not provide an absolute measure of the condensate material properties, rather they provide simple tools to measure the relative changes in droplet material properties at different conditions. Advanced methods such as active microrheology using optical tweezers may provide critical and much-needed tools to measure the viscoelastic properties of biomolecular condensates.

Acknowledgments

The authors acknowledge Benjamin Cammett, Dr. Wei Wang, and other members of Banerjee Laboratory for helpful discussions at various stages of manuscript preparation. **Funding**: We gratefully acknowledge support for this work from University at Buffalo, SUNY, College of Arts and Sciences to P.R.B. and funding from the National Institute on Aging (NIA) of the National Institutes of Health (R21 AG064258) to P.R.B.

References

Aguzzi, A., & Altmeyer, M. (2016). Phase separation: Linking cellular compartmentalization to disease. *Trends in Cell Biology*, *26*(7), 547–558.

Alberti, S. (2017a). Phase separation in biology. *Current Biology*, *27*(20), R1097–R1102.

Alberti, S. (2017b). The wisdom of crowds: Regulating cell function through condensed states of living matter. *Journal of Cell Science*, *130*(17), 2789–2796.

Alberti, S., & Hyman, A. A. (2016). Are aberrant phase transitions a driver of cellular aging? *BioEssays*, *38*(10), 959–968.

Alshareedah, I., Kaur, T., Ngo, J., Seppala, H., Kounatse, L.-A. D., Wang, W., et al. (2019). Interplay between short-range attraction and long-range repulsion controls reentrant liquid condensation of ribonucleoprotein–RNA complexes. *Journal of the American Chemical Society*, *141*(37), 14593–14602. https://doi.org/10.1021/jacs.9b03689.

Axelrod, D., Koppel, D., Schlessinger, J., Elson, E., & Webb, W. W. (1976). Mobility measurement by analysis of fluorescence photobleaching recovery kinetics. *Biophysical Journal*, *16*(9), 1055.

Banani, S. F., Lee, H. O., Hyman, A. A., & Rosen, M. K. (2017). Biomolecular condensates: Organizers of cellular biochemistry. *Nature Reviews Molecular Cell Biology*, *18*(5), 285–298.

Banani, S. F., Rice, A. M., Peeples, W. B., Lin, Y., Jain, S., Parker, R., et al. (2016). Compositional control of phase-separated cellular bodies. *Cell*, *166*(3), 651–663.

Banerjee, P. R., & Deniz, A. A. (2014). Shedding light on protein folding landscapes by single-molecule fluorescence. *Chemical Society Reviews*, *43*(4), 1172–1188.

Banerjee, P. R., Milin, A. N., Moosa, M. M., Onuchic, P. L., & Deniz, A. A. (2017). Reentrant phase transition drives dynamic substructure formation in ribonucleoprotein droplets. *Angewandte Chemie International Edition*, *56*(38), 11354–11359.

Berthier, J., & Brakke, K. A. (2012). *The physics of microdroplets*. John Wiley & Sons.

Boeynaems, S., Holehouse, A. S., Weinhardt, V., Kovacs, D., Van Lindt, J., Larabell, C., et al. (2019). Spontaneous driving forces give rise to protein−RNA condensates with coexisting phases and complex material properties. *Proceedings of the National Academy of Sciences of the United States of America*, *116*(16), 7889–7898.

Braeckmans, K., Peeters, L., Sanders, N. N., De Smedt, S. C., & Demeester, J. (2003). Three-dimensional fluorescence recovery after photobleaching with the confocal scanning laser microscope. *Biophysical Journal*, *85*(4), 2240–2252.

Brangwynne, C. P., Eckmann, C. R., Courson, D. S., Rybarska, A., Hoege, C., Gharakhani, J., et al. (2009). Germline P granules are liquid droplets that localize by controlled dissolution/condensation. *Science*, *324*(5935), 1729–1732.

Brangwynne, C. P., Mitchison, T. J., & Hyman, A. A. (2011). Active liquid-like behavior of nucleoli determines their size and shape in Xenopus laevis oocytes. *Proceedings of the National Academy of Sciences of the United States of America*, *108*(11), 4334–4339.

Brangwynne, C. P., Tompa, P., & Pappu, R. V. (2015). Polymer physics of intracellular phase transitions. *Nature Physics*, *11*(11), 899–904.

Ceballos, A. V., McDonald, C. J., & Elbaum-Garfinkle, S. (2018). Methods and strategies to quantify phase separation of disordered proteins. *Methods in Enzymology*, *611*, 31–50. Elsevier.

Chalupníková, K., Lattmann, S., Selak, N., Iwamoto, F., Fujiki, Y., & Nagamine, Y. (2008). Recruitment of the RNA helicase RHAU to stress granules via a unique RNA-binding domain. *Journal of Biological Chemistry*, *283*(50), 35186–35198.

Crocker, J. C., & Grier, D. G. (1996). Methods of digital video microscopy for colloidal studies. *Journal of Colloid and Interface Science*, *179*(1), 298–310.

Dahiya, P., Caggioni, M., & Spicer, P. T. (2016). Arrested coalescence of viscoelastic droplets: Polydisperse doublets. *Philosophical Transactions of the Royal Society A: Mathematical, Physical and Engineering Sciences*, *374*(2072)20150132.

Day, C. A., Kraft, L. J., Kang, M., & Kenworthy, A. K. (2012). Analysis of protein and lipid dynamics using confocal fluorescence recovery after photobleaching (FRAP). *Current Protocols in Cytometry*, *62*(1), 2.19. 11–12.19. 29.

Ditlev, J. A., Case, L. B., & Rosen, M. K. (2018). Who's in and who's out—Compositional control of biomolecular condensates. *Journal of Molecular Biology*, *430*(23), 4666–4684.

Eggers, J., Lister, J. R., & Stone, H. A. (1999). Coalescence of liquid drops. *Journal of Fluid Mechanics*, *401*, 293–310.

Elbaum-Garfinkle, S., Kim, Y., Szczepaniak, K., Chen, C. C.-H., Eckmann, C. R., Myong, S., et al. (2015). The disordered P granule protein LAF-1 drives phase separation into droplets with tunable viscosity and dynamics. *Proceedings of the National Academy of Sciences of the United States of America*, *112*(23), 7189–7194.

Feric, M., & Brangwynne, C. P. (2013). A nuclear F-actin scaffold stabilizes ribonucleoprotein droplets against gravity in large cells. *Nature Cell Biology*, *15*(10), 1253–1259.

Feric, M., Vaidya, N., Harmon, T. S., Mitrea, D. M., Zhu, L., Richardson, T. M., et al. (2016). Coexisting liquid phases underlie nucleolar subcompartments. *Cell*, *165*(7), 1686–1697.

Forman-Kay, J. D., Kriwacki, R. W., & Seydoux, G. (2018). Phase separation in biology and disease. *Journal of Molecular Biology*, *430*(23), 4603.

Franzmann, T. M., Jahnel, M., Pozniakovsky, A., et al. (2018). Phase separation of a yeast prion protein promotes, cellular fitness. *Science*, *359*(6371), eaao5654. 2. https://doi.org/10.1126/science.aao5654.

Gitler, A. D., Dhillon, P., & Shorter, J. (2017). *Neurodegenerative disease: Models, mechanisms, and a new hope*. The Company of Biologists Ltd.

Hansen, A. S., Woringer, M., Grimm, J. B., Lavis, L. D., Tjian, R., & Darzacq, X. (2018). Robust model-based analysis of single-particle tracking experiments with Spot-On. *eLife*, 7, e33125.

Hyman, A. A., Weber, C. A., & Jülicher, F. (2014). Liquid-liquid phase separation in biology. *Annual Review of Cell and Developmental Biology*, *30*, 39–58.

Jahnel, M., Behrndt, M., Jannasch, A., Schäffer, E., & Grill, S. W. (2011). Measuring the complete force field of an optical trap. *Optics Letters*, *36*(7), 1260–1262.

Jain, S., Wheeler, J. R., Walters, R. W., Agrawal, A., Barsic, A., & Parker, R. (2016). ATPase-modulated stress granules contain a diverse proteome and substructure. *Cell*, *164*(3), 487–498.

Jawerth, L. M., Ijavi, M., Ruer, M., Saha, S., Jahnel, M., Hyman, A. A., et al. (2018). Salt-dependent rheology and surface tension of protein condensates using optical traps. *Physical Review Letters*, *121*(25), 258101.

Jeon, B.-J., Nguyen, D. T., Abraham, G. R., Conrad, N., Fygenson, D. K., & Saleh, O. A. (2018). Salt-dependent properties of a coacervate-like, self-assembled DNA liquid. *Soft Matter*, *14*(34), 7009–7015.

Kang, M., Day, C. A., Kenworthy, A. K., & DiBenedetto, E. (2012). Simplified equation to extract diffusion coefficients from confocal FRAP data. *Traffic*, *13*(12), 1589–1600.

Kang, M., & Kenworthy, A. K. (2008). A closed-form analytic expression for FRAP formula for the binding diffusion model. *Biophysical Journal*, *95*(2), L13–L15.

Kaur, T., Alshareedah, I., Wang, W., Ngo, J., Moosa, M. M., & Banerjee, P. R. (2019). Molecular crowding tunes material states of ribonucleoprotein condensates. *Biomolecules*, *9*(2), 71.

King, O. D., Gitler, A. D., & Shorter, J. (2012). The tip of the iceberg: RNA-binding proteins with prion-like domains in neurodegenerative disease. *Brain Research*, *1462*, 61–80.

Kroschwald, S., Maharana, S., Mateju, D., Malinovska, L., Nüske, E., Poser, I., et al. (2015). Promiscuous interactions and protein disaggregases determine the material state of stress-inducible RNP granules. *eLife*, 4e06807.

Kubitscheck, U., & Peters, R. (2013). *Fluorescence microscopy*. Wiley Online Library.

Leal, L. G. (2007). Advanced transport phenomena: Fluid mechanics and convective transport processes. In (Vol. 7)Cambridge University Press.

Li, P., Banjade, S., Cheng, H.-C., Kim, S., Chen, B., Guo, L., et al. (2012). Phase transitions in the assembly of multivalent signalling proteins. *Nature*, *483*(7389), 336–340.

Mazza, D., Braeckmans, K., Cella, F., Testa, I., Vercauteren, D., Demeester, J., et al. (2008). A new FRAP/FRAPa method for three-dimensional diffusion measurements based on multiphoton excitation microscopy. *Biophysical Journal*, *95*(7), 3457–3469.

Metzler, R., Jeon, J.-H., Cherstvy, A. G., & Barkai, E. (2014). Anomalous diffusion models and their properties: Non-stationarity, non-ergodicity, and ageing at the centenary of single particle tracking. *Physical Chemistry Chemical Physics*, *16*(44), 24128–24164.

Mitrea, D. M., & Kriwacki, R. W. (2016). Phase separation in biology; functional organization of a higher order. *Cell Communication and Signaling: CCS*, *14*(1), 1.

Molliex, A., Temirov, J., Lee, J., Coughlin, M., Kanagaraj, A. P., Kim, H. J., et al. (2015). Phase separation by low complexity domains promotes stress granule assembly and drives pathological fibrillization. *Cell*, *163*(1), 123–133.

Nott, T. J., Petsalaki, E., Farber, P., Jervis, D., Fussner, E., Plochowietz, A., et al. (2015). Phase transition of a disordered nuage protein generates environmentally responsive membraneless organelles. *Molecular Cell, 57*(5), 936–947.

Onuchic, P. L., Milin, A. N., Alshareedah, I., Deniz, A. A., & Banerjee, P. R. (2019). Divalent cations can control a switch-like behavior in heterotypic and homotypic RNA coacervates. *Scientific Reports, 9*(1), 1–10. https://doi.org/10.1038/s41598-019-48457-x.

Patel, A., Lee, H. O., Jawerth, L., Maharana, S., Jahnel, M., Hein, M. Y., et al. (2015). A liquid-to-solid phase transition of the ALS protein FUS accelerated by disease mutation. *Cell, 162*(5), 1066–1077.

Paulsen, J. D., Carmigniani, R., Kannan, A., Burton, J. C., & Nagel, S. R. (2014). Coalescence of bubbles and drops in an outer fluid. *Nature Communications, 5*(1), 1–7.

Pawar, A. B., Caggioni, M., Hartel, R. W., & Spicer, P. T. (2012). Arrested coalescence of viscoelastic droplets with internal microstructure. *Faraday Discussions, 158*(1), 341–350.

Ramaswami, M., Taylor, J. P., & Parker, R. (2013). Altered ribostasis: RNA-protein granules in degenerative disorders. *Cell, 154*(4), 727–736.

Rowlinson, J. S., & Widom, B. (2013). *Molecular theory of capillarity.* Courier Corporation.

Rubinstein, M., & Colby, R. H. (2003). Polymer physics. In (Vol. 23), New York: Oxford University Press.

Rueden, C. T., Schindelin, J., Hiner, M. C., DeZonia, B. E., Walter, A. E., Arena, E. T., et al. (2017). ImageJ2: ImageJ for the next generation of scientific image data. *BMC Bioinformatics, 18*(1), 529.

Shen, H., Tauzin, L. J., Baiyasi, R., Wang, W., Moringo, N., Shuang, B., et al. (2017). Single particle tracking: From theory to biophysical applications. *Chemical Reviews, 117*(11), 7331–7376.

Shin, Y., & Brangwynne, C. P. (2017). Liquid phase condensation in cell physiology and disease. *Science, 357*(6357), eaaf4382.

Soumpasis, D. (1983). Theoretical analysis of fluorescence photobleaching recovery experiments. *Biophysical Journal, 41*(1), 95–97.

Sprague, B. L., Pego, R. L., Stavreva, D. A., & McNally, J. G. (2004). Analysis of binding reactions by fluorescence recovery after photobleaching. *Biophysical Journal, 86*(6), 3473–3495.

Taylor, N., Elbaum-Garfinkle, S., Vaidya, N., Zhang, H., Stone, H. A., & Brangwynne, C. P. (2016). Biophysical characterization of organelle-based RNA/protein liquid phases using microfluidics. *Soft Matter, 12*(45), 9142–9150.

Taylor, N. O., Wei, M.-T., Stone, H. A., & Brangwynne, C. P. (2019). Quantifying dynamics in phase-separated condensates using fluorescence recovery after photobleaching. *Biophysical Journal, 117*(7), 1285–1300.

Wang, J., Choi, J.-M., Holehouse, A. S., Lee, H. O., Zhang, X., Jahnel, M., et al. (2018). A molecular grammar governing the driving forces for phase separation of prion-like RNA binding proteins. *Cell, 174*(3), 688–699.e616.

Weidtkamp-Peters, S., Lenser, T., Negorev, D., Gerstner, N., Hofmann, T. G., Schwanitz, G., et al. (2008). Dynamics of component exchange at PML nuclear bodies. *Journal of Cell Science, 121*(16), 2731–2743.

Zhang, H., Elbaum-Garfinkle, S., Langdon, E. M., Taylor, N., Occhipinti, P., Bridges, A. A., et al. (2015). RNA controls PolyQ protein phase transitions. *Molecular Cell, 60*(2), 220–230.

Zhou, H., Song, Z., Zhong, S., Zuo, L., Qi, Z., Qu, L. J., et al. (2019). Mechanism of DNA-induced phase separation for transcriptional repressor VRN1. *Angewandte Chemie, 131*(15), 4912–4916.

CHAPTER SEVEN

Small-angle X-ray scattering experiments of monodisperse intrinsically disordered protein samples close to the solubility limit

Erik W. Martin[a], Jesse B. Hopkins[b], and Tanja Mittag[a,*]

[a]Department of Structural Biology, St. Jude Children's Research Hospital, Memphis, TN, United States
[b]The Biophysics Collaborative Access Team (BioCAT), Department of Biological Sciences, Illinois Institute of Technology, Chicago, IL, United States
*Corresponding author: e-mail address: tanja.mittag@stjude.org

Contents

Abstract

The condensation of biomolecules into biomolecular condensates via liquid-liquid phase separation (LLPS) is a ubiquitous mechanism that drives cellular organization. To enable these functions, biomolecules have evolved to drive LLPS and facilitate

Methods in Enzymology, Volume 646
ISSN 0076-6879
https://doi.org/10.1016/bs.mie.2020.07.002

partitioning into biomolecular condensates. Determining the molecular features of proteins that encode LLPS will provide critical insights into a plethora of biological processes. Problematically, probing biomolecular dense phases directly is often technologically difficult or impossible. By capitalizing on the symmetry between the conformational behavior of biomolecules in dilute solution and dense phases, it is possible to infer details critical to phase separation by precise measurements of the dilute phase thus circumventing complicated characterization of dense phases. The symmetry between dilute and dense phases is found in the size and shape of the conformational ensemble of a biomolecule—parameters that small-angle X-ray scattering (SAXS) is ideally suited to probe. Recent technological advances have made it possible to accurately characterize samples of intrinsically disordered protein regions at low enough concentration to avoid interference from intermolecular attraction, oligomerization or aggregation, all of which were previously roadblocks to characterizing self-assembling proteins. Herein, we describe the pitfalls inherent to measuring such samples, the experimental details required for circumventing these issues and analysis methods that place the results of SAXS measurements into the theoretical framework of LLPS.

1. Introduction

Intrinsically disordered protein regions (IDRs) are a class of proteins which do not adopt stable secondary or tertiary structure (Oldfield & Dunker, 2014; Tompa, 2012; van der Lee et al., 2014). This lack of structure can be thought of as arising from the flattening of the conformational energy landscape which results in a large ensemble of interconverting conformations with similar free energies. Thus, the characterization of the structure of IDRs must account for the ensemble nature of IDRs (Mittag & Forman-Kay, 2007). Many traditional structural techniques fail to provide information on IDRs; those that do, including NMR spectroscopy and single molecule fluorescence, provide largely ensemble-average or time-average observables.

Important measures of the structural features of IDRs are parameters that report on their global dimensions—quantities that small-angle X-ray scattering (SAXS) is ideally suited to measuring. In a homogeneous, dilute solution, SAXS directly probes the spatial distribution of atoms in the measurement volume thus providing direct access to the size and shape of the ensemble of conformations. The utility of SAXS in probing the ensemble properties of IDRs is well established and has been reviewed several times (Bernado & Svergun, 2012; Cordeiro et al., 2017; Kachala, Valentini, & Svergun, 2015; Rambo & Tainer, 2011).

Recent developments in cell biology have drawn attention to the physical process of liquid-liquid phase separation (LLPS) as a relevant driver of non-stoichiometric molecular assembly and cellular compartmentalization (Banani, Lee, Hyman, & Rosen, 2017; Boeynaems et al., 2018; Shin & Brangwynne, 2017). IDRs were suspected early on as being able to provide the driving force for biomolecular phase separation (Kato et al., 2012; Molliex et al., 2015), perhaps due to their qualitative similarity to well characterized aqueous multi-phase systems involving synthetic polymers (i.e., polyethylene glycol and dextran). Indeed, disordered regions have been shown to be sufficient for driving the phase transitions of many proteins (Conicella, Zerze, Mittal, & Fawzi, 2016; Kato et al., 2012; Molliex et al., 2015; Nott et al., 2015; Patel et al., 2015). Determining what elements within IDRs are providing the adhesive interactions that drive phase transitions has spurred increased interest in the conformational behavior of IDRs (Brady et al., 2017; Burke, Janke, Rhine, & Fawzi, 2015; Martin et al., 2020; Ryan, Dignon, et al., 2018).

Polymer phase transitions occur if the sum of three attractive potentials, i.e., polymer-polymer, solvent-solvent and polymer-solvent potentials, favors the solvation of polymer molecules by themselves rather than by solvent and is able to overcome the favorable entropy of mixing. This is captured in the Flory-Huggins mean field theory describing phase-separating polymers (Flory, 1942; Huggins, 1942). If a polymer is of sufficient length (often termed the infinite chain limit), Flory-Huggins theory predicts that the emergent properties of a solution of many molecules is recapitulated internally in a single polymer; the affinity between monomers in the same polymer is identical to that across polymers (Fig. 1A). While IDRs are heteropolymers and of finite length, homopolymer theory often describes their behavior well suggesting that heterogeneous, long-range interactions within IDRs are sufficiently dynamic that they may cancel on the ensemble level (Hofmann et al., 2012; Martin et al., 2016; Schuler, Soranno, Hofmann, & Nettels, 2016). Current research indeed supports the prediction from homopolymer theory that the dimensions of IDRs are predictive of their phase behavior (Dignon, Zheng, Best, Kim, & Mittal, 2018; Lin & Chan, 2017; Martin et al., 2020). Therefore, the details of molecular interactions that can be inferred from precise measurements of dilute IDR samples provides insight into the properties of dense IDR phases. For example, perturbations to the chemical nature of an IDR either via mutation/permutation of the amino acid sequence or via posttranslational modification (phosphorylation, methylation, etc.) impact the single chain

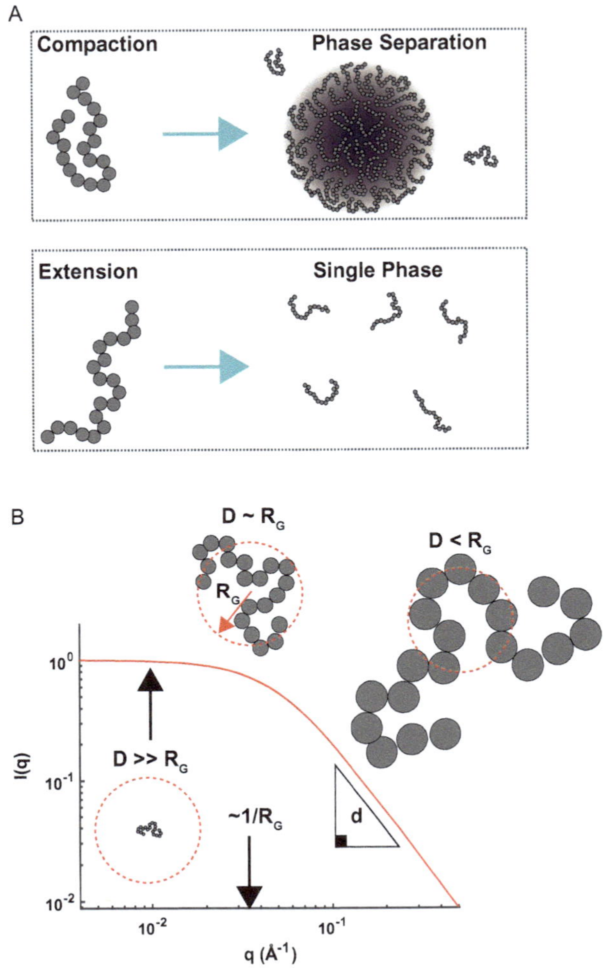

Fig. 1 Single chain IDR dimensions report on emergent properties. (A) Conditions that lead to dilute chain compaction similarly drive phase separation, while conditions that result in expansion maintain disperse protein. (B) The length scale information inherent in a SAXS experiment. The parameter D, with $D = 2\pi/q$, represents the resolution or distances within the protein that are primarily contributing to the SAXS curve at a defined point. The correlation length, $\sim 1/R_G$, signifies the length scale at which the small-angle regime gives way to the power-law regime. The fractal dimension, d, is defined as the slope of log(I(q)) vs log(q) in the higher angles.

protein properties and can be measured by SAXS; these changes are reflective of changes in the shape of the coexistence curve, or binodal, which quantifies the free energy of concentrated protein solutions. As a consequence of this symmetry, quantitative measurements informative of phase behavior can be made in technologically accessible dilute regimes. In the context of SAXS, this means that IDR shape information can be measured without convolution with interparticle interactions inherent in semi-dilute or concentrated solutions.

2. Resolution of structural information from SAXS

The minimum distance accessible to a SAXS experiment (Fig. 1B) is related to the momentum transfer vector, q.

$$q = \frac{4\pi \ sin(\theta)}{\lambda}, \tag{1}$$

where λ is the X-ray wavelength (1–2 Å at synchrotron sources) and 2θ is the scattering angle. The resolution increases with increasing angle such that its length scale is $D = \frac{2\pi}{q}$. The typical SAXS experiment provides a transfer vector range of $0.005\ \text{Å}^{-1} < q < 0.5\ \text{Å}^{-1}$. Therefore, the highest obtainable resolution is ~10 Å. However, this cutoff is not strictly analogous to the resolution in high-resolution structural techniques such as X-ray crystallography or cryo-electron microscopy (Tuukkanen, Kleywegt, & Svergun, 2016). While it is true that SAXS does not provide details at the amino acid level, parameters such as the radius of gyration (R_G) can be calculated at precision far exceeding this resolution (Svergun & Feĭgin, 1986). This apparent dichotomy is rooted in the fact that SAXS is a "low information" technique where only a limited number of parameters can be extracted precisely (Koch, Vachette, & Svergun, 2003).

In addition to the available q range, the resolution of protein SAXS data is limited by the scattering intensity, which is related to the electron scattering length density (SLD) difference between the protein and the solvent. The SLD quantifies the scattering power of an object. X-rays interact with electrons and thus the SLD for SAXS is related to electron density. Generally, the SLD of a particular atom scales with the number of electrons and therefore with atomic number (Z). Due to absorption, the relationship between Z and SLD is not completely monotonic—a property exploited in anomalous X-ray scattering experiments. Due to the presence of only atoms with low Z, protein samples suffer from low scattering contrast that becomes vanishingly small at higher angles. This is particularly problematic in IDRs where the electron density is spread over a wider distribution of distances than in globular proteins. Therefore, measurements at wider angles provide little additional information.

The scattering intensity as a function of the scattering angle, I(q), from completely disordered IDRs is typically characterized by two distinct resolution regimes. At the smallest angles, the scattering intensity approaches a finite value, I_0, that, in the absence of correlations between molecules, is dependent only on concentration and molecular mass. At wider angles, the scattering is approximately a power law, $I(q) \propto q^{-d}$. The exponent d is the fractal dimension which quantifies self-similarity and is a measure of how the mass increases relative to volume (where, $mass \propto distance^{d}$). For folded proteins, the mass roughly scales directly with volume and d approaches 3. However, inhomogeneities in electron density caused by the non-random atom distribution in folded domain structure result in departures from pure power law behavior—a feature exploited to model protein shape. In contrast, the diverse ensembles of conformations that are characteristic of IDRs average out many inhomogeneities and the scattering intensity at higher angles is often well characterized by $I(q) \propto q^{-d}$.

The fractal dimension, d, thus reports on the ensemble-average density in the sample. At the extremes, $d \approx 3$ for a collapsed globule, similar to folded proteins, and for a fully extended rod, $d = 1$. For a homopolymer near the infinite length limit, values of d are bounded between ~1.6 and 2 which represent the limits of a self-avoiding random walk and a Gaussian coil, respectively (Rubinstein & Colby, 2003). The self-avoiding limit applies to a polymer with a net repulsion between monomers whereas the Gaussian coil limit applies at the specific condition, designated the "theta state," where repulsive and attractive forces cancel. IDRs most often have a fractal dimension that falls within this window (Bernado & Blackledge, 2009; Riback et al., 2017). Due to the finite length of IDRs, regions of local or long-range attraction or repulsion could drive the conformational ensemble outside of the theoretical limits of a homopolymer toward either a disordered globule (Hofmann et al., 2012; Martin et al., 2020) or rod-like shape (Muller-Spath et al., 2010). For example, short polyelectrolytes are more extended than a self-avoiding polymer and have values of d less than 1.6. Values of d between 2 and 3 could suggest partial folding or collapse into a molten globule state (Ptitsyn, 1995; Uversky, 2002). In short, the fractal dimension reports on the packing of the IDR and hence is diagnostic of the shape of the ensemble. The crossover between fractal regime into the small-angle regime is defined by a correlation length past which the IDR no longer exhibits fractal-like behavior. This characteristic length is directly related to the R_G of the IDR (Fig. 1B) (Hammouda, 1993).

Both d and R_G are properties determined by the complete distribution of atomic distances in the sample and thus ensemble-averaged quantities that report on the entire accessible conformational space of the IDR. Assuming that the sample is monodisperse and sufficiently dilute, R_G can be determined precisely and reflects the mean distance of all protein atoms from the center of mass (Koch et al., 2003). Determination of the fractal dimension, d, can be less straightforward. If the IDR truly behaves like a homopolymer in the sense that the amino acid positions can be represented by a smooth distribution (i.e., all heterogeneity is averaged out), the fractal dimension can be accurately obtained and should be coupled to R_G via the scaling relation $R_G \sim N^{\frac{1}{d}}$, where N is the number of amino acids in the IDR. The inverse of the fractal dimension is the Flory scaling exponent (Flory, 1953; Rubinstein & Colby, 2003), $\nu = \frac{1}{d}$. Strictly speaking, the scaling relation applies to a homopolymer of infinite length—a condition not met by finite length IDRs. Therefore, the measured value of ν can be thought of as an "apparent scaling exponent" comparing the IDR to a similar homopolymer and allowing values outside the theoretical limits.

If the distribution of monomers in the IDR deviate from statistical randomness as a result of long-range correlations between protein regions, which can arise due to strong binary interactions between distant regions, R_G and ν are decoupled (Banks, Qin, Weiss, Stanley, & Zhou, 2018; Fuertes et al., 2017; Riback et al., 2019). The case of an IDR that is linked to a folded domain is an extreme example of long-range correlations and could have a scattering profile with additional correlation lengths at short distances that reflect the atomic distance distribution in the folded domain. In short, due to the limited resolution of SAXS experiments, the parameters R_G and ν contain the majority of the information content for samples of completely disordered IDRs. The goal of a SAXS experiment is to accurately extract these parameters to obtain insight into the size and shape of the conformational ensemble and quantify deviations from ideal, random behavior.

3. Complications of SAXS measurements of self-assembling proteins

SAXS measurements on soluble biomolecules are straightforward and often done rapidly and effectively in high-throughput facilities (Classen et al., 2013). However, the very features that allow some IDRs to phase separate limit our ability to make high-quality SAXS measurements. The primary issues fall into two categories. (1) Some phase-separating proteins form soluble oligomers via oligomerization domains which enhance phase separation (Mitrea et al., 2018; Powers et al., 2019; Wang et al., 2018); they may also form off-pathway aggregates (Conicella et al., 2016; Molliex et al., 2015; Patel et al., 2015; Schmidt, Barreau, & Rohatgi, 2019). The SAXS curve of a heterogeneous ensemble is the mass average, not the number average, of components, and small populations of oligomers or aggregates contribute disproportionately to the scattering. Additionally, the contribution of large species is particularly high at the small angles which are required to precisely determine R_G. While SAXS can be used to monitor both oligomerization (Williamson, Craig, Kondrashkina, Bailey-Kellogg, & Friedman, 2008) and aggregation (Herranz-Trillo et al., 2017), both processes severely impair analysis of the ensemble of monomers. (2) The saturation concentration (c_{sat}), i.e., the concentration above which the protein forms a dense phase, determines a concentration limit above which SAXS data on monomeric samples is not accessible. However, the effective limit is often lower because the attractive interaction potential between proteins that mediates phase separation can lead to an upturn in intensity at small angles even below c_{sat}. This feature is indicative of interference between molecules arising from intermolecule distances that are close enough in space to lie within the SAXS resolution. These features can be modeled by a function called the structure factor which decays to unity at high angles and is discussed further in Section 7. Measurements at these low concentrations limit the signal-to-noise of IDR SAXS data with low scattering contrast. These issues are so pervasive that SAXS measurements have historically favored highly soluble proteins. This point is well illustrated by the fact that the overwhelming majority of IDRs measured by SAXS are more expanded than the polymer theta state (i.e., $\nu > 0.5$), which implies self-avoidance (Bernado & Blackledge, 2009; Cordeiro et al., 2017). These observations have fostered the view that IDRs are generally highly soluble and expanded (Riback et al., 2017). However, phase-separating IDRs—under conditions that promote phase separation—have been reported to have $\nu < 0.5$ (Martin et al., 2020) highlighting the ability of adhesive elements in the chain to act inter- as well as intramolecularly.

In order to obtain high-quality data on challenging IDRs characterized by limited solubility, contrast and high aggregation potential, we recommend size-exclusion chromatography (SEC)-coupled measurements at high-flux synchrotron radiation sources. SEC-coupled SAXS (SEC-SAXS) eliminates small aggregates and provides flexibility in sample buffer conditions, superior baseline subtraction and rigorous analysis of interparticle interference which we will discuss below.

4. IDR sample measurement in SEC-SAXS mode

A critical consideration when developing a SAXS experiment on an associative IDR is to map the conditions under which the IDR is soluble, forms aggregates and phase separates. These conditions will set the practical limits for the experiment. In the ideal SEC-SAXS setup, the SEC column will be plumbed to minimize the dead volume between the elution from the column and the X-ray scattering measurement. This minimizes the possibility of sample aggregation or phase separation allowing measurements close to the solubility limit. To have the highest obtainable concentration at the X-ray beam, samples need to be loaded onto the SEC column several fold higher in concentration to account for the on-column dilution factor. A practical way to approach this is to load the samples onto the column in a buffer in which the protein is soluble to high concentrations (Riback et al., 2017). For IDRs, this buffer can contain a denaturant such as guanidinium hydrochloride because there is no risk of perturbing important structural features in the protein in an irreversible manner. Alternatively, buffers that include high (or very low) salt concentrations can be effective in maintaining solubility. The experiment then relies on the buffer-exchanging ability of SEC to place the sample in the correct solution conditions at the X-ray beam.

As a practical consideration, the associative properties of phase-separating proteins can slow their passage through a SEC column. The solubilizing additives in the sample loading buffer can "push" the protein through the column. This will manifest as one protein peak eluting where expected based on the protein's hydrodynamic radius, followed by a second peak that contains more protein as well as the additives. For this reason, it is useful if the loading buffer has a significantly different conductance than the running buffer to easily distinguish elution frames containing the sample in the desired buffer.

SEC columns are selected such that resin sufficiently separates the protein from small molecules. SAXS beamlines differ in their preference for silica- or dextran-based SEC resins. Superdex Increase resins from GE Healthcare have been used successfully at the BioCAT beamline at the Advanced Photon Source at Argonne National Lab.

An important consideration in column selection is the maximum flowrate. Radiation damage is mitigated by the sample flow and, in a traditional system, the minimum flow rate will be determined by the rate of X-ray damage to the sample—faster flow results in shorter exposure of any given sample volume resulting in less damage. The coflow systems implemented at the Australian National Synchrotron and BioCAT, which will be discussed in detail later, allow for the use of smaller columns and slower flow. A GE Healthcare Superdex Increase 5/150 column with 3 mL volume, which has a maximum flow rate of 0.45 mL/min, is typically sufficient to avoid radiation damage. The minimally allowable flowrate is beamline dependent. While it is impossible to exceed the solubility limit of a particular protein, longer columns allow for larger volumes to be injected; they spread the elution over a larger volume and therefore more frames that can be averaged. The experimental conditions thus represent a tradeoff that has to take into account protein solubility, flux, column flowrate and sample availability.

The SAXS data is collected as individual exposures; their length is a balance between flow rate, the volume in which the sample elutes and the X-ray flux. With a 3 mL SEC column eluting at 0.4 mL/min into a coflow sample chamber, 0.5 s exposures are a good compromise. The resulting data is a series of detector images that can be analyzed in a similar fashion to all SAXS data with freely available software in packages such as BioXTAS RAW (Hopkins, Gillilan, & Skou, 2017) and CHROMIX in ATSAS (Panjkovich & Svergun, 2018).

The obvious benefit of SEC-SAXS is the elimination of oligomers and aggregates (Fig. 2). Equally important for the analysis of adhesive proteins is an ideal baseline subtraction (Fig. 2) and a built-in continuous concentration gradient (Fig. 3). SEC-SAXS provides the best possible baseline subtraction via the selection of buffer frames that are as close to the protein elution peak as possible. As a result, the composition of the buffer is identical to that in the sample. While this is often a minor consideration when dealing with concentrated samples with high relative signal to background, the quality of the buffer match becomes critically important as the sample concentration decreases. The continuous concentration gradient that is intrinsic in the elution peak from a SEC column (Fig. 3) enables the critical examination of the data for interparticle interference. Interparticle interference will appear as a dome when the R_G is plotted as a function of elution frame (Fig. 3). If associative IDRs oligomerize in a concentration dependent manner on the column, the R_G plot will have a negative slope, because larger species elute earlier from SEC columns (Fig. 3). If the association happens with a lag time

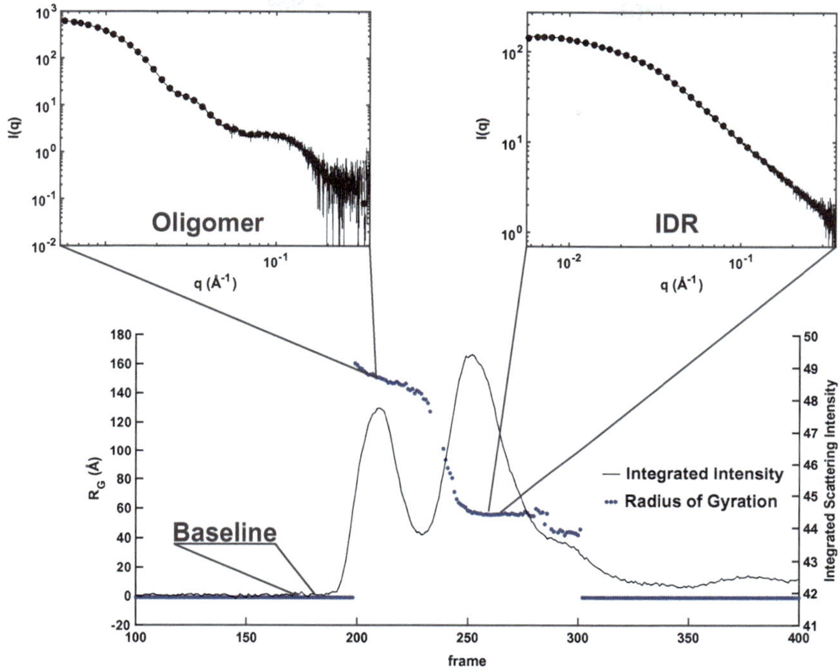

Fig. 2 SEC-SAXS enables the separation of monomeric IDR from unwanted oligomers and aggregates. The baseline can be chosen from statistically similar frames, where no part of the curve deviates from expected random noise, near the sample elution. The desired sample data is averaged from a region where the R_G is not concentration dependent.

but is faster than the deadtime between SEC elution and data collection, the R_G plot will also appear as a dome because self-association is strongest at the highest protein concentrations in the middle of the elution peak (Fig. 3). Even in the presence of such effects, regions of the SEC elution with the lowest concentration may be selectively averaged to ensure elimination of these artifacts.

5. Primary data analysis

The goal of the SAXS experiment is to extract the ensemble-average R_G of the IDR and the shape of the ensemble of conformations. The primary analysis of data on adhesive IDRs proceeds through a series of relatively simple transformations of the scattering profile (Fig. 4) which inform on the size and shape.

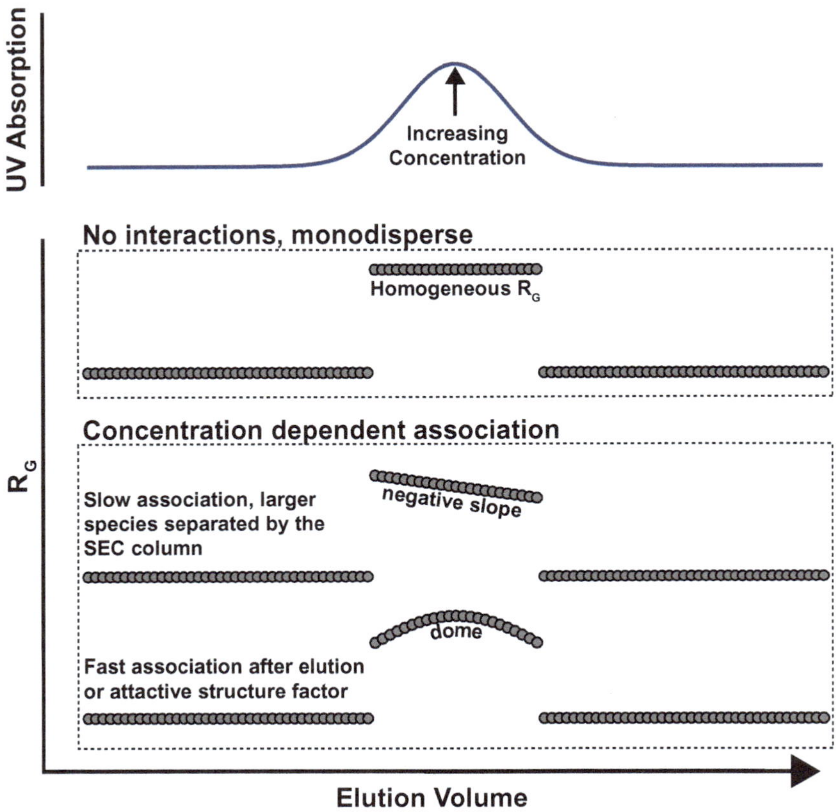

Fig. 3 The R_G as a function of elution time is diagnostic of intermolecular interactions. The elution from the SEC column is monitored by UV absorption. The magnitude of the absorption is directly related to the protein concentration. The R_G is calculated from a rolling average along the protein elution. If the R_G is flat across all elution concentrations, there are no inter-monomer interactions and the sample is monodispersed. A negative slope indicates that larger species are separated by the column and are eluting first. A domed elution profile indicates that the protein self-associates in a concentration dependent manner after elution from the column.

5.1 Guinier analysis

The easiest way to obtain size information from a SAXS profile is a linear fit to the Guinier transform of small-angle scattering data. The Guinier equation results from a first order expansion of the Debye scattering equation in the limit of very small angles.

$$I(q) = I_0 e^{\frac{-q^2 R_G^2}{3}} \tag{2}$$

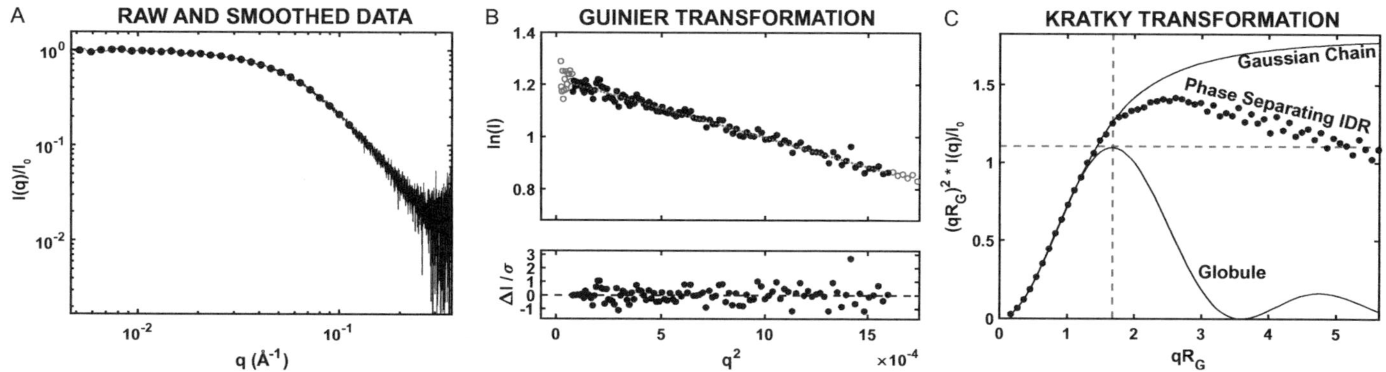

Fig. 4 Primary SAXS data transformations. (A) Raw SAXS data on IDRs presented in log-log format highlights the low q data, power law regimes and correlation length. Binned data is shown as circles. (B) The Guinier transformation of low q ($qR_G < 1$) data is used to calculate R_G. Lower plot shows the residuals. (C) The Kratky transformation enables visual inspection of the fractal dimension as indicated by the slope of the data at high q relative to Gaussian chain and globule references. On a dimensionless Kratky plot, the intersection of red lines indicates the location of the maximum for a globule. Deviation of the maximum from this point indicates flexibility.

I_0, the zero-angle scattering, is determined by protein concentration and molecular weight. In a Guinier plot, data is plotted as the natural logarithm of the intensity as a function of q^2 (Fig. 4B). Thus, the resulting linear slope is proportional to the square of R_G. Due to the fact that the Guinier equation discards all higher-order terms in the expansion, the equation is only valid at very small angles, which for IDRs is taken to be approximately $q < \frac{1}{R_G}$ (Svergun, Feĭgin, & Taylor, 1987). Guinier analysis provides a good first approximation of the R_G and information on sample quality. The pattern of the residuals in a linear fit of ln(I) vs q^2 (Fig. 4B) can indicate aggregation or interparticle interference, both of which manifest as an upturn at the smallest angles. While these data points can be excluded, the effects impact the whole scattering profile. When analyzing low-concentration data on IDRs, Guinier analysis can be of limited utility. The noise combined with the limited utilizable q range can result in a high uncertainty of the R_G, and these issues are magnified if there is the slightest concern that the sample is not monodisperse.

5.2 Kratky transformation

The Kratky transformation scales the intensity by q^2 (Fig. 4C). In folded proteins, a Kratky plot can inform on the flexibility of the system. The Kratky plots of well-folded proteins have bell curves with well-defined maxima and converge to zero at higher q (Fig. 4C). In contrast, Kratky plots of unfolded proteins appear hyperbolic and serve as a visual indicator of d. An IDR at the theta point, with $d = 2$, reaches a plateau that is flat until the end of the fractal region of the curve at high angles (a limit which is rarely experimentally accessible and not discussed here). For self-avoiding IDRs with $d < 2$, the Kratky plot at high q has a positive slope. For self-interacting IDRs with $d > 2$, the slope is negative (Fig. 4C). The Kratky plot can be normalized and made dimensionless by scaling the intensity by the zero-angle scattering and the scattering angle by R_G. By normalizing the intensity by I_0, samples of different concentrations or molecular weights can be compared. Multiplying q by R_G normalizes the distance resolution by the protein radius. In the context of a Kratky plot, these normalizations allow for direct comparison of d, and therefore the scaling exponent ν, across samples (Durand et al., 2010).

5.3 Comments on indirect Fourier transform

In X-ray crystallography (XRC), molecules are present in regular orientation in the crystal lattice and the resulting diffraction pattern represents the position of individual atoms in inverse space. With knowledge of phases, Fourier transform of this data yields the position of atoms in 3D space. Similarly, the SAXS pattern is also determined by the distribution of atoms in space. However, unlike in XRC, the particles in a SAXS sample are typically randomly distributed resulting in spherical averaging of the particle electron density. In the context of IDRs and highly flexible systems, the data represents an additional average over the ensemble of conformations. The quantity derived from a SAXS experiment that is analogous to 3D coordinates in XRC is the atomic pair distribution function, P(r). The SAXS intensity can be directly calculated from the Fourier transform of P(r).

$$I(q) = 4\pi \int_0^\infty P(r) \frac{\sin(qr)}{qr} dr \qquad (3)$$

In principle, recovering the P(r) distribution from the SAXS intensities should be possible by the inverse Fourier transform. In practice, experimental limitations in measurable scattering angles would require extrapolation to zero angle and wide angles. To circumvent this issue, use of the indirect Fourier transform (IFT), in which the data is represented by a series of basis functions, was proposed by Glatter (1977).

Similar methods are implemented in most SAXS data analysis software and generally are effective in determining parameters such as the R_G and the shape within the distance resolution of the measurement. This information is used for ab initio shape determination for well-folded systems (Svergun, 1999). However, IFT is problematic for SAXS data analysis of highly flexible, disordered systems. The issue is inherent to finding the solution to the IFT which is an ill-posed inverse problem. The fit therefore relies heavily on regularization to ensure that the solution is smooth. Further, the IFT requires the probability density to converge to zero at zero distance and at a fixed maximum dimension (D_{max}). Finding an appropriate value of D_{max} for proteins with well-defined shape can be accomplished by sampling values around an estimated solution while monitoring the quality of fit. In the case of IDRs, the D_{max} is poorly defined. Conformations with a D_{max} that approaches the contour length of the IDR may exist but with vanishingly small populations—a problem that is magnified by the weaker contrast for very extended conformations. Given that the solution for R_G from IFT depends on the choice of D_{max}, the R_G from IFT for IDRs can, at best, have a large error and, at worst, be inaccurate.

The P(r) distribution has some utility in analyzing SAXS data on phase-separating proteins. The shape of the P(r) distribution is likely similar at intermediate distances for all possible solutions and could thus be used to assess changes in shape. If phase separation in a particular protein is initiated by folding or changes in relative domain orientations, the shape of the P(r) distribution will be diagnostic of these changes. However, it is important to keep in mind that the information content of the P(r) distribution is limited by the resolution of the experiment. For small IDRs, high signal beyond $q = 0.25 - 0.3\ \text{Å}^{-1}$ is rare. Changes in P(r) that suggest shape changes at intermediate distances may thus be caused by artifacts from low signal or suspect baseline subtraction—a problem equally likely for larger proteins if higher resolution angles are included in the fit but are not free from issues of low signal or baseline subtraction artifacts. IFT is often advertised as a more precise method for obtaining radii of gyration compared to Guinier analysis because it uses scattering data from the full q range, but for IDRs, the problems resulting from IFT may outweigh the benefits.

6. Synchrotron SAXS beamline hardware

Making high-quality SAXS measurements of IDRs with low inherent contrast and concentration mandates attention to maximizing signal-to-noise. This requires the use of high-flux synchrotron radiation sources that are specifically designed for biological SAXS measurements.

The Advanced Photon Source (APS) at Argonne National Lab is a high brilliance, third generation synchrotron radiation source ideally suited for measurement of demanding biological samples. The Biophysical Collaborative Access Team (BioCAT) beamline at Sector 18 is dedicated to biological SAXS, both of fibers and muscles (fiber diffraction) and solutions and has optimized X-ray optics and detection hardware. The BioCAT beamline is a good model for discussing hardware and will thus be the template for this section. Similar beamlines at synchrotrons around the world will be discussed at the end of this section. The overall layout and primary X-ray optics of the beamline are well described (Fischetti et al., 2004). In brief, BioCAT uses both vertical and horizontal focusing optics to run monochromic SAXS experiments at 12 keV with $\sim 2 \times 10^{13}$ ph/s in the full $30 \times 140\ \mu m^2$ focused beam. Collimation for solution SAXS experiments results in an available experimental flux of $\sim 6 \times 10^{12}$ ph/s. A Pilatus3 × 1M detector (Dectris) is used to measure the resulting scattering.

The signal-to-noise in SAXS measurements can be improved in three ways via the beamline hardware: improving X-ray detection, reducing background scattering, and increasing X-ray flux. X-ray detector technology is already mostly optimized. In the last ~10 years, X-ray pixel array detectors (PADs) (Barna et al., 1995) have become the most common detector type for SAXS. These are single photon counting detectors with zero readout noise, high dynamic range, and detection efficiencies near 100% for X-ray energies in the range of 10–12 keV. Therefore, the noise level is only limited by the shot noise (Poisson noise) inherent in a counting experiment. The previous and current generation of PADs, most commonly the Pilatus and Eiger detectors made by Dectris (Kraft et al., 2009), provide large detection areas (up to $\sim 300 \times 300\,mm^2$), fast readouts (up to 2000 Hz), and minimal (~1 ms) or no deadtime between images. While improvements are possible for certain applications, these detectors are already ideal for SAXS. Beamlines typically have the largest area detector they can afford to maximize the percentage of scattered photons captured. Charge-coupled device (CCD)-based detectors may still be used at some beamlines but are non-ideal for SAXS due to generally slow readouts, non-zero readout noise, and lower detection efficiencies.

SAXS measures X-rays coming from three sources: the sample, the solution, and the SAXS instrument itself. Minimizing the number of X-rays from the instrument can significantly improve the signal-to-noise of a measurement. Instrumental X-ray scattering, also called background or parasitic scattering, is the dominant component of measured X-rays from the instrument, though there may also be a contribution from X-ray fluorescence. Instrumental scattering comes from anything in the beam path, including X-ray optics, X-ray windows, beam monitoring devices, and most significantly air. All SAXS beamlines use at least a partial vacuum environment, which removes air scattering from the measurement. Some SAXS beamlines use in-air sample cells, which increases flexibility for doing experiments with different types of sample holders at the cost of adding scattering from two X-ray windows and a small amount of air. BioCAT and other SAXS beamlines run experiments in a full vacuum environment, including using an in-vacuum sample cell. This removes as much scattering as possible from the X-ray flight path.

Some things are required to be in the beam path, for example, X-ray optical elements used to clean and shape the beam before measuring the sample. The scattering from these items is generally divergent from the incident X-ray beam. By placing these items far away from the sample and using one or more sets of slits or pinholes carefully positioned in the beam path, the excess scattering from these items can be almost entirely removed from the measurement. The exact type and layout of these slits or pinholes varies from beamline to beamline, depending on available space. BioCAT uses three sets of single crystal bladed scatterless X-ray slits with both horizontal and vertical blades (Xenocs and JJ X-ray) to collimate the beam and remove excess instrumental scattering. This, combined with minimizing the number of scattering sources by using an in-vacuum sample cell, keeps the intensity of instrumental scattering ~100-fold lower than the solution/sample scattering.

Signal-to-noise levels for a measurement can also be increased by increasing the X-ray flux on the sample. The maximum flux available at a beamline is limited by the X-ray source and beamline design and cannot be easily improved without expensive and intrusive upgrades. However, an additional, practical limitation on flux is defined by the susceptibility of the sample to radiation damage (Hopkins & Thorne, 2016; Jeffries, Graewert, Svergun, & Blanchet, 2015). Thus, strategies which minimize radiation damage without compromising flux are crucial to maximizing signal-to-noise. Historically, strategies for reducing radiation damage have involved a combination of continuously flowing the sample through the X-ray beam to spread the incident flux over more sample and the addition of buffer components that act as radical scavengers. While these methods are partially effective, the majority of high-flux beamlines like BioCAT still require attenuation of the incident beam to prevent damage.

Recently, a novel sample cell design, called a coflow cell (Fig. 5A and B), was proposed to mitigate issues surrounding radiation damage (Kirby et al., 2016). The coflow cell provides an exterior sheath of buffer "coflowing" with a central core of sample solution in a laminar flow regime. This flow geometry both minimizes the likelihood of damage to the sample while simultaneously preventing spurious background scattering that can occur from damaged samples adhering to the cell. Using this design, Kirby et al. were able to use an order of magnitude more X-ray flux without damage to the sample, and were able to take full advantage of the high-flux beamline (Kirby et al., 2016). Versions of the coflow cell are used at the SAXS/WAXS beamline at the Australian Synchrotron and BioCAT and allow the beamlines to use all available flux without causing radiation damage, which increases signal-to-noise. The increase in flux coupled to decrease in noise allows for measurement at concentrations as low as 0.4 mg/mL (prior to injection onto an SEC column) for a 12 kDa IDR (Fig. 5C). This advance has been critical to acquiring interpretable SAXS data on IDRs with low c_{sat}.

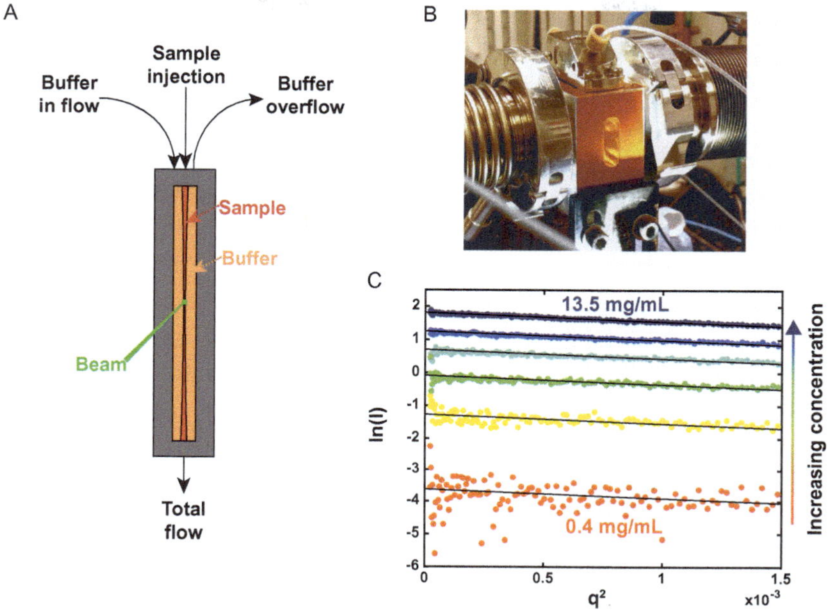

Fig. 5 The coflow sample chamber enables maximizing signal-to-noise while minimizing radiation damage. (A) A schematic view of the coflow sample chamber. The top of the chamber contains a buffer inlet and an outlet to remove excess buffer. The sample is injected into the middle of the capillary where a constant flow is maintained by the total flow outlet pump. The result is the sample "coflowing" with the buffer in the laminar regime and not mixing. (B) The coflow sample chamber installed at the BioCAT beamline (18ID-D) at the Advanced Photon Source. (C) Low q data shown in Guinier format indicating that identical R_Gs can be calculated from 0.05 mL samples injected onto a 3 mL SEC column at concentrations ranging from 13.5 to 0.4 mg/mL (12.5 kDa protein).

In the next ~3 years, the APS will be undergoing an upgrade (APS-U), which will increase flux at the BioCAT beamline. BioCAT will also be adding a new multilayer monochromator, which will further increase available flux. While not all of this flux may be useable for SAXS experiments, any portion that is will further improve signal-to-noise.

There are a number of X-ray beamlines around the world with dedicated biological solution SAXS setups similar to that at BioCAT. While details often differ, depending on the specifics of the X-ray source and layout, all beamlines with a dedicated biological SAXS program provide broadly similar resources. These include in-line SEC-SAXS, and increasingly SEC-MALS-SAXS, batch mode SAXS measurements, and often automated sample changing and sample cell cleaning. All beamlines also provide PAD detectors, usually Dectris Pilatus or Eiger detectors with detecting areas equal to or greater than the Pilatus 1M.

Despite the similarities, different beamlines have different specialties. For example, the SIBYLS beamline at the Advanced Light Source (Classen et al., 2013) is highly specialized for high-throughput batch mode SAXS measurements, using a commercial pipetting robot with custom built sample cells to measure a full 96 well plate every few hours. However, a tradeoff to the focus on high throughput is that the experiments must be done in an air gap, which reduces the signal-to-noise of the measurement. Some BioSAXS beamlines are limited by the source, such as BM29 at the ESRF (Pernot et al., 2013) which is on a bending magnet, limiting the total flux at the beamline to around 10^{12} ph/s. Additionally, the adoption of the coflow technology has been somewhat slow, and only two beamlines worldwide currently use the coflow cell (though others are either in the process of or are planning to implement it), which limits the number of high-flux beamlines that can fully utilize all available photons. Taken together, there are only a few beamlines particularly suitable for experiments attempting to measure extremely low signal-to-noise systems.

In addition to BioCAT, the other optimal beamline for these measurements is the SAXS/WAXS beamline at the Australian Synchrotron where the coflow system was designed (Ryan, Trewhella, et al., 2018). It has a full beam flux similar to that at BioCAT ($\sim 5 \times 10^{12}$ ph/s), uses an in-vacuum sample cell, and numerous other improvements to minimize excess scattering from beamline components. The P12 beamline at Petra III (Blanchet et al., 2015) and the SWING beamline at Synchrotron SOLEIL (David & Perez, 2009) both provide high flux (up to $\sim 10^{13}$ and 5×10^{12} ph/s respectively), and a fully optimized system with an in-vacuum sample cell and minimal parasitic scattering. However, neither beamline uses a coflow cell, which limits the useable flux and thus the signal to noise achievable. Most other SAXS beamlines around the world lack either the flux or the coflow cell, while several also do not use an in-vacuum sample cell.

7. Model-dependent analysis

Phase separation of synthetic polymers which are analogous to IDRs has been theoretically modeled for decades. Homopolymer theories either treat the interaction between monomers with a mean field approach (i.e., Flory-Huggins theory) (Flory, 1942; Huggins, 1942) or explicitly consider the networking of monomer units (i.e., Flory-Stockmayer theory) (Flory, 1941; Stockmayer, 1944). In either case, the phase-separating system is considered to be composed of identical monomers distributed in space based on connectivity and entropy. Percolation and networking models can be extended to heterogeneous "stickers-and-spacers" systems (Choi, Holehouse, & Pappu, 2020) wherein the adhesive elements within the polymer—Flory-Stockmayer monomers—are the "stickers" and they are linked by "spacers" which are similar to non-interacting monomers with positive excluded volume in Flory-Huggins theory. Deducing the adhesive elements in IDRs via careful SAXS experiments can enable the parameterization of an IDR system such that it can be treated with an appropriate polymer model with the level of complexity relevant to the system. This section will discuss analytic form factors that can model X-ray scattering from polymer systems. The degree of complexity required to model the single chain system hints at the complexity needed to model phase separation.

When polymer theories are applied to IDRs, it is assumed that IDRs can be treated similarly to ideal polymers or with an additional level of complexity. In this context, it can be useful to interpret SAXS data from IDRs through the lens of polymer models which define the probability density of monomers in the polymer chain based on their covalent connections and non-covalent contacts. Different types of polymer chain models result in different form factors and therefore ensemble size distributions. The form factor P(q) is related to the scattering intensity normalized by I_0:

$$I(q) = I_0 P(q) S(q) \tag{4}$$

This form is apparent in Eq. (2), where the first order expansion of the Debye scattering equation is an estimate of the form factor at small angles. It is important to note that the structure factor S(q) should always equal 1 in dilute solutions. In the presence of interparticle interference, S(q) can take on values below 1 (indicating repulsion) and above 1 (indicating attraction) at small angles; it converges to 1 at wide angles. Procedures to model S(q) are well established for a number of colloidal systems (Lindner & Zemb, 2002) but are beyond the scope of this chapter. If these features appear in dilute phase data, it is best to further dilute the sample and only analyze low concentration fractions.

Given the nature of SAXS resolution, all variations of analytical form factors based on the Debye scattering equation (shown in Eq. 2) should converge on the same R_G at very small angles. Up to which angle—relative to the IDR R_G—the form factor fits the data, varies between models. Understanding the assumptions inherent in a model and determining systematic deviations from it, can provide valuable information about the ensemble of conformations of the IDR vs an ideal homopolymer.

7.1 Gaussian chain

The simplest polymer form factor describes a random chain in which the monomers obey Gaussian statistics, i.e., the orientation of each monomer is not correlated with any other monomer other than due to connectivity in the chain. The Gaussian chain model results in a relationship between the end-to-end distance R_{ee}, R_G and number of residues N.

$$\langle R_{ee}^2 \rangle = Nb^2 \tag{5}$$

$$\langle R_G^2 \rangle = \frac{\langle R_{ee}^2 \rangle}{6} \tag{6}$$

This relationship leads to the Flory scaling exponent $\nu = 0.5$ (Flory, 1953). The Kuhn length, b, is the renormalized monomer length that is required so that the IDR can follow Gaussian statistics. Classically, this model would be constructed such that an IDR with N residues is renormalized to N′ residues each with a length b such that the model holds. The Kuhn length would then be a function of features of the IDR that result in departure from Gaussian statistics such as steric clashes between sidechains and restrictions to bond rotation. Experimentally, and conveniently, the Kuhn length for IDRs is often ~1 residue (Kohn et al., 2004; Wilkins et al., 1999). The Gaussian chain model is physically unrealistic in that it allows multiple monomers to occupy the same space. Although this aspect of the model is non-realistic and does not allow the extraction of ensembles of conformations, if the form factor fits the data, the size distribution of the real IDR ensemble will be statistically identical to the non-realistic model. Gaussian chain statistics have the following form factor:

$$P(q) = \frac{2}{q^4 R_G^4} \left[e^{-q^2 R_G^2} - 1 + q^2 R_G^2 \right]. \tag{7}$$

This form factor was derived by Peter Debye and is a special case of the Debye scattering function where the explicit sum over all atom pairs is replaced by an integral over the probability density (or pair distribution function) of interatom distances in a Gaussian chain. The Gaussian chain form factor has identical limiting behavior as the Debye scattering equation at small angles:

$$P(q) \cong \frac{2}{q^4R_G^4}\left[-1 + q^2R_G^2 + 1 - q^2R_G^2 + \frac{\left(q^2R_G^2\right)^2}{2} - \frac{\left(q^2R_G^2\right)^3}{6} + \ldots\right]$$
$$\cong 1 - \frac{q^2R_G^2}{3} \tag{8}$$

At wide angles, the equations can be simplified to:

$$P(q) \cong \frac{2}{q^2R_G^2} \sim q^{-2} \tag{9}$$

The fractal dimension of the Gaussian chain, $d = 2$, is characteristic of the theta state of a polymer where attractive interactions between monomers exactly balance with excluded volume, i.e., repulsive interactions. The theta state can also be thought of as a tipping point above which the polymer is well solvated and below which attractive interactions between monomers are dominant and the polymer collapses. This collapse below the theta state is analogous to phase separation in polymer dense solutions. A consequence of this connection is that an IDR, under conditions where it will phase separate, will likely be near the theta state and therefore a simple Gaussian chain model may be a good fit to experimental data (Fig. 6A and B). The R_G is the only free parameter determining the shape of the Gaussian chain form factor (Eq. 7). Given that the small angles are sufficient to determine the R_G (Eq. 8), the Gaussian chain form factor should fit a wide range of data up to $q = \frac{1}{R_G}$. How far experimental data deviates from the model in the crossover regime and the fractal scaling regime is diagnostic of how similar the conformational ensemble of the IDR is to that of the theta state of a homopolymer (Fig. 6A and B).

7.2 Swollen Gaussian chain

Real polymer chains cannot intersect with themselves, which results in self-avoidance and can be quantified by a parameter called "excluded volume." Excluded volume is added as an exponent to the Gaussian chain statistics from Eqs. 5 and 6:

$$\left\langle R_{ee}{}^2\right\rangle = N^{2\nu}b^2 \tag{10}$$

$$\left\langle R_G{}^2\right\rangle = b^2\left[\frac{N^{2\nu}}{(2\nu + 1)(2\nu + 2)}\right] \tag{11}$$

where ν is the Flory scaling exponent discussed in Section 2 which is generally between $\frac{1}{3}$ for collapsed polymers and globules and $\frac{3}{5}$ for self-avoiding

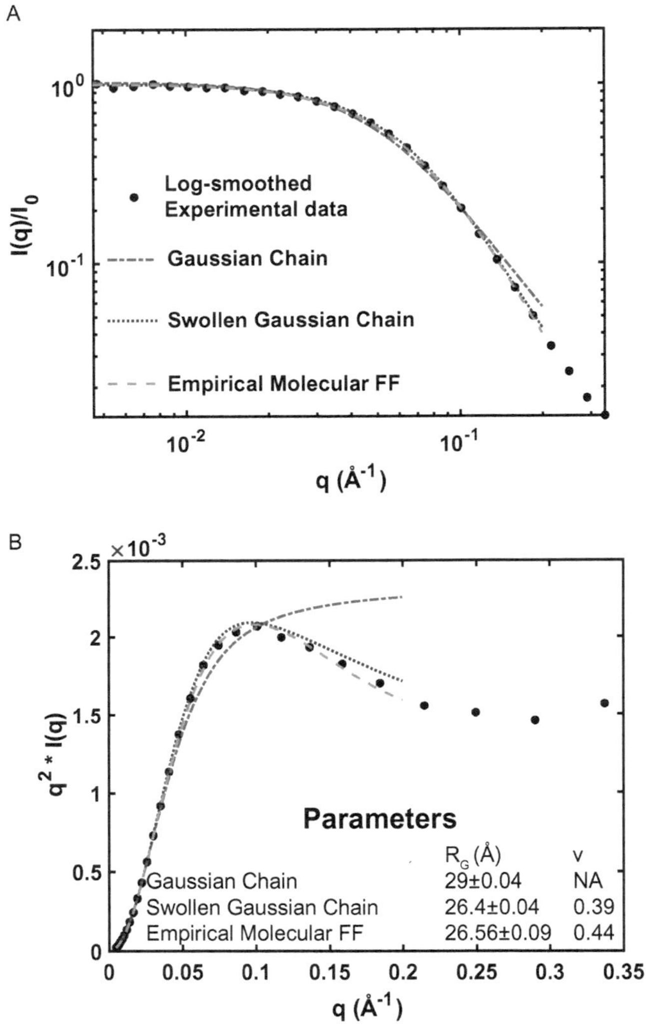

Fig. 6 SAXS data on a phase-separating IDR fit to different polymer models. Fits using the Gaussian Chain, Swollen Gaussian Chain and the Empirical Molecular FF to experimental data on a phase-separating IDR are shown in (A) raw and (B) Kratky format. The Swollen Gaussian Chain and Empirical Molecular FF provide good fits to the data and near equal R_G. However, the scaling exponents differ significantly.

polymers. The scaling exponent of $\frac{1}{2}$ represents the theta state and restores the Gaussian statistics in the previous section—a special case where the excluded volume is effectively zero. Unlike for the Gaussian chain, there is no analytic solution to the form factor that includes self-avoidance for any value of excluded volume other than $\frac{1}{2}$. The solution was originally put into integral form by Henry Benoit in 1957 and was more recently put in a more tractable "semi analytic form" by Hammouda (1993) using a combination of incomplete gamma functions.

$$P(q) = \frac{1}{\nu U^{\frac{1}{2\nu}}}\left[\gamma\left(\frac{1}{2\nu}, U\right) - \frac{1}{U^{\frac{1}{2\nu}}}\gamma\left(\frac{1}{2\nu}, U\right)\right] \tag{12}$$

$$U = \frac{q^2 R_G^2 (2\nu + 1)(2\nu + 2)}{6} \tag{13}$$

$$\gamma(x, U) = \int_0^U e^{-t} t^{x-1}\, dt \tag{14}$$

This form factor can be useful for obtaining R_G because it is valid to wider angles than the Guinier approximation or the fit to a Gaussian chain form factor due to the variable parameter ν (Fig. 6A and B). ν itself is a useful parameter for the characterization of IDR size scaling and can even hint at a driving force for phase separation in some proteins (Martin et al., 2020). However, the absolute accuracy of the parameter ν extracted by a fit to this form factor has been called into question (Fig. 6B) (Riback et al., 2017). Specifically, the swollen Gaussian chain form factor can fail to reproduce parameters of limiting cases such as self-avoiding walks generated by molecular simulations. Given these limitations, it is important to be cautious when interpreting fitted values. Differences in values in response to changes in condition or IDR sequence are likely meaningful, while the absolute values with respect to theoretical limiting cases should be viewed with skepticism.

7.3 Empirical derivations

Instead of deriving a form factor based on an ideal pair distribution function, an alternate strategy is to empirically derive a form factor based on molecular simulations. This strategy was effectively implemented using Monte Carlo simulations of polystyrene by Pedersen and Schurtenberger (1996). In this application, simulations were used to parameterize numerical solutions to the form factor by linear combination of the form factor of a Gaussian chain and an infinitely thin rod. Riback et al. used a similar approach to parameterize a form factor specific to IDRs. They used a series of molecular dynamics simulations in which the excluded volume was effectively titrated by adjusting the attractive potential between beta sidechain carbons in a poly-alanine chain (Riback et al., 2017). The simulations were used to parameterize an empirical function that fits scattering data as a function of

R_G and ν (Fig. 6A and B). The empirical molecular form factor (FF) has proven useful in analyzing SAXS data on IDRs (Banks et al., 2018; Martin et al., 2020; Riback et al., 2017, 2019). The IDR-specific parameterization aids in the ability of this model to reproduce values of ν across sequences (Fig. 6B). Further, using a model derived from IDRs allows for rapid evaluation of SAXS data relative to an ideal, homopolymer IDR sequence. Specifically, by evaluating the prefactor, obtained from the form factor fit for a given IDR sequence length, R_G and ν, relative to the prefactor in the poly-alanine model, it is possible to diagnose deviations from the expected statistics. For example, local stiffness or transient structuring could result in deviations in the Kuhn length which would be manifest as an atypical prefactor. Additionally, long-range correlations between monomers in an IDR will impact the shape of the SAXS curve outside of the small-angle regime. In extreme cases where there are significant long-range correlations, Riback et al. parameterized an extension of the empirical form factor (Riback et al., 2019). It is important to note that high signal-to-noise and ideal baseline subtraction is required to quantitatively compare a model that includes heterogeneity with a homogeneous ensemble.

7.4 Simulation-based analysis

Typical samples of synthetic polymers contain a distribution of molecular weights. Assuming that the IDR sample is pure (a feature that should be controlled for and can be ensured by using SEC-SAXS), knowledge of the chemical identity of the IDR can be used to improve SAXS data interpretation over that with the polymer models from above. Knowledge of the sequence length and amino acid identity can provide details aiding in analysis.

Incorporating information about protein sequence into the analysis usually involves generating an explicit ensemble of conformations that fits the experimental data. A number of different methods exist, but all rely on the concept that the data can be represented by a linear combination of representative conformations. First, an ensemble of conformations is generated. This can be done by sampling from amino-acid specific torsion angle libraries, creating the starting distribution of conformations. The ensemble can also be generated by Molecular Dynamics or Monte Carlo simulations with a physics-based force field. Second, the fit of the data to the ensemble is inspected. In the best-case scenario (e.g., in the case of a well-performing force field and sampling method), the starting ensemble already gives good agreement with the data. If the fit is deemed suboptimal, the conformations in the ensemble can be reweighted or a sub-ensemble can be selected. This process has been accomplished via Bayesian inference optimization of the complete prior distribution (Antonov, Olsson, Boomsma, & Hamelryck, 2016), genetic algorithms which subsample and reweight the prior (Bernado, Mylonas, Petoukhov, Blackledge, & Svergun, 2007; Pelikan, Hura, & Hammel, 2009), clustering followed by maximum entropy reweighting (Rozycki, Kim, & Hummer, 2011), and reweighting the prior based on minimization of a pseudo-energy through simulated annealing (Krzeminski, Marsh, Neale, Choy, & Forman-Kay, 2013; Marsh & Forman-Kay, 2012).

The quality of the reweighted ensemble strongly depends on the starting ensemble; if the distribution of properties in the starting ensemble is smooth and already recapitulates the experimental data closely (e.g., because the force field performs well), then reweighting has been shown to result in ensembles with physically reasonable properties such as smooth size distributions (Tria, Mertens, Kachala, & Svergun, 2015). If the starting ensemble is in poor agreement with the experimental data, reweighting can result in ensembles with rough distributions of properties that are physically unreasonable. Hence, realistic and extensive sampling of the conformational space of the sequence in question is key to ensuring the extracted model is minimally biased by the prior distribution (Hummer & Kofinger, 2015; Larsen et al., 2020).

The low information content of SAXS data results in a danger of overfitting. This is particularly true if the number of conformations in a minimal ensemble are similar to the information content in the SAXS curve, which may erroneously result in suggesting heterogeneity in the ensemble. A solution to preventing overfitting is implemented in the MultiFoXS algorithm which increases the size of the ensemble until it no longer significantly improves the quality of the fit (Schneidman-Duhovny, Hammel, Tainer, & Sali, 2016). This can also be accomplished through Bayesian inference optimization of minimal ensembles (Yang, Parisien, Major, & Roux, 2010). Intriguingly, the majority of SAXS curves of IDRs can be well represented with fewer than five conformations. These results stress the low information content in SAXS data and the fact that useful data analysis should be testing compatibility with a defined physical model. The molecular details of conformations in minimal ensembles, including the contacts that result in their size and shape and correlations between such interactions, all of which are sought when trying to understand how self-association is encoded in IDRs, cannot be extracted from minimal ensembles. Therefore, the only reliable parameter in minimal models may be the ensemble-average level of compaction—a parameter that is rapidly accessible from analytic models—and not features like intramolecular contacts.

All-atom or coarse-grained simulations using physics-based force fields can, in favorable cases, be used for the molecular interpretation of SAXS data without fitting. The SAXS form factor calculated from an ensemble that represents the conformational space of the IDR in the simulation can be directly compared with experimental data. This type of approach has been effectively applied using Monte Carlo simulations with the ABSINTH implicit solvent model (Martin et al., 2016, 2020). In case the form factor from simulations and experiment are in agreement, the simulations are likely capturing physically realistic conformational behavior of the IDR. Comparison to additional data, e.g., from NMR experiments, helps to establish whether the simulations capture additional features of the conformational behavior of the IDR such as formation of transient contacts or transient secondary structure.

Molecular-detail insight from simulations is particularly useful when evaluating the conformations of phase-separating IDRs because intra-chain contacts can be assumed to also exist intermolecularly in semi-dilute and dense solutions and drive phase transitions. Thus, SAXS-validated simulations can aid in identifying the adhesive elements in IDRs. The synergy between simulation and experiment can be exploited to devise a mutational strategy in which proposed adhesive elements in an IDR are rapidly probed via simulation and their impact verified by SAXS and characterization of the phase behavior (Martin et al., 2020).

A well-known difficulty in this approach is finding a simulation paradigm that accurately reflects experimental data—a problem particularly relevant to IDRs for which simulations often result in overly compact ensembles (Best & Mittal, 2010). Continued force field development with particular attention being paid to the interaction between solvent and protein has helped ameliorate these shortcomings in recent years (Best, Zheng, & Mittal, 2014; Robustelli, Piana, & Shaw, 2018). If current force fields that are optimized for IDRs are insufficient to match experimental data, ad hoc modification of the force field or simulation parameters may help. This can often be accomplished by tuning the simulation temperature (Das, Huang, Phillips, Kriwacki, & Pappu, 2016; Francis, Lindorff-Larsen, Best, & Vendruscolo, 2006; Martin et al., 2020) or by subtle modifications to the interactions between solvent and protein (Best et al., 2014; Piana, Donchev, Robustelli, & Shaw, 2015). Tuning force field parameters and reweighting ensembles can even be mathematically equivalent. In the context of a mutation-based experimental design, the force field could be tuned to recreate experimental data of an experimentally well-characterized IDR sequence. This same force field is used to simulate sequence variants where proposed adhesive elements are modified. If the force field modifications succeed in capturing physically realistic behavior, the simulation results should be a good fit to experimental data on all variants without any further modification.

An alternative approach is bottom-up coarse-grained simulations in which intraresidue interaction potentials are parameterized to match experimental data (Norgaard, Ferkinghoff-Borg, & Lindorff-Larsen, 2008). These types of simulations can be implemented in generalized coarse-grained engines such as HOOMD-Blue (Anderson, Lorenz, & Travesset, 2008). This analysis is particularly attractive for studying phase-separating systems. Parameterizing a coarse-grained simulation enables testing of models in which the affinities between particular residues or groups of residues, so-called "stickers," are titrated and the resulting global dimensions are compared to experimental data. The profound advantage of coarse-grained systems is that the same force field can be transferred to simulations of dense IDR solutions to assay how the interaction potentials translate into emergent properties. An example of this type of simulation are coarse-grained on-lattice Monte Carlo simulations that allow for rapid sampling and convergence in simulations containing many IDR molecules (Choi, Dar, & Pappu, 2019; Holehouse & Pappu, 2020). The PIMMS and LASSI software packages are designed specifically to implement this workflow. PIMMS was used recently to translate the interaction potential between individual amino acids parameterized based on experimental SAXS data into the calculation of complete coexistence curves for a phase-separating IDR (Martin et al., 2020). An alternate approach relies on "slab geometry" molecular dynamics simulations (Dignon, Zheng, Best, et al., 2018; Dignon, Zheng, Kim, Best, & Mittal, 2018). These methods also allow the direct transfer of a coarse-grained force field from single chain to dense solutions and have the advantage that they are off-lattice and could contain more information about dense phase structure. However, this comes at the cost of higher simulation time.

8. Implementation of SAXS measurements of single chain IDR behavior for characterizing phase behavior

Careful analysis of dilute protein data has the potential to reveal vital information about the nature of sequence features that contribute to higher-order assembly and phase separation. E.g., data on a series of IDRs with progressively higher aromatic amino acid content displayed an increasing drive to compact (Martin et al., 2020) (Fig. 7). The compaction was quantified by the decreasing scaling exponent (Fig. 7B) which correctly predicted the phase behavior of these sequences. This example suggests a coherent

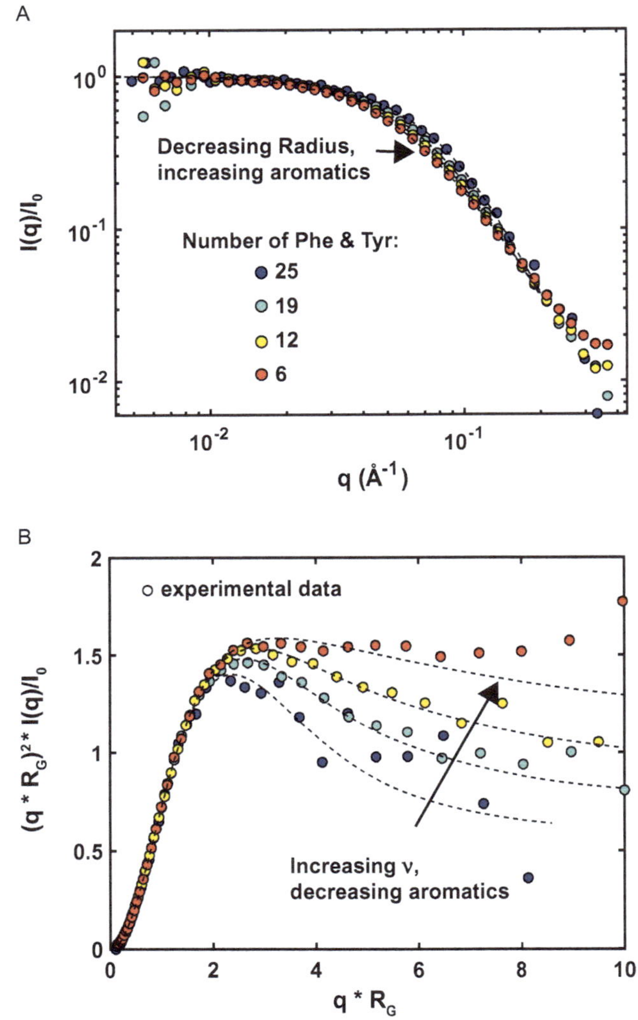

Fig. 7 Changing sequence features that influence the driving force for phase separation are reflected in the protein dimensions under dilute conditions. (A) Raw data on a series of IDR variants with different numbers of aromatic amino acids shows the predicted decrease in radius with increased propensity to phase separate. (B) The scaling exponent increases with the removal of aromatic amino acids indicating more favorable solvation. Dashed lines are fits to the Empirical Molecular FF.

workflow where IDR sequence or solution conditions can be systematically titrated, and the results directly mapped to protein phase behavior. If the IDR behaves like a homopolymer on the length scales measurable by SAXS, this connection will likely exist. However, as IDRs are heteropolymers of finite length, cases will exist with decoupling and symmetry breaking between single chain and emergent properties. This behavior can be predicted within the "stickers-and-spacers" formalism recently described by Choi et al. (2020). If the "stickers" are strong enough and the "spacers" have sufficient excluded volume, it is possible that the "stickers" on a single chain may not interact with each other. Importantly, even if R_G and ν do not directly report on c_{sat}, the dilute protein ensemble still informs on the characteristics of the dense phase via the excluded volume of spacers and the cumulative strength of interactions between stickers—parameters that can be inferred from the scaling exponent and c_{sat} respectively. Systems where R_G and ν do not, a priori, suggest a low c_{sat}, but the stickers are strong enough to drive assembly, could result in a lower dense phase concentration or even a percolated, gelled, network in the absence of phase separation (Harmon, Holehouse, Rosen, & Pappu, 2017).

9. Conclusions

SAXS has always been a valuable tool for characterizing IDRs because it provides information on their ensemble-average sizes and size distributions. However, these measurements have been largely restricted to highly soluble IDRs due to the requirement for aggregate-/oligomer-free samples of high enough concentration to achieve sufficient signal-to-noise. We propose that these limitations resulted in a bias toward IDR sequences that behave as if they are in a good solvent ($\nu > 0.5$) (Bernado & Blackledge, 2009; Cordeiro et al., 2017; Riback et al., 2017). While the oligomerization of self-associating IDRs has been monitored by SAXS in individual, interesting examples (Herranz-Trillo et al., 2017), examination of this category of protein in the dilute regime has largely been ignored. Recent advancements in hardware at synchrotron radiation sources along with refined sample preparation has opened the door to SAXS characterization of IDRs that have traditionally been inaccessible.

Acknowledgments

The authors would like to thank Thomas Boothby for providing the sample used in Fig. 2. T.M. acknowledges funding by NIH grant R01GM112846, the St. Jude Children's

Research Hospital Research Collaborative on Membrane-less Organelles in Health and Disease and the American Lebanese Syrian Associated Charities (ALSAC). This research used resources of the Advanced Photon Source, a U.S. Department of Energy (DOE) Office of Science User Facility operated for the DOE Office of Science by Argonne National Laboratory under Contract No. DE-AC02-06CH11357. This project was supported by grant 9 P41 GM103622 from the National Institute of General Medical Sciences of the National Institutes of Health. Use of the Pilatus3 × 1M detector was provided by grant 1S10OD018090-01 from NIGMS. The content is solely the responsibility of the authors and does not necessarily reflect the official views of the National Institute of General Medical Sciences or the National Institutes of Health.

References

Anderson, J. A., Lorenz, C. D., & Travesset, A. (2008). General purpose molecular dynamics simulations fully implemented on graphics processing units. *Journal of Computational Physics*, *227*(10), 5342–5359. https://doi.org/10.1016/j.jcp.2008.01.047.

Antonov, L. D., Olsson, S., Boomsma, W., & Hamelryck, T. (2016). Bayesian inference of protein ensembles from SAXS data. *Physical Chemistry Chemical Physics*, *18*(8), 5832–5838. https://doi.org/10.1039/c5cp04886a.

Banani, S. F., Lee, H. O., Hyman, A. A., & Rosen, M. K. (2017). Biomolecular condensates: Organizers of cellular biochemistry. *Nature Reviews Molecular Cell Biology*, *18*(5), 285–298. https://doi.org/10.1038/nrm.2017.7.

Banks, A., Qin, S., Weiss, K. L., Stanley, C. B., & Zhou, H. X. (2018). Intrinsically disordered protein exhibits both compaction and expansion under macromolecular crowding. *Biophysical Journal*, *114*(5), 1067–1079. https://doi.org/10.1016/j.bpj.2018.01.011.

Barna, S., Shepherd, J., Wixted, R., Tate, M., Rodricks, B., & Gruner, S. (1995). Development of a fast pixel array detector for use in microsecond time-resolved X-ray diffraction. In (Vol. 2521)SPIE.

Bernado, P., & Blackledge, M. (2009). A self-consistent description of the conformational behavior of chemically denatured proteins from NMR and small angle scattering. *Biophysical Journal*, *97*(10), 2839–2845. https://doi.org/10.1016/j.bpj.2009.08.044.

Bernado, P., Mylonas, E., Petoukhov, M. V., Blackledge, M., & Svergun, D. I. (2007). Structural characterization of flexible proteins using small-angle X-ray scattering. *Journal of the American Chemical Society*, *129*(17), 5656–5664. https://doi.org/10.1021/ja069124n.

Bernado, P., & Svergun, D. I. (2012). Analysis of intrinsically disordered proteins by small-angle X-ray scattering. *Methods in Molecular Biology*, *896*, 107–122. https://doi.org/10.1007/978-1-4614-3704-8_7.

Best, R. B., & Mittal, J. (2010). Protein simulations with an optimized water model: Cooperative helix formation and temperature-induced unfolded state collapse. *The Journal of Physical Chemistry B*, *114*(46), 14916–14923. https://doi.org/10.1021/jp108618d.

Best, R. B., Zheng, W., & Mittal, J. (2014). Balanced protein-water interactions improve properties of disordered proteins and non-specific protein association. *Journal of Chemical Theory and Computation*, *10*(11), 5113–5124. https://doi.org/10.1021/ct500569b.

Blanchet, C. E., Spilotros, A., Schwemmer, F., Graewert, M. A., Kikhney, A., Jeffries, C. M., et al. (2015). Versatile sample environments and automation for biological solution X-ray scattering experiments at the P12 beamline (PETRA III, DESY). *Journal of Applied Crystallography*, *48*(Pt. 2), 431–443. https://doi.org/10.1107/S160057671500254X.

Boeynaems, S., Alberti, S., Fawzi, N. L., Mittag, T., Polymenidou, M., Rousseau, F., et al. (2018). Protein phase separation: A new phase in cell biology. *Trends in Cell Biology, 28*(6), 420–435. https://doi.org/10.1016/j.tcb.2018.02.004.

Brady, J. P., Farber, P. J., Sekhar, A., Lin, Y. H., Huang, R., Bah, A., et al. (2017). Structural and hydrodynamic properties of an intrinsically disordered region of a germ cell-specific protein on phase separation. *Proceedings of the National Academy of Sciences of the United States of America, 114*(39), E8194–E8203. https://doi.org/10.1073/pnas.1706197114.

Burke, K. A., Janke, A. M., Rhine, C. L., & Fawzi, N. L. (2015). Residue-by-residue view of in vitro FUS granules that bind the C-terminal domain of RNA polymerase II. *Molecular Cell, 60*(2), 231–241. https://doi.org/10.1016/j.molcel.2015.09.006.

Choi, J. M., Dar, F., & Pappu, R. V. (2019). LASSI: A lattice model for simulating phase transitions of multivalent proteins. *PLoS Computational Biology, 15*(10), e1007028. https://doi.org/10.1371/journal.pcbi.1007028.

Choi, J. M., Holehouse, A. S., & Pappu, R. V. (2020). Physical principles underlying the complex biology of intracellular phase transitions. *Annual Review of Biophysics, 49*, 107–133. https://doi.org/10.1146/annurev-biophys-121219-081629.

Classen, S., Hura, G. L., Holton, J. M., Rambo, R. P., Rodic, I., McGuire, P. J., et al. (2013). Implementation and performance of SIBYLS: A dual endstation small-angle X-ray scattering and macromolecular crystallography beamline at the advanced light source. *Journal of Applied Crystallography, 46*(Pt. 1), 1–13. https://doi.org/10.1107/S0021889812048698.

Conicella, A. E., Zerze, G. H., Mittal, J., & Fawzi, N. L. (2016). ALS mutations disrupt phase separation mediated by alpha-helical structure in the TDP-43 low-complexity C-terminal domain. *Structure, 24*(9), 1537–1549. https://doi.org/10.1016/j.str.2016.07.007.

Cordeiro, T. N., Herranz-Trillo, F., Urbanek, A., Estaña, A., Cortés, J., Sibille, N., et al. (2017). Structural characterization of highly flexible proteins by small-angle scattering. In B. Chaudhuri, I. G. Muñoz, S. Qian, & V. S. Urban (Eds.), *Biological small angle scattering: Techniques, strategies and tips* (pp. 107–129). Singapore: Springer Singapore.

Das, R. K., Huang, Y., Phillips, A. H., Kriwacki, R. W., & Pappu, R. V. (2016). Cryptic sequence features within the disordered protein p27Kip1 regulate cell cycle signaling. *Proceedings of the National Academy of Sciences of the United States of America, 113*(20), 5616–5621. https://doi.org/10.1073/pnas.1516277113.

David, G., & Perez, J. (2009). Combined sampler robot and high-performance liquid chromatography: A fully automated system for biological small-angle X-ray scattering experiments at the synchrotron SOLEIL SWING beamline. *Journal of Applied Crystallography, 42*(5), 892–900. https://doi.org/10.1107/S0021889809029288.

Dignon, G. L., Zheng, W., Best, R. B., Kim, Y. C., & Mittal, J. (2018). Relation between single-molecule properties and phase behavior of intrinsically disordered proteins. *Proceedings of the National Academy of Sciences of the United States of America, 115*(40), 9929–9934. https://doi.org/10.1073/pnas.1804177115.

Dignon, G. L., Zheng, W., Kim, Y. C., Best, R. B., & Mittal, J. (2018). Sequence determinants of protein phase behavior from a coarse-grained model. *PLoS Computational Biology, 14*(1), e1005941. https://doi.org/10.1371/journal.pcbi.1005941.

Durand, D., Vivès, C., Cannella, D., Pérez, J., Pebay-Peyroula, E., Vachette, P., et al. (2010). NADPH oxidase activator p67phox behaves in solution as a multidomain protein with semi-flexible linkers. *Journal of Structural Biology, 169*(1), 45–53. https://doi.org/10.1016/j.jsb.2009.08.009.

Fischetti, R., Stepanov, S., Rosenbaum, G., Barrea, R., Black, E., Gore, D., et al. (2004). The BioCAT undulator beamline 18ID: A facility for biological non-crystalline diffraction and X-ray absorption spectroscopy at the advanced photon source. *Journal of Synchrotron Radiation, 11*(Pt. 5), 399–405. https://doi.org/10.1107/S0909049504016760.

Flory, P. J. (1941). Molecular size distribution in three dimensional polymers. I. Gelation. *Journal of the American Chemical Society, 63*(11), 3083–3090. https://doi.org/10.1021/ja01856a061.

Flory, P. J. (1942). Thermodynamics of high polymer solutions. *The Journal of Chemical Physics, 10*(1), 51–61. https://doi.org/10.1063/1.1723621.

Flory, P. J. (1953). *Principles of polymer chemistry*. Ithaca: Cornell University Press.

Francis, C. J., Lindorff-Larsen, K., Best, R. B., & Vendruscolo, M. (2006). Characterization of the residual structure in the unfolded state of the Δ131Δ fragment of staphylococcal nuclease. *Proteins: Structure, Function, and Bioinformatics, 65*(1), 145–152. https://doi.org/10.1002/prot.21077.

Fuertes, G., Banterle, N., Ruff, K. M., Chowdhury, A., Mercadante, D., Koehler, C., et al. (2017). Decoupling of size and shape fluctuations in heteropolymeric sequences reconciles discrepancies in SAXS vs. FRET measurements. *PNAS, 114*(31), E6342–E6351. https://doi.org/10.1073/pnas.1704692114.

Glatter, O. (1977). A new method for the evaluation of small-angle scattering data. *Journal of Applied Crystallography, 10*(5), 415–421. https://doi.org/10.1107/S0021889877013879.

Hammouda, B. (1993). SANS from homogeneous polymer mixtures: A unified overview. In *Polymer characteristics* (pp. 87–133). Berlin, Heidelberg: Springer Berlin Heidelberg.

Harmon, T. S., Holehouse, A. S., Rosen, M. K., & Pappu, R. V. (2017). Intrinsically disordered linkers determine the interplay between phase separation and gelation in multivalent proteins. *eLife, 6*, e30294. https://doi.org/10.7554/eLife.30294.

Herranz-Trillo, F., Groenning, M., van Maarschalkerweerd, A., Tauler, R., Vestergaard, B., & Bernado, P. (2017). Structural analysis of multi-component amyloid systems by chemometric SAXS data decomposition. *Structure, 25*(1), 5–15. https://doi.org/10.1016/j.str.2016.10.013.

Hofmann, H., Soranno, A., Borgia, A., Gast, K., Nettels, D., & Schuler, B. (2012). Polymer scaling laws of unfolded and intrinsically disordered proteins quantified with single-molecule spectroscopy. *Proceedings of the National Academy of Sciences of the United States of America, 109*(40), 16155–16160. https://doi.org/10.1073/pnas.1207719109.

Holehouse, A. S., & Pappu, R. V. (2020). *PIMMS (0.24 pre-beta)*. Zenodo. https://doi.org/10.5281/zenodo.3588456.

Hopkins, J. B., Gillilan, R. E., & Skou, S. (2017). BioXTAS RAW: Improvements to a free open-source program for small-angle X-ray scattering data reduction and analysis. *Journal of Applied Crystallography, 50*(Pt. 5), 1545–1553. https://doi.org/10.1107/S1600576717011438.

Hopkins, J. B., & Thorne, R. E. (2016). Quantifying radiation damage in biomolecular small-angle X-ray scattering. *Journal of Applied Crystallography, 49*(Pt. 3), 880–890. https://doi.org/10.1107/S1600576716005136.

Huggins, M. L. (1942). Some properties of solutions of long-chain compounds. *The Journal of Physical Chemistry, 46*(1), 151–158. https://doi.org/10.1021/j150415a018.

Hummer, G., & Kofinger, J. (2015). Bayesian ensemble refinement by replica simulations and reweighting. *The Journal of Chemical Physics, 143*(24), 243150. https://doi.org/10.1063/1.4937786.

Jeffries, C. M., Graewert, M. A., Svergun, D. I., & Blanchet, C. E. (2015). Limiting radiation damage for high-brilliance biological solution scattering: Practical experience at the EMBL P12 beamline PETRAIII. *Journal of Synchrotron Radiation, 22*(2), 273–279. https://doi.org/10.1107/S1600577515000375.

Kachala, M., Valentini, E., & Svergun, D. I. (2015). Application of SAXS for the structural characterization of IDPs. *Advances in Experimental Medicine and Biology, 870*, 261–289. https://doi.org/10.1007/978-3-319-20164-1_8.

Kato, M., Han, T. W., Xie, S., Shi, K., Du, X., Wu, L. C., et al. (2012). Cell-free formation of RNA granules: Low complexity sequence domains form dynamic fibers within hydrogels. *Cell, 149*(4), 753–767. https://doi.org/10.1016/j.cell.2012.04.017.

Kirby, N., Cowieson, N., Hawley, A. M., Mudie, S. T., McGillivray, D. J., Kusel, M., et al. (2016). Improved radiation dose efficiency in solution SAXS using a sheath flow sample environment. *Acta Crystallographica Section D, Structural Biology*, *72*(Pt. 12), 1254–1266. https://doi.org/10.1107/S2059798316017174.

Koch, M. H., Vachette, P., & Svergun, D. I. (2003). Small-angle scattering: A view on the properties, structures and structural changes of biological macromolecules in solution. *Quarterly Reviews of Biophysics*, *36*(2), 147–227. https://doi.org/10.1017/s0033583503003871.

Kohn, J. E., Millett, I. S., Jacob, J., Zagrovic, B., Dillon, T. M., Cingel, N., et al. (2004). Random-coil behavior and the dimensions of chemically unfolded proteins. *Proceedings of the National Academy of Sciences of the United States of America*, *101*(34), 12491. https://doi.org/10.1073/pnas.0403643101.

Kraft, P., Bergamaschi, A., Broennimann, C., Dinapoli, R., Eikenberry, E. F., Henrich, B., et al. (2009). Performance of single-photon-counting PILATUS detector modules. *Journal of Synchrotron Radiation*, *16*(Pt. 3), 368–375. https://doi.org/10.1107/S0909049509009911.

Krzeminski, M., Marsh, J. A., Neale, C., Choy, W. Y., & Forman-Kay, J. D. (2013). Characterization of disordered proteins with ENSEMBLE. *Bioinformatics*, *29*(3), 398–399. https://doi.org/10.1093/bioinformatics/bts701.

Larsen, A. H., Wang, Y., Bottaro, S., Grudinin, S., Arleth, L., & Lindorff-Larsen, K. (2020). Combining molecular dynamics simulations with small-angle X-ray and neutron scattering data to study multi-domain proteins in solution. *PLoS Computational Biology*, *16*(4), e1007870. https://doi.org/10.1371/journal.pcbi.1007870.

Lin, Y. H., & Chan, H. S. (2017). Phase separation and single-chain compactness of charged disordered proteins are strongly correlated. *Biophysical Journal*, *112*(10), 2043–2046. https://doi.org/10.1016/j.bpj.2017.04.021.

Lindner, P., & Zemb, T. (2002). *Neutrons, X-rays, and light: Scattering methods applied to soft condensed matter* (1st ed.). Amsterdam; Boston: Elsevier.

Marsh, J. A., & Forman-Kay, J. D. (2012). Ensemble modeling of protein disordered states: Experimental restraint contributions and validation. *Proteins*, *80*(2), 556–572. https://doi.org/10.1002/prot.23220.

Martin, E. W., Holehouse, A. S., Grace, C. R., Hughes, A., Pappu, R. V., & Mittag, T. (2016). Sequence determinants of the conformational properties of an intrinsically disordered protein prior to and upon multisite phosphorylation. *Journal of the American Chemical Society*, *138*(47), 15323–15335. https://doi.org/10.1021/jacs.6b10272.

Martin, E. W., Holehouse, A. S., Peran, I., Farag, M., Incicco, J. J., Bremer, A., et al. (2020). Valence and patterning of aromatic residues determine the phase behavior of prion-like domains. *Science*, *367*(6478), 694–699. https://doi.org/10.1126/science.aaw8653.

Mitrea, D. M., Cika, J. A., Stanley, C. B., Nourse, A., Onuchic, P. L., Banerjee, P. R., et al. (2018). Self-interaction of NPM1 modulates multiple mechanisms of liquid-liquid phase separation. *Nature Communications*, *9*(1), 842. https://doi.org/10.1038/s41467-018-03255-3.

Mittag, T., & Forman-Kay, J. D. (2007). Atomic-level characterization of disordered protein ensembles. *Current Opinion in Structural Biology*, *17*(1), 3–14. https://doi.org/10.1016/j.sbi.2007.01.009.

Molliex, A., Temirov, J., Lee, J., Coughlin, M., Kanagaraj, A. P., Kim, H. J., et al. (2015). Phase separation by low complexity domains promotes stress granule assembly and drives pathological fibrillization. *Cell*, *163*(1), 123–133. https://doi.org/10.1016/j.cell.2015.09.015.

Muller-Spath, S., Soranno, A., Hirschfeld, V., Hofmann, H., Ruegger, S., Reymond, L., et al. (2010). From the cover: Charge interactions can dominate the dimensions of intrinsically disordered proteins. *Proceedings of the National Academy of Sciences of the United States of America*, *107*(33), 14609–14614. https://doi.org/10.1073/pnas.1001743107.

Norgaard, A. B., Ferkinghoff-Borg, J., & Lindorff-Larsen, K. (2008). Experimental parameterization of an energy function for the simulation of unfolded proteins. *Biophysical Journal, 94*(1), 182–192. https://doi.org/10.1529/biophysj.107.108241.

Nott, T. J., Petsalaki, E., Farber, P., Jervis, D., Fussner, E., Plochowietz, A., et al. (2015). Phase transition of a disordered nuage protein generates environmentally responsive membraneless organelles. *Molecular Cell, 57*(5), 936–947. https://doi.org/10.1016/j.molcel.2015.01.013.

Oldfield, C. J., & Dunker, A. K. (2014). Intrinsically disordered proteins and intrinsically disordered protein regions. *Annual Review of Biochemistry, 83*, 553–584. https://doi.org/10.1146/annurev-biochem-072711-164947.

Panjkovich, A., & Svergun, D. I. (2018). CHROMIXS: Automatic and interactive analysis of chromatography-coupled small-angle X-ray scattering data. *Bioinformatics, 34*(11), 1944–1946. https://doi.org/10.1093/bioinformatics/btx846.

Patel, A., Lee, H. O., Jawerth, L., Maharana, S., Jahnel, M., Hein, M. Y., et al. (2015). A liquid-to-solid phase transition of the ALS protein FUS accelerated by disease mutation. *Cell, 162*(5), 1066–1077. https://doi.org/10.1016/j.cell.2015.07.047.

Pedersen, J. S., & Schurtenberger, P. (1996). Scattering functions of semiflexible polymers with and without excluded volume effects. *Macromolecules, 29*(23), 7602–7612. https://doi.org/10.1021/ma9607630.

Pelikan, M., Hura, G. L., & Hammel, M. (2009). Structure and flexibility within proteins as identified through small angle X-ray scattering. *General Physiology and Biophysics, 28*(2), 174–189. https://doi.org/10.4149/gpb_2009_02_174.

Pernot, P., Round, A., Barrett, R., De Maria Antolinos, A., Gobbo, A., Gordon, E., et al. (2013). Upgraded ESRF BM29 beamline for SAXS on macromolecules in solution. *Journal of Synchrotron Radiation, 20*(Pt. 4), 660–664. https://doi.org/10.1107/S0909049513010431.

Piana, S., Donchev, A. G., Robustelli, P., & Shaw, D. E. (2015). Water dispersion interactions strongly influence simulated structural properties of disordered protein states. *The Journal of Physical Chemistry B, 119*(16), 5113–5123. https://doi.org/10.1021/jp508971m.

Powers, S. K., Holehouse, A. S., Korasick, D. A., Schreiber, K. H., Clark, N. M., Jing, H., et al. (2019). Nucleo-cytoplasmic partitioning of ARF proteins controls auxin responses in *Arabidopsis thaliana*. *Molecular Cell, 76*(1), 177–190.e175. https://doi.org/10.1016/j.molcel.2019.06.044.

Ptitsyn, O. B. (1995). Molten globule and protein folding. *Advances in Protein Chemistry, 47*, 83–229. https://doi.org/10.1016/s0065-3233(08)60546-x.

Rambo, R. P., & Tainer, J. A. (2011). Characterizing flexible and intrinsically unstructured biological macromolecules by SAS using the Porod-Debye law. *Biopolymers, 95*(8), 559–571. https://doi.org/10.1002/bip.21638.

Riback, J. A., Bowman, M. A., Zmyslowski, A. M., Knoverek, C. R., Jumper, J. M., Hinshaw, J. R., et al. (2017). Innovative scattering analysis shows that hydrophobic disordered proteins are expanded in water. *Science, 358*(6360), 238–241. https://doi.org/10.1126/science.aan5774.

Riback, J. A., Bowman, M. A., Zmyslowski, A. M., Plaxco, K. W., Clark, P. L., & Sosnick, T. R. (2019). Commonly used FRET fluorophores promote collapse of an otherwise disordered protein. *Proceedings of the National Academy of Sciences of the United States of America, 116*(18), 8889–8894. https://doi.org/10.1073/pnas.1813038116.

Robustelli, P., Piana, S., & Shaw, D. E. (2018). Developing a molecular dynamics force field for both folded and disordered protein states. *Proceedings of the National Academy of Sciences of the United States of America, 115*(21), E4758–E4766. https://doi.org/10.1073/pnas.1800690115.

Rozycki, B., Kim, Y. C., & Hummer, G. (2011). SAXS ensemble refinement of ESCRT-III CHMP3 conformational transitions. *Structure, 19*(1), 109–116. https://doi.org/10.1016/j.str.2010.10.006.

Rubinstein, M., & Colby, R. H. (2003). *Polymer physics*. Oxford; New York: Oxford University Press.

Ryan, V. H., Dignon, G. L., Zerze, G. H., Chabata, C. V., Silva, R., Conicella, A. E., et al. (2018). Mechanistic view of hnRNPA2 low-complexity domain structure, interactions, and phase separation altered by mutation and arginine methylation. *Molecular Cell, 69*(3), 465–479.e467. https://doi.org/10.1016/j.molcel.2017.12.022.

Ryan, T. M., Trewhella, J., Murphy, J. M., Keown, J. R., Casey, L., Pearce, F. G., et al. (2018). An optimized SEC-SAXS system enabling high X-ray dose for rapid SAXS assessment with correlated UV measurements for biomolecular structure analysis. *Journal of Applied Crystallography, 51*(1), 97–111. https://doi.org/10.1107/S1600576717017101.

Schmidt, H. B., Barreau, A., & Rohatgi, R. (2019). Phase separation-deficient TDP43 remains functional in splicing. *Nature Communications, 10*(1), 4890. https://doi.org/10.1038/s41467-019-12740-2.

Schneidman-Duhovny, D., Hammel, M., Tainer, J. A., & Sali, A. (2016). FoXS, FoXSDock and MultiFoXS: Single-state and multi-state structural modeling of proteins and their complexes based on SAXS profiles. *Nucleic Acids Research, 44*(W1), W424–W429. https://doi.org/10.1093/nar/gkw389.

Schuler, B., Soranno, A., Hofmann, H., & Nettels, D. (2016). Single-molecule FRET spectroscopy and the polymer physics of unfolded and intrinsically disordered proteins. *Annual Review of Biophysics, 45*, 207–231. https://doi.org/10.1146/annurev-biophys-062215-010915.

Shin, Y., & Brangwynne, C. P. (2017). Liquid phase condensation in cell physiology and disease. *Science, 357*(6357), eaaf4382. https://doi.org/10.1126/science.aaf4382.

Stockmayer, W. H. (1944). Theory of molecular size distribution and gel formation in branched polymers II. General cross linking. *The Journal of Chemical Physics, 12*(4), 125–131. https://doi.org/10.1063/1.1723922.

Svergun, D. I. (1999). Restoring low resolution structure of biological macromolecules from solution scattering using simulated annealing. *Biophysical Journal, 76*(6), 2879–2886. https://doi.org/10.1016/S0006-3495(99)77443-6.

Svergun, D. I., & Feĭgin, L. A. (1986). *Rentgenovskoe i neĭtronnoe malouglovoe rasseîa͡nie*. Moskva: "Nauka," Glav. red. fiziko-matematicheskoĭ lit-ry.

Svergun, D. I., Feĭgin, L. A., & Taylor, G. W. (1987). *Structure analysis by small-angle X-ray and neutron scattering*. New York: Plenum Press.

Tompa, P. (2012). Intrinsically disordered proteins: A 10-year recap. *Trends in Biochemical Sciences, 37*(12), 509–516. https://doi.org/10.1016/j.tibs.2012.08.004.

Tria, G., Mertens, H. D., Kachala, M., & Svergun, D. I. (2015). Advanced ensemble modelling of flexible macromolecules using X-ray solution scattering. *IUCrJ, 2*(Pt. 2), 207–217. https://doi.org/10.1107/S205225251500202X.

Tuukkanen, A. T., Kleywegt, G. J., & Svergun, D. I. (2016). Resolution of ab initio shapes determined from small-angle scattering. *IUCrJ, 3*(6), 440–447. https://doi.org/10.1107/S2052252516016018.

Uversky, V. N. (2002). Natively unfolded proteins: A point where biology waits for physics. *Protein Science, 11*(4), 739–756. https://doi.org/10.1110/ps.4210102.

van der Lee, R., Buljan, M., Lang, B., Weatheritt, R. J., Daughdrill, G. W., Dunker, A. K., et al. (2014). Classification of intrinsically disordered regions and proteins. *Chemical Reviews, 114*(13), 6589–6631. https://doi.org/10.1021/cr400525m.

Wang, A., Conicella, A. E., Schmidt, H. B., Martin, E. W., Rhoads, S. N., Reeb, A. N., et al. (2018). A single N-terminal phosphomimic disrupts TDP-43 polymerization, phase separation, and RNA splicing. *The EMBO Journal*, *37*(5), e97452. https://doi.org/10.15252/embj.201797452.

Wilkins, D. K., Grimshaw, S. B., Receveur, V., Dobson, C. M., Jones, J. A., & Smith, L. J. (1999). Hydrodynamic radii of native and denatured proteins measured by pulse field gradient NMR techniques. *Biochemistry*, *38*(50), 16424–16431. https://doi.org/10.1021/bi991765q.

Williamson, T. E., Craig, B. A., Kondrashkina, E., Bailey-Kellogg, C., & Friedman, A. M. (2008). Analysis of self-associating proteins by singular value decomposition of solution scattering data. *Biophysical Journal*, *94*(12), 4906–4923. https://doi.org/10.1529/biophysj.107.113167.

Yang, S., Parisien, M., Major, F., & Roux, B. (2010). RNA structure determination using SAXS data. *The Journal of Physical Chemistry B*, *114*(31), 10039–10048. https://doi.org/10.1021/jp1057308.

CHAPTER EIGHT

SAXS methods for investigating macromolecular and self-assembled polyelectrolyte complexes

Amanda B. Marciel[a], **Samanvaya Srivastava**[b], **Jeffrey M. Ting**[c,d,†], **and Matthew V. Tirrell**[c,d,*]

[a]Department of Chemical and Biomolecular Engineering, Rice University, Houston, TX, United States
[b]Department of Chemical and Biomolecular Engineering, University of California, Los Angeles, Los Angeles, CA, United States
[c]Pritzker School of Molecular Engineering, University of Chicago, Chicago, IL, United States
[d]Center for Molecular Engineering and Materials Science Division, Argonne National Laboratory, Lemont, IL, United States
*Corresponding author: e-mail address: mtirrell@uchicago.edu

Contents

† Current address: 3M Company, 3M Center, Saint Paul, MN 55144, United States.

Methods in Enzymology, Volume 646
ISSN 0076-6879
https://doi.org/10.1016/bs.mie.2020.09.013

Abstract

Polyelectrolyte complexation is driven by associative interactions between oppositely charged polyelectrolytes, resulting in formation of a macroscopic polymer dense phase and a polymer dilute phase with applications in coatings, adhesives, and purification membranes. Beyond macroscale phase separation, precision polymer synthesis has enabled further development of polyelectrolyte complex (PEC)-based self-assembled micelles and hydrogels with applications in biotechnology. Interestingly, it has been suggested that mechanisms similar to polyelectrolyte complexation drive formation of biological condensates that play an indispensable role in cellular biogenesis. The formation pathways and functionality of these complex materials is dependent on the physical properties that are built into polymer structure and the resulting physical conformation in the dilute and dense phase. Scattering techniques have enabled *in situ* investigation of structure-function relationships in PEC materials that may address unresolved biophysical questions in cellular processes as well as catalyze the development of novel materials for diverse applications. We describe preparation of PEC materials with controlled polymer characteristics (length, blockiness, charge density), small-angle X-ray scattering (SAXS) techniques employed to probe appropriate length scales, and the data analysis routines from a practical standpoint for new users. This article deals with bulk complexes and not with the related, important and interesting area of non-equilibrium layer-by-layer assembly of polyelectrolytes.

1. Introduction

Oppositely charged polyelectrolytes undergo complexation when mixed under appropriate conditions, resulting in formation of a polymer-rich polyelectrolyte complex (PEC) phase and a polymer-poor supernatant phase (Srivastava & Tirrell, 2016; Stewart, Wang, & Shao, 2011; van der Gucht, Spruijt, Lemmers, & Cohen Stuart, 2011). This phase separation is driven by attractive electrostatic inter-chain interactions and entropically favorable release of bound counterions and water of hydration (Schlenoff, Rmaile, & Bucur, 2008; Srivastava & Tirrell, 2016). Therefore, complexation is influenced not only by the density and association strength of ionizable groups along the chains, but also external solution conditions, including salt concentration, pH, and temperature (Nakashima, André, & Spruijt, 2020; Prabhu, Ali, Bleuel, Mao, & Ma, 2020). These parameters can be tuned to produce PECs with materials properties that span from viscous liquids to elastic solids (Perry et al., 2015), which are of interest in applications ranging from adhesives and coatings to membranes and encapsulation agents. More recently, liquid-like PECs have been used as model systems to investigate the underlying physics that drive *in vitro* liquid-liquid phase separation (LLPS) and dissolution of biological condensates (Perry, 2019). Biological condensates are dynamic liquid-like droplets found in all kingdoms of life (except archaebacteria, thus far) that play critical roles in cellular biogenesis (Banani, Lee, Hyman, & Rosen, 2017; Vernon & Forman-Kay, 2019). Many studies have shown that LLPS of biological condensates is dependent on the electrostatic interactions of unstructured protein sequences called intrinsically disordered regions (IDRs) or intrinsically disordered proteins (IDPs), therefore *in vitro* LLPS resembles complex coacervation of synthetic polyelectrolytes (Brangwynne, 2013; Brangwynne, Tompa, & Pappu, 2015; Chong & Forman-Kay, 2016; Darling, Liu, Oldfield, & Uversky, 2017; Keating, 2012; Marin, Hopkins, & Mittag, 2020). A major question in the field of intracellular physics is how IDR sequence and conformation leads to LLPS and influences the functionality of the biological condensate. This question, along with interest in better understanding the PEC structure-function continuum, has spurred investigations on polymer conformation in the dense and dilute phases. In this context, scattering methods have been particularly useful in elucidating the evolution of polymer chain conformations upon complexation and phase separation (Fig. 1A).

Advances in precision polymer synthesis have also enabled the controlled synthesis of block polyelectrolytes comprising charged and neutral blocks. Complexation of such oppositely charged block polyelectrolytes has been harnessed to create PEC-based self-assembled materials, including spherical and rod-like micelles, as well as hydrogels (Pergushov, Müller, & Schacher, 2012). These materials have been employed for diverse biomedical and biotechnological applications such as drug and gene delivery, 3-D bioprinting, and bioadhesion (Zhou, Shi, Li, & He, 2020). This additional layer of complexity may enable LLPS of biological molecules to be readily compartmentalized and released from these modular carrier materials. Expanding the realm of applications for these PEC self-assemblies requires an intimate control over the nanoscale morphologies—the size, shape, and distance between assembled structures. Again, scattering methods have been instrumental in revealing the structural characteristics of PEC self-assemblies, both of PEC micelles and PEC hydrogels (Fig. 1B and C, respectively).

In general, polymer conformation is investigated using scattering techniques, such as dynamic/static light scattering (DLS/SLS), small-angle

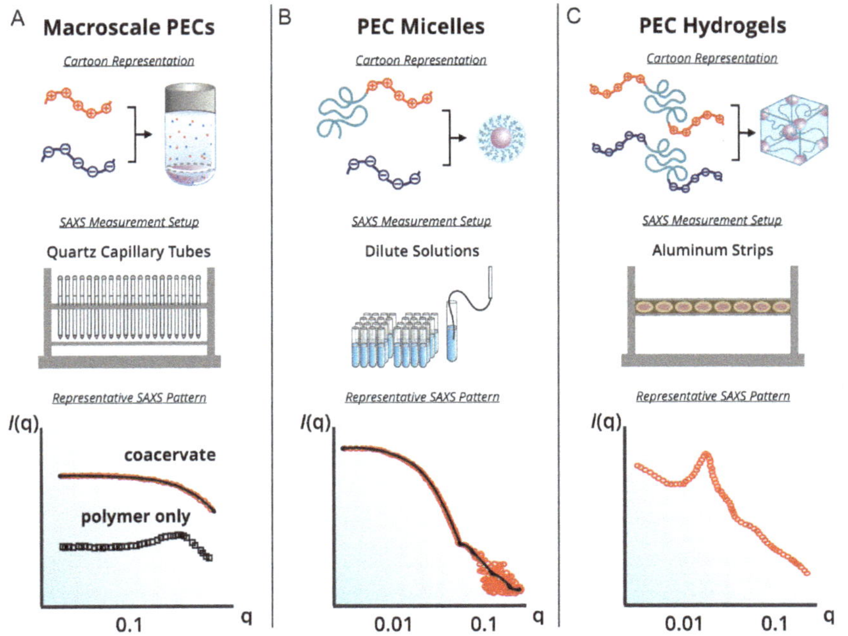

Fig. 1 Overview of SAXS experiments and analyses for oppositely charged (A) macroscale PECs, (B) PEC micelles, and (C) PEC hydrogels. For each PEC material from top to bottom, a cartoon representation of the complexation and assembly of positively- and negatively-charged polymers (not drawn to scale), the SAXS measurement setup of various instrumentations, and a representative SAXS pattern are illustrated.

X-ray scattering (SAXS), small-angle neutron scattering (SANS), and their advanced variations (Prabhu et al., 2020; Sztucki, Di Cola, & Narayanan, 2012; Zhang & Ilavsky, 2010). Synchrotron SAXS at national facilities (e.g., Argonne National Laboratory, Advanced Photon Source) is a high-flux technique that enables investigation of polymer conformation over three-orders of magnitude in length scale (~1–100 nm). In polyelectrolyte solutions and PECs, SAXS intensity curves, $I(q)$ describe the conformation of individual chains (Fig. 1A), whereas for self-assembled PECs the assembled structures dominate $I(q)$ (Fig. 1B and C). In this chapter, we describe how SAXS can be used to investigate polyelectrolyte conformation in PECs and the self-assembled structure of PEC micelles and hydrogels. For each material type, we describe sample preparation, measurement considerations and expectations, as well as data analysis routines (Fig. 2).

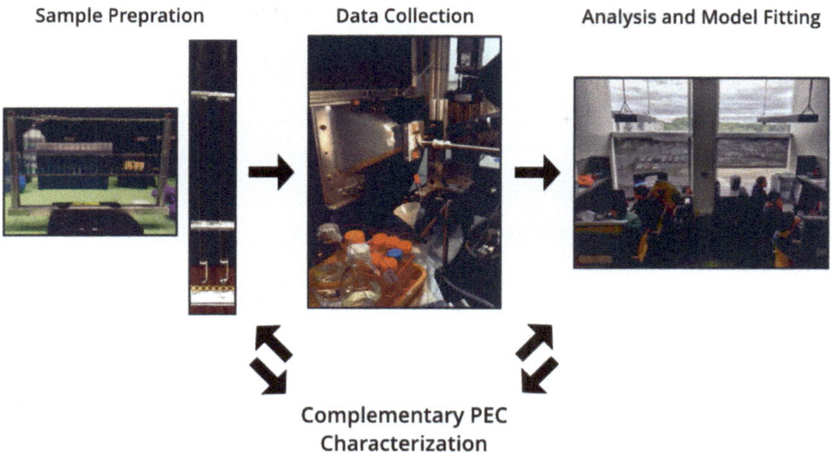

Fig. 2 Workflow of SAXS sample preparation, data collection, analysis and model fitting, and complementary PEC characterization.

2. Macromolecular polyelectrolyte complexes (PECs)

2.1 Materials

The following polypeptide-based polyelectrolytes (Alamanda™ Polymers) were used to investigate chain conformation before and after complexation: poly(L-lysine hydrochloride) ($N_c = 400$, 65,466.98 g/mol), poly(L-glutamic acid sodium salt), and poly(D,L-glutamic acid sodium salt) ($N_c = 400$, 60,859.5 g/mol). N_c denotes the degree of polymerization. Polypeptides are excellent systems to analyze PEC physics due to their hydrophilicity and identical backbone structure. In addition, polypeptide-based PECs can be tuned from viscous liquids to solid precipitates by varying the chirality of the constituent polypeptides (Marciel, Chung, Brettmann, & Leon, 2017; Perry et al., 2015). In our 2018 study, this strategy was implemented to investigate polyelectrolyte conformation in the liquid phase throughout the entirety of two-phase region (0–1000 mM NaCl) (Marciel, Srivastava, & Tirrell, 2018).

2.1.1 APS beamline 12-ID-B

SAXS measurements were made at beamline 12-ID-B of the Advanced Photon Source (APS) at Argonne National Laboratory (Lemont, IL). PEC samples were irradiated in 1 mm quartz thin-walled capillaries

(Charles Supper Company) at 13 keV at a detector distance of 4 m (q range of 0.002–0.5 $Å^{-1}$). Glassy carbon was used as the reference for absolute intensity calibration.

2.2 Sample preparation

1. Stock solutions of polyelectrolyte (20 wt%), NaCl (5 M), and HCl or NaOH (2 M) were prepared in Milli-Q ultrapure water (resistivity > 18 MΩ cm, Millipore, USA). To prepare stock polyelectrolyte, add weighed powder to half of the calculated Milli-Q ultrapure water. Polypeptides are weak polyelectrolytes and must be pH adjusted with NaOH or HCl (add 2 μL at 2 M and measure pH with pH meter or pH paper) to the desired charge state (in this example poly-lysine and poly-glutamic acid are fully charged at pH 7). Add the remaining water to reach final wt%. Rotate on a tube revolver for 1.5 or 10 mL tubes or roll on a tube roller for 50 mL tubes overnight. Note: Sonication and vortex mixing may be used, but it must be determined if this damages the polyelectrolytes via SEC, MALDI, and/or NMR characterization.
2. Characterize individual polyelectrolytes prior to complexation experiments. We recommend assessing polyelectrolyte dispersity using size-exclusion chromatography (SEC), charge state by determining the pH titration curve, solubility at all experimental conditions (e.g., added salt, temperature, pH, etc.) using optical microscopy, hydrogen-bonding character via circular dichroism (CD) and Fourier-transform infrared spectroscopy (FTIR), and lastly overlap concentration (c^*) and radius of gyration (R_g) via viscometry measurements. Note: see the 2018 publication from our group in *Soft Matter* for experimental details (Marciel et al., 2018). After individual polyelectrolyte characterization, proceed to characterization of the phase behavior of your oppositely charged polyelectrolytes. We recommend assessing phase behavior in varying salt, temperature, pH, and charge-matching conditions using optical microscopy and/or thermogravimetric analysis (TGA). Note: see publications from our group in *Macromolecules* for experimental details (Li et al., 2018; Li, Srivastava, Meng, Ting, & Tirrell, 2020).
3. Prepare backgrounds, polyelectrolytes, and PEC samples for SAXS analysis of polyelectrolyte conformation before and after complexation. In this example, PECs are prepared at 1:1 monomer charge-matched ratios (2 wt% total polyelectrolyte) with varied added NaCl. First, anionic polyelectrolyte stock solution is added to Milli-Q ultrapure water and

NaCl (0–1000 mM final concentration) in a 1.5 mL Eppendorf Tube™ and vortexed for 20 s. Next, cationic polyelectrolyte is added and vortexed for 20 s, where upon complexation the solution becomes turbid. The total polyelectrolyte wt% for a PEC sample needs to result in a visible coacervate phase at the bottom of the 1 mm capillary for a successful SAXS measurement (see capillary photo in Fig. 2). The volume for PEC sample preparation is determined by the maximum volume of the 1 mm capillary (~200 μL). Lastly, background samples of water and corresponding water + NaCl solutions are prepared to isolate polyelectrolyte scattering intensities.

4. Turbid solutions are then transferred to the capillaries and centrifuged at 5000 × *g* for 10 min resulting in separation of the polymer-rich phase and polymer-poor phase. To prevent the capillary from breaking during centrifugation we recommend wrapping it in padding (e.g., Kimwipes, Kimtech™) and placing into a 10 mL Falcon tube. Backgrounds (water or water + NaCl) and individual polyelectrolyte solutions may be prepared and directly loaded into capillaries. We recommend preparing polyelectrolyte solutions at the total added wt% (in this example 1 wt % for each polyelectrolyte was mixed to form PECs). If possible, we also recommend preparing each polyelectrolyte at the wt% in the coacervate phase (in this example each polyelectrolyte is at 20 wt% in the coacervate phase, as measured by TGA). Knowledge of the solution conditions (dilute, semidilute, entangled) in the coacervate phase enables appropriate polymer models to be fit. In addition, it allows separation of effects of concentration on conformation vs effects from complexation. Lastly, seal the top of the capillaries with wax or glue and store at room temperature.

2.3 Scattering methods

1. SAXS enables investigation of polyelectrolyte conformation in both the polymer-rich and polymer-poor phases. Prior to the experiment, calculate the expected polyelectrolyte length scales, including radius of gyration, R_g and correlation length, ξ (see Marciel et al., 2018 for more details). This will determine the appropriate q range, although most SAXS instruments cover three decades in q and are thus sufficient to study internal polymer length scales. In this example, we use the 4 m detector distance (q range of 0.002–0.5 $Å^{-1}$).
2. Load backgrounds, polyelectrolytes, and PECs into a multi capillary holder (Fig. 2) after calibration with glassy carbon and establishing an

appropriate beam mask. First, determine radiation exposure time. Sufficient exposure results in significant scattering intensity over the background and throughout the desired q range with no visible radiation damage (increase in scattering intensity at low q values between shots). For new samples, vary polyelectrolyte sample concentration (dilute to semidilute) to verify presence of correlation peak behavior at high q and determine if you have decent signal-to-noise. In this example, three shots at 0.1 s exposure with 1 s between shots was sufficient. Note: If beam damage is observed, we recommend decreasing exposure time, moving positions between shots, and/or increasing time between exposures.

3. Measure the polymer-rich PEC phase and the polymer-poor supernatant phase of each LLPS sample. First, vertically align the capillary via the motorized stage to the phase of interest. Sufficient volume in the polymer-rich phase (~15 μL in the 1 mm capillaries) is necessary for vertical alignment. Next, horizontally align the capillary to radiate through the center of the capillary. Note: when moving the stage be aware that you are not irradiating the edge of the capillary or the metal capillary holder. If no beam damage is detected, azimuthally average the scattering intensities and move to next sample.

4. The devil is in the details. Choice of ion (counterion or added salt) matters for phase behavior and also may interfere with scattering intensity at a particular X-ray wavelength. For instance, the K-edge of Br has an X-ray absorption ~13 keV. Thus, the SAXS signature of the polymer chains from PEC samples with KBr, including the correlation peak and the characteristic q^{-2} decay of $I(q)$ at high q may be occluded if the experiments are carried out at 13 keV.

2.4 Data analysis

SAXS data collected from semidilute polyelectrolyte solutions (above overlap concentration c^*) exhibit distinct features compared to neutral polymer solutions, namely a peaked behavior at high q values (Barrat & Joanny, 1996; Colby, 2009; Dobrynin, Colby, & Rubinstein, 1995; Dobrynin & Rubinstein, 2005; Muthukumar, 2017) (Fig. 3A). This peak is attributed to a new length scale that emerges when excluded volume effects dominate called the correlation length, ξ and scales with concentration approximately as $c^{-1/2}$. The correlation length defines the distance between two polyelectrolyte chains set by the repulsion of correlation volumes due to the high

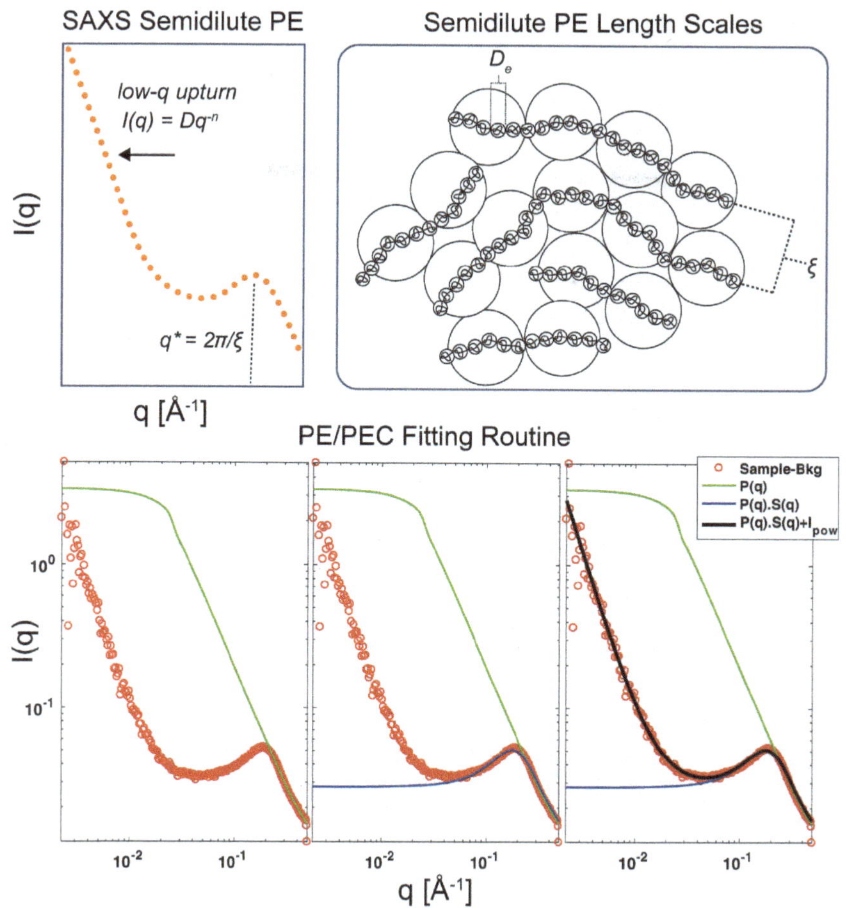

Fig. 3 (A) Features of semidilute polyelectrolyte SAXS curves. (B) Schematic of length scales associated with semidilute polyelectrolyte solutions. (C) Fitting routine used to analyze SAXS curves of semidilute polyelectrolyte solutions and PEC. Fitting is accomplished in a piecewise fashion where background-subtracted samples are first fit to the form factor, followed by the structure factor and lastly a power law to fit the low-q upturn. *Adapted from Marciel, A.B., Srivastava, S., Tirrell, M.V. (2018). Structure and rheology of polyelectrolyte complex coacervates.* Soft Matter, 14*(13), 2454–2464. https://doi.org/10.1039/C7SM02041D.*

osmotic pressure of the counterions that suppress long-range polyelectrolyte fluctuations (Dobrynin et al., 1995; Muthukumar, 2017, 2016). Semidilute polyelectrolyte solutions have three distinct length scales that may be probed with SAXS, (i) $r > \xi$ electrostatics are screened, and the chain behaves like a random walk of correlation volumes, (ii) $D_e < r < \xi$ inside the correlation volume there is strong intrachain electrostatic repulsion, and the chain behaves like a directed random walk of electrostatic volumes, D_e, and (iii) $r < D_e$ inside the electrostatic volume where electrostatics are weaker than thermal motion, and the chain behaves like a neutral polymer with bare Kuhn length, b (Fig. 3B). Three big questions arise when modeling the semidilute polyelectrolyte scattering intensity, (i) what form factor, $P(q)$ describes the chain conformation inside the correlation volume, $D_e < r < \xi$ and (ii) what is the physical meaning of the large scattering intensity at low q? Lastly, (iii) can we treat chain conformations in PECs as semidilute solutions of polyelectrolytes?

2.4.1 Form factor

The form factor, $P(q)$ describes the shape of individual scatterers and in this case the conformation of individual polyelectrolyte chains. In semidilute solutions of polyelectrolytes, there is no designated form factor that describes the directed random walk inside the correlation volume making conformation predictions difficult. However, the worm-like chain model and the semi-flexible chain model with excluded volume have been implemented with success (Josef & Bianco-Peled, 2012; Lopez, Rogers, Colby, Graham, & Cabral, 2014; Murnen, Rosales, Dobrynin, Zuckermann, & Segalman, 2013).

The Kratky and Porod worm-like chain model transitions from a Gaussian chain to a rigid rod at length scales below the persistence length, (l_p) of the chain and is characterized by the Sharp and Bloomfield function (Sharp & Bloomfield, 1968):

$$P(q, l_p, L) = 2\frac{e^{-x} + x - 1}{x^2} + \frac{2l_p}{15L}\left[4 + \frac{7}{x} - \left(11 + \frac{7}{x}\right)e^{-x}\right] \quad (1)$$

Here P is the form factor, L is the contour length of the chain, and $x = q^2 L l_p / 3$. This relation may be utilized for finite chain lengths $L > 10b$ (upper limit $qb < 3.1$ *or* 4). To expand the q range Cloizeaux (1973) calculated an exact expression for infinitely long chains (can be combined with Eq. (1) and used $qb > 3$) (Buhler & Boué, 2004; Ragnetti & Oberthur, 1986):

$$P(q, l_p, L) = \frac{\pi}{qL} + \frac{2}{3q^2 l_p L} \quad (2)$$

The Pedersen-Schurtenberger model for semi-flexible chains with excluded volume has been utilized in several studies to describe single-chain polyelectrolyte conformation (Pedersen & Schurtenberger, 1996). This approach added excluded volume effects into a parametrized worm-like chain model. The form factor assumes that the polyelectrolyte is semi-flexible with an effective $l_p = 0.5b$, where below the chain acts like a rigid rod. The following expressions include modifications suggested by Chen, Butler, and Magid (2006):

$$P(q, L, b) = \begin{cases} P_1(q, L, b), qb < 3.1 \\ P_2(q, L, b), qb \geq 3.1 \end{cases} \tag{3}$$

With

$$\begin{aligned} P_1(q, L, b) = {} & (1 - w(qR_G))\frac{2}{u_1^2}(e^{-u_1} + u_1 - 1) + f_{corr}(q)w(qR_G) \\ & \left[1.22(qR_G)^{-\frac{1}{0.585}} + 0.4288(qR_G)^{-\frac{2}{0.585}} - 1.651(qR_G)^{-\frac{3}{0.585}}\right] \\ & + \frac{C(n_b)}{n_b}\left[\frac{4}{15} + \frac{7}{15u_2} - \left(\frac{11}{15} + \frac{7}{15u_2}\right)e^{-u_2}\right] \end{aligned} \tag{4}$$

where

$$w(x) = 0.5\left(1 + \tanh\frac{x - 1.523}{0.1477}\right) \tag{5}$$

$$u_1 = \frac{Lb}{6}\left(1 - \frac{3}{2n_b} + \frac{3}{2n_b^2} - \frac{3}{4n_b^3}\left(1 - e^{-2n_b}\right)\right)q^2 \tag{6}$$

$$R_G^2 = (\alpha(n_b))^2\frac{Lb}{6} \tag{7}$$

$$u_2 = (\alpha(n_b))^2\frac{q^2Lb}{6} \tag{8}$$

$$\alpha(x) = \left(1 + \left(\frac{x}{3.12}\right)^2 + \left(\frac{x}{8.67}\right)^3\right)^{\frac{0.176}{6}} \tag{9}$$

$$n_b = \frac{L}{b} \tag{10}$$

$$C(n_b) = \begin{cases} 3.06n_b^{-0.44}, L > 10b \\ 1, L \leq 10b \end{cases} \tag{11}$$

$$f_{corr}(q) = \begin{cases} 1, \dfrac{dP_1}{dq} \leq 0 \\ 0, \dfrac{dP_1}{dq} > 0 \end{cases} \tag{12}$$

$$P_2(q, L, b) = \frac{a_1(q, L, b)}{(qb)^{4.95}} + \frac{a_2(q, L, b)}{(qb)^{5.29}} + \frac{\pi}{qL} \tag{13}$$

The expressions for a_1 *and* a_2 are determined by imposing continuity and differentiability at $qb = 3.1$.

2.4.2 Structure factor

The structure factor $S(q)$ describes the contribution of interactions between scatterers and in the case of semidilute polyelectrolyte solutions and PECs describes the electrostatic interactions between chains. Assemblies of chains can be described by $I(q) \sim KP(q)S(q)$, where K is a sample specific pre-factor describing the number of scatters and their scattering contrast. The polymer version of the reference interaction site model (PRISM) theory is commonly used to describe inter-chain interactions of polyelectrolytes.

$$I(q) = k \frac{P(q, L, b)}{1 + \beta c(q, \xi) P(q, L, b)} \tag{14}$$

where $c(q, \xi)$ is the Fourier transform of the direct correlation function $c(q, \xi) = \exp - q^2 \xi^2$ and β is inversely related to the osmotic pressure through $\beta = (S(q=0) - 1)^{-1}$ (Dobrynin & Rubinstein, 2005; Marciel et al., 2018). PRISM successfully predicts the peaked behavior at high q. Note: PRISM theory assumes chains follow a self-avoiding random walk conformation (*i.e.*, chains are in good solvent).

2.4.3 Model independent analysis

Semidilute solutions of polyelectrolytes have large scattering intensity at low q. The origin of this *upturn* is unknown but suggests that there are structures on length scales larger than the chain size (Ermi & Amis, 1998). However, this directly conflicts with the suppression of long-range concentration fluctuation caused by the high osmotic pressure. These structures may be related to the slow modes observed in dynamic light scattering measurements, although they seem to not influence macroscopic rheology or solution thermodynamics (Ermi & Amis, 1998). Scattering intensity at low q is often ignored, or the solution is filtered immediately prior to X-ray exposure. Several studies include an additional power law $I_{PL}(q) = Dq^{-n}$ to describe the upturn, where D and n (2–4) are concentration dependent (Lopez et al., 2014; Marciel et al., 2018).

2.4.4 Fitting routine

Scattering intensity from polyelectrolyte solutions and PECs were fit in a piecewise fashion using the Pederson-Schurtenberger form factor with excluded volume, PRISM theory structure factor, and a low q power law (Fig. 3C) (Marciel et al., 2018). The SAXS intensity $I(q)$ is first fit with the form factor $P(q) + I_{bkg}$ for $I(q) > q^*$ where $q^* = 2\pi/\xi$ and I_{bkg} is the background scattering. The three fitting parameters from the form factor fit k_1, b

and α were then used to fit the structure factor, $P(q)S(q) + I_{bkg}$ with three fitting parameters k_2, β and ξ. Lastly, the power law fit was added $P(q) S(q) + I_{PL}(q) + I_{bkg}$ to achieve the final fit throughout $I(q)$. Using our polypeptide system as an example, the bare Kuhn length b is set at 7Å (Marciel et al., 2018), therefore each Kuhn monomer consists of two amino acid monomers where the total number of Kuhn monomers, $N = N_c/2$, 200. The contour length may be calculated by $L = Nb$ resulting in 1400Å. Note: if estimating the overlap concentration using $c^* \approx b^{-3}u^{-1}f^{-2}N^{-2}$, where f is the fraction of charged monomer on the backbone and u is the interaction parameter defined as the ratio of the Bjerrum length l_b to b use the fraction per Kuhn length, in this case 2.

3. Self-assembled PEC micelles

3.1 Materials and methods

3.1.1 Representative polycations

The following polycations are discussed in this section. Poly(ethylene oxide)-*block*-poly(vinyl benzyl trimethylammonium chloride) (PEO-*b*-PVBTMA, synthesized as previously described (Ting, Wu, Herzog-Arbeitman, Srivastava, & Tirrell, 2018)), PEG-*block*-poly(L-lysine) (PEG-*b*-PLK, Alamanda Polymers), poly(2-methacryloyloxyethyl phosphorylcholine methacrylate)-*block*-poly(vinyl benzyl trimethylammonium chloride) (PMPC-*b*-PVBTMA, synthesized as previously described (Ting, Marras, Mitchell, Campagna, & Tirrell, 2020)).

3.1.2 Representative polyanions

The following polyanions are discussed in this section. Poly(acrylic acid sodium salt) (PAA, Polymer Source, 14,800g/mol), Poly(ethylene oxide)-*block*-poly(sodium 4-vinylbenzenesulfonate) (PEO-*b*-PSS, synthesized as previously described (Ting et al., 2018)), various DNA oligonucleotides (Integrated DNA Technologies, described in greater detail as previously described (Lueckheide, Vieregg, Bologna, Leon, & Tirrell, 2018; Marras, Vieregg, Ting, Rubien, & Tirrell, 2019)), and sheared salmon sperm DNA (Invitrogen).

3.1.3 APS beamline 12-ID-B

SAXS measurements were made at beamline 12-ID-B of the Advanced Photon Source (APS) at Argonne National Laboratory. Micelle samples were irradiated in a thin-wall glass capillary flow cell with a photon energy of 13keV.

3.1.4 SSRL beamline 4–2

SAXS experiments of the pre-formed micelles were performed at the Stanford Synchrotron Radiation Lightsource (SSRL) Beamline 4–2 (Smolsky et al., 2007). In-house autosampler robotics enabled high-throughput data collection for temperature-controlled 96-well plate datasets (average rate of 3–5 min/sample, or 96 samples in 5–8 h). Briefly, approximately 10–30 μL sample aliquots were pumped to a 1.5 mm quartz capillary cell, which underwent sample oscillations upon exposure to minimize radiation damage. In between sample runs, extensive washing of the capillary, needle, and tubing was programmed to be performed by dispensing water and detergent. Data acquisition was conducted using Blu-Ice software with a customized Bio-Logic 4 syringe stopped-flow mixer (SFM400). Although the following optional features were not used for the examples shown below, an *in situ* fiber-optic UV-vis spectrophotometer and in-line size-exclusion chromatography upstream of the SAXS detector are available for users, which may be particularly important for biological applications.

3.2 Sample preparation

1. Samples of PEC micelles for SAXS are prepared in aqueous solvent at charge matched and dilute conditions. As a sample calculation, consider a PEO_{10k}-$PVBTMA_{100}$/PAA_{158} system with 100 mM added salt in water for SAXS prepared at a total polymer concentration of 1 mg/mL (0.1 wt%). For PEC micelles containing nucleic acids, selected buffer should be used instead of water.
2. Directly dissolve dried PEO_{10k}-$PVBTMA_{100}$ and PAA_{158} polymers into filtered Milli-Q water (resistivity > 18 MΩ cm, Millipore, USA) at 10 mg/mL stock solutions. Adjust pH with NaOH or HCl, if needed. Prepare a 5 M salt stock solution, if needed. Vortex all stock solutions for at least 1 min to ensure full dissolution, indicated by visual transparency and homogeneity.
3. Calculate the added volumes from each stock solution to maintain (i) the total polymer concentration of 1 mg/mL, (ii) equimolar charge between polycation and polyanion, and (iii) a total desired sample volume of 500 μL (so that the formation of micelles in solution can be checked by DLS and, if necessary, multiple repeats can be examined by SAXS). In this example, these constraints result in 440 μL water, 10 μL of the salt stock solution, 38.2 μL of the PEO_{10k}-$PVBTMA_{100}$ stock solution, and 11.8 μL of the PAA_{158} stock solution to maintain 1.2 μmol of cationic and anionic charge.

4. To a clean, dried borosilicate glass tube, sequentially add the predetermined volumetric amounts of water, salt, polycation, and polyanion at room temperature. Immediately after the polyanion is added, seal the tube and vortex the solution for at least 30 s. Set the solution aside for at least 1 h to allow for equilibration.
5. To prevent potential beam radiation damage to the dilute PEC micelle samples, add 1 wt% glycerol to each sample. Glycerol acts as a scavenger for free radicals in preventing polymer degradation. Pure glycerol is viscous, so dilution with solvent (e.g., to 50% stock solution) is recommended for practical purposes. In addition, prepare at least 100 mL of filtered Milli-Q water with 1 wt% glycerol (and/or at least 100 mL of solvent matching the sample's buffer or salt content with 1 wt% glycerol) for background SAXS measurements.
6. Examine the quality of PEC micelle formation by DLS. Comprehensive guides for DLS analysis of nanoparticles are available elsewhere (Stetefeld, McKenna, & Patel, 2016). While in most cases this has not been reported to be an issue, it is recommended to check whether the order of polyelectrolyte addition results in different micelle sizes or size distributions; if so, the micelles may be kinetically trapped. In this case, thermal or salt annealing protocols can be used to reduce the colloidal polydispersity of systems like oligonucleotide DNA PEC micelles (Lueckheide et al., 2018; Marras et al., 2019; Marras, Vieregg, & Tirrell, 2020).
7. Store PEC micelle solutions at 10 °C overnight prior to SAXS.

3.3 Scattering methods

1. Prepare the synchrotron beamline flow cell setup. As a continuation of the model example above, the protocol for the PEO_{10k}-$PVBTMA_{100}$/PAA_{158} system at APS Beamline 12-ID-B involved a 3 mm diameter quartz capillary in line with a syringe pump, where a minimum volume of approximately 140 μL was required per sample.
2. Determine the PEC micelle sample radiation exposure time; for PEO_{10k}-$PVBTMA_{100}$/PAA_{158} micelles, 1.0 s exposure was sufficient for achieving reasonable scattering intensity over the desired q range. In general, an appropriate radiation dose per unit volume of sample will depend on beam intensity, detector sensitivity, and solution conditions (e.g., total polymer concentration, scattering length density, fidelity of sample to potential radiation damage). For new samples, this should be examined by setting a series of exposure times and monitoring

evidence of beam damage to the samples as a function of total polymer concentration. Sample signals should be reasonably higher than background SAXS patterns over the desired q range with sufficiently small error bars in intensity. Note that the scaled scattering intensity should be independent of total polymer concentration in measuring dilute micelle solutions; evidence of interparticle interactions at low q may result in erroneous data analysis. Additional information on interparticle effects in PEC micelles can be found elsewhere (Wu, Ting, Weiss, & Tirrell, 2019).

3. After preparing the physical sample setup inside the synchrotron beamline, acquire data of the background solution and PEC micelles. In most setups, samples can be recovered if there is no evidence of radiation damage.
4. Azimuthally average collected data across all exposures that are within statistical error. Gradual increase of intensity usually denotes beam damage; in this case, it is recommended to wash the capillary and redo the sample run with decreased exposure time. Extremely large intensity fluctuations can be a sign of air bubbles and may be addressed by decreasing the syringe pump flow rate.
5. It is recommended to collect background data often (e.g., every 5th sample run) and overlay the SAXS patterns. Gradual increase of intensity over time indicates that the capillary may be contaminated with damaged polymer samples, which may result in erroneous data analysis. If this is observed, wash the capillary with bleach, reduce exposure time, and repeat recent SAXS measurements.

3.4 Data analysis

SAXS is a powerful solution technique for characterizing dilute nanoscale PEC micelles. Structural modeling of the scattering intensity as a function of the scattering angle (represented in the scattering wavevector q) can allow quantitative determination of micelle size, shape, and internal structures after appropriate data reduction and analysis. Although the following is not a comprehensive collection of all existing models that can potentially describe macroscopically phase-separated multicompartment assemblies, we describe here two model cases that users typically can first try fitting for an unknown sample, based on common assembly morphologies reported in the literature. In general, the slope at low q ($\lesssim 0.1\,\text{Å}^{-1}$) approximates the nanoparticle shape as spheroids (q^0), cylindrical rods (q^{-1}), or flexible cylinders (q^{-2}).

The Irena software package in Igor Pro (Ilavsky & Jemian, 2009) will be discussed in fittings below. For more sophisticated models, a particularly important limitation to keep in mind is potential overfitting data with an excess of physically unmeaningful fitting parameters. Further discussion of advanced SAXS fitting is provided in the Section 4.

3.4.1 Spheroidal and cylindrical PEC micelles

For this subsection, we focus attention to DNA oligonucleotide-containing PEC micelles that form spheroidal and cylindrical morphologies as reported by Marras et al. (2019). In general, both the micelle and background data are imported and subtracted, in which the high q values ($q > 0.5$) are usually used to scale the matching process. It is important to match the environmental conditions of samples to appropriate background (e.g., PEC micelles in PBS containing 100 mM salt to PBS with 100 mM salt). Fig. 4 shows representative background-subtracted SAXS data of individual polyelectrolytes (PEO-PVBTMA/pVBTMA-PEG polycation and 88 nt DNA polyanion) and the PEC micelles. In this case, the low q shows a $\sim q^{-1}$ dependency, indicating that the nanoparticles resemble rigid cylinders. By varying the block lengths of both polyelectrolytes, targeted PEC micelle size and morphologies were outlined using SAXS accompanied by cryo-electron imaging. Furthermore, the hybridization state of the oligonucleotide is important for controlling the properties of the PEC micelles. Single- and double-stranded DNA are polyanions with inherently different charge density and persistence length values—we compared this difference directly and outlined structure-property relationships for both PEC micellization and macroscale polyelectrolyte complexation. Further modifications, such as examining how neutral PEO compares to zwitterionic polymer shells have also been examined in follow up studies (Ting et al., 2020).

Fitting was performed using the multi-level modeling macros distributed with the Irena software package, consisting of a hard-surface form factor, power law, and diffraction peak models. Fig. 5 shows an example of applying a flexible cylinder form factor, unified level power law, and diffraction peak fit to SAXS data using a routine nonlinear least-squares method. At low q ($<10^{-2}$ Å^{-1}), the intensity showed an approximately q^{-2} dependence, a signature of worm-like micelles. Detailed discussions on the form factors are available online (Ilavsky & Jemian, 2009). Eq. (15) shows the calculation of the polydispersity of PEC micelles using the fitted radius and width values:

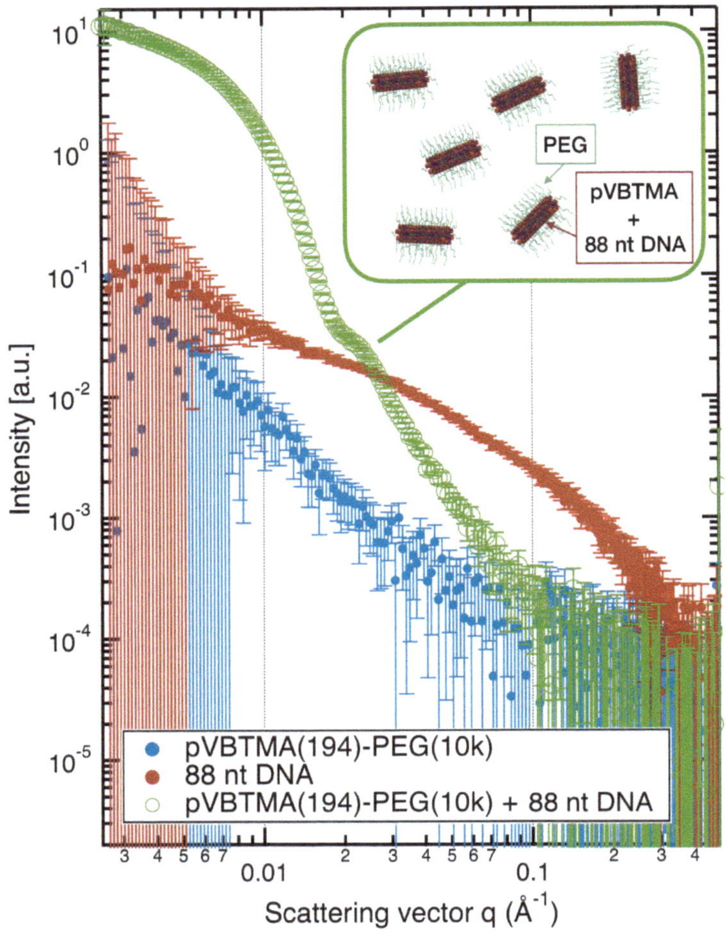

Fig. 4 SAXS data for individual polyelectrolytes pVBTMA(194)-PEG(10*k*) (blue filled circle) and 88 nt DNA (red filled circle) show no structure formation compared to the PEC micelle (green open circle). Both polyelectrolytes are at 2 mM charge concentration in 1xPBS. Intensity is not scaled or offset in this plot. *This is adapted from Marras, A.E., Vieregg, J.R., Ting, J.M., Rubien, J.D., Tirrell, M.V. (20190. Polyelectrolyte complexation of oligonucleotides by charged hydrophobic—Neutral hydrophilic block copolymers.* Polymers, 11*(1), 83. https://doi.org/10.3390/polym11010083.*

$$PDI = \left(\frac{standard\ deviation}{radius}\right)^2 = \frac{(width)^2/4}{(radius)^2} \tag{15}$$

Scattering from the power law regime results from the polymers, which should follow q^{-2} and $q^{-5/3}$ dependency for ideal chains and swollen chains, respectively. The diffraction peak at $q \sim 0.23\,\text{Å}^{-1}$ arises from parallel,

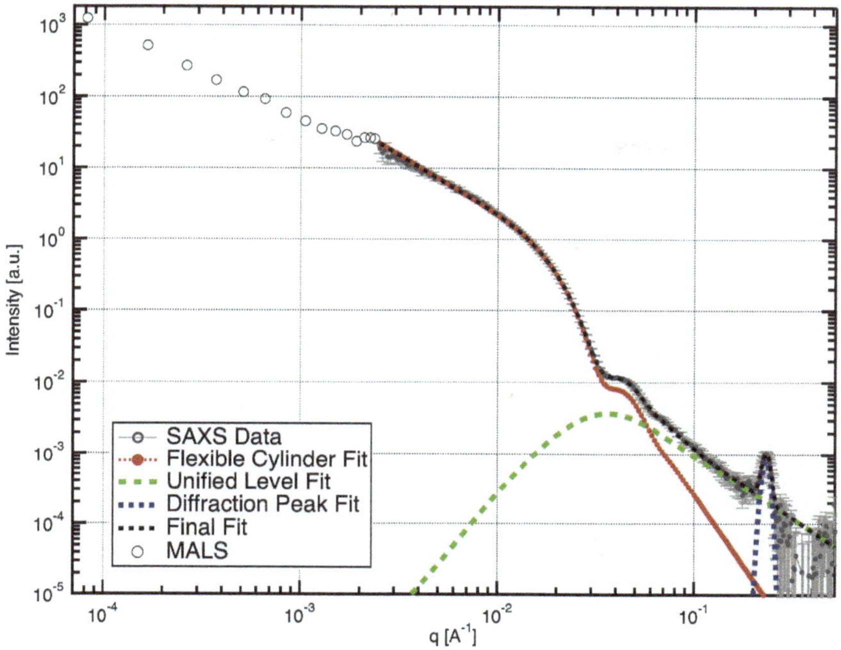

Fig. 5 Representative SAXS fitting for a cylindrical micelle of PEG-*b*-PLK/DNA, consisting of SAXS data (gray circles), multi-angle light scattering data (open black circles), multicomponent fitting model (overall black line consists of flexible cylinder model in red; power law in green; quasi-Bragg diffraction peak in blue). The inset shows a cartoon of the cylindrical PEC micelles comprising the PLK/DNA core and the PEG corona. Fitting results gave a PEC micelle population with mean radius of 11.0 nm and colloidal polydispersity value of 0.03. *This is adapted from Marras, A.E., Vieregg, J.R., Tirrell, M.V. (2020). Assembly and characterization of polyelectrolyte complex micelles.* Journal of Visualized Experiments, 157, *e60894. https://doi.org/10.3791/60894.*

hexagonal packing of dsDNA helices in the PEC micelle core. In general, after applying the fitting it is good practice to check if the final values are physically meaningful. Other solution characterization tools such as electron imaging or light scattering can complement SAXS results. We direct interested readers to a review by Patterson et al. for more information (Patterson, Robin, Chassenieux, Colombani, & O'Reilly, 2014).

3.4.2 Kinetically trapped PEC micelles

Depending on the selection of polyelectrolytes and the physicochemical attributes, PEC micelle assemblies can form polydisperse, kinetically trapped aggregates. However, structure-property relationships such as the effect of salt on loosening or breaking up assembled structures can be probed with

SAXS. As an example, complexation of synthesized PEO-*b*-PSS and PEO-*b*-PVBTMA forms non-equilibrium aggregates (Wu, Ting, Werba, Meng, & Tirrell, 2018). Using a two-level Beaucage model shown in Eq. (16) for collected SAXS data, Wu and coworkers captured general trends of how kinetically trapped aggregates change over increased NaCl salt levels:

$$I(q) = \sum_{i=1}^{2}\left[G_i \exp\left(-\frac{q^2 R_{g,i}^2}{3}\right) + \frac{B_i\left[\mathrm{erf}\left(\frac{qR_{g,i}}{\sqrt{6}}\right)\right]^{3P_i}}{q^{P_i}}\right] \quad (16)$$

Here, G is the Guinier scaling factor, R_g is the radius of gyration, and P is the Porod exponential corresponding to the ith level fit, with G and B being scaling constants. In the Irena software package, this is also referred to as the "Unified Fit" option. By examining the fitting of mass fractals of the low q regime, the authors observed a change in P from 2.0 to 1.63 corresponding to signatures of Gaussian chains ($P\sim 2$) to full swollen coils ($P\sim 5/3$). Combining this analysis with the characteristic -4 slope at mid q ($0.03\,\text{Å}^{-1} < q < 0.1\,\text{Å}^{-1}$), the PEO-*b*-PSS/PEO-*b*-PVBTMA complexes physically form globular assemblies that are locally aggregated and interconnected at no added salt; upon the incorporation of salt, the complexes swell, expand, and eventually break up as observed with complementary light scattering and cryo-transmission electron microscopy imaging. Further detailed analysis on the breakup of uniform, spherical PEC micelles using time-resolved dynamic light scattering can be found in other work from our group (Wu, Ting, Meng, & Tirrell, 2020; Wu, Ting, & Tirrell, 2020).

4. Self-assembled PEC hydrogels

Self-assembly of oppositely charged triblock polyelectrolytes with ionizable endblocks results in complexation of the oppositely charged blocks, leading to microphase separated assemblies of polyelectrolyte complex (PEC) domains containing the oppositely charged blocks and connected to each other by the neutral midblocks (Audus et al., 2015; Audus & Fredrickson, 2016; Hunt et al., 2011; Krogstad, Choi, et al., 2014; Krogstad et al., 2013; Krogstad, Lynd, et al., 2014; Lemmers et al., 2012; Lemmers, Sprakel, Voets, van der Gucht, & Cohen Stuart, 2010; Lemmers, Voets, Cohen Stuart, & van der Gucht, 2011; Ortony et al.,

2014; Rahalkar et al., 2018; Srivastava, Levi, Goldfeld, & Tirrell, 2020). At total polymer concentrations $\Phi \geq 5$ wt%, such self-assemblies form three-dimensional hydrogels (Audus et al., 2015; Hunt et al., 2011; Krogstad et al., 2013; Lemmers et al., 2010; Srivastava et al., 2020). At low polymer concentrations, these assemblies may form networks that cannot occupy the entire solution volume and therefore phase separate (Srivastava et al., 2017). Experimental mappings of the PEC morphology and ordering in hydrogels composed of mixtures of oppositely charged symmetric *ABA* triblock polyelectrolytes with varying composition of the triblock polyelectrolytes and total polymer concentration, along with schematics representing the PEC assemblies are presented in Fig. 6 (Srivastava et al., 2020). At low Φ (≤ 20 wt%), the hydrogels are expected to contain spherical PEC domains not arranged in any specific manner. Upon increasing Φ, an ordering transition of PEC domains from a disordered arrangement to an ordering in a body centered cubic (BCC) arrangement, followed by successive morphological transitions from BCC spheres to hexagonally-close-packed (HCP)

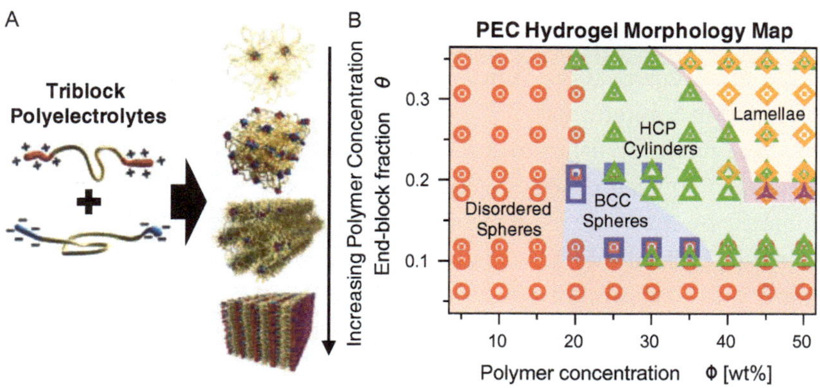

Fig. 6 (A) Schematic depicting the formation of PEC hydrogel upon mixing of oppositely charged polyelectrolytes. The morphology and ordering of the PEC domains evolve with increasing total polymer concentration from disordered spheres to BCC spheres to HCP cylinders to parallel lamellae. (B) A morphology map for PEC hydrogels depicting the domains morphology and the (dis)order state as a function of end-block fraction in the block polyelectrolytes and the total polymer concentration. Red circles, blue squares, green triangles, pink three-pointed stars and orange diamonds denote disordered spheres, BCC spheres, HCP cylinders, gyroid, parallel lamellae, respectively. *Adapted with permission from Srivastava, S., Levi, A.E., Goldfeld, D.J., Tirrell, M.V. (2020). Structure, morphology, and rheology of polyelectrolyte complex hydrogels formed by self-assembly of oppositely charged triblock polyelectrolytes.* Macromolecules, 53*(14), 5763–5774. https://doi.org/10.1021/acs.macromol.0c00847. Copyright 2020 American Chemical Society.*

cylinders to parallelly-stacked lamellae has been reported, as revealed from scattering and electron microscopy measurements. These mappings of PEC domain morphologies also enabled direct comparison with the corresponding predictions from self-consistent field theory calculations (Audus et al., 2015; Audus & Fredrickson, 2016).

4.1 Model block polyelectrolytes: Synthesis and characterization

Model *ABA* triblock polyelectrolytes, with ionizable functionalized poly(allyl glycidyl ether) endblocks on either ends of neutral poly(ethylene oxide) midblocks, were synthesized using established two-step protocol (Hunt et al., 2011; Krogstad et al., 2013; Srivastava et al., 2020). In the first step, poly(allyl glycidyl ether) blocks were polymerized on each end of a hydroxy-terminated poly(ethylene oxide) chains, of known length distributions, through ring opening anionic polymerization. The polymerization reaction, carried out in tetrahydrofuran solutions under dry nitrogen atmosphere and catalyzed by potassium naphthalenide, led to formation of symmetric poly(allyl glycidyl ether)$_n$-poly(ethylene oxide)$_m$-poly(allyl glycidyl ether)$_n$ (PAGE$_n$-PEO$_m$-PAGE$_n$) *ABA* triblock polymers. Here, *n* and *m* denote the degree of polymerization of the PAGE and the PEO blocks, respectively. The reaction was terminated by addition of methanol. The block polymers were precipitated in *n*-hexane, filtered and dried under high vacuum. In the second step, ionizable groups were attached to each monomer in the PAGE endblocks by thiol-ene click reactions. The block polymers were dissolved in dimethyl formamide (DMF) along with functionalized thiols (4 equiv. per alkene) and azobisisobutyronitrile (AIBN) (0.75 equiv. per alkene). The mixture was sparged with dry nitrogen and then heated at 75 °C for 12 h while stirring. Subsequently, the solution was dialyzed repeatedly to remove the unreacted thiols and AIBN; and lyophilized to obtain the functionalized PAGE$_n$-PEO$_m$-PAGE$_n$ polyelectrolytes (*f*PAGE$_n$-PEO$_m$-PAGE$_n$). Ammonium and sulfonate functionalized block polyelectrolytes (*a*PAGE$_n$-PEO$_m$-PAGE$_n$ and *s*PAGE$_n$-PEO$_m$-PAGE$_n$) were synthesized by employing cysteamine and sodium 3-mercapto-1-propanesulfonate, respectively. Guanidinium functionalized block polyelectrolytes (*g*PAGE$_n$-PEO$_m$-PAGE$_n$) were synthesized by functionalization of the ammonium groups of *a*PAGE$_n$-PEO$_m$-PAGE$_n$ in pH 10 aqueous solutions by 1H-pyrazole-1-carboxamidine (4 equiv. per -NH_2 group), followed by dialysis and lyophilization. The triblock polymer precursors were characterized by nuclear magnetic resonance (NMR) spectroscopy and size-exclusion chromatography to determine their molar mass. In the former, the degree of polymerization was estimated by comparing the area under the peaks corresponding to the allyl groups with the area under the peaks corresponding to the methyl end groups. In the latter, measurements were carried out with tetrahydrofuran as solvent, and the elution times of the polymers were compared against those of PEO standards with known molecular weights. NMR spectroscopy was also employed to characterize the extent of the thiol-ene coupling reaction by monitoring the disappearance of the peaks corresponding to the allyl groups.

4.2 PEC hydrogel preparation

1. Concentrated stock solutions of the *f*$PAGE_n$-PEO_m-$PAGE_n$ polyelectrolytes were prepared in 18 MΩ cm Milli-Q water at 50 wt% polymer concentration. 4 M NaCl solutions were also prepared in 18 MΩ cm Milli-Q water.
2. In a typical hydrogel preparation, 0.1 mL of polyelectrolyte complex (PEC) hydrogels were prepared in 0.5 mL Eppendorf tubes by sequential mixing the block polycation stock solution, Milli-Q water, NaCl stock solution and block polycation stock solution, in that order, in stoichiometric proportions. Between each addition step, the mixture was vortex-mixed for 5 s.
3. The hydrogel samples, obtained after all the additions were made, were again vortex-mixed for 30 s to ensure homogenous mixing of the oppositely charged block polyelectrolytes.

PEC hydrogels composed of equimolar mixtures of *ABA* block polyelectrolytes of matching lengths (by a virtue of being derived from the same triblock polymer precursor) were investigated to eliminate the influence of macromolecular charge mismatch on the hydrogel structure and properties. Thus, systematic studies of PEC hydrogel structure (investigated through scattering measurements) were facilitated by a series of hydrogels with methodically varying sizes of the charged (DP: 12–75) and the neutral (DP: 112–455) blocks, and total polymer concentration ($\Phi = 5$–50 wt%) (Srivastava et al., 2020).

4.3 Sample preparation for SAXS measurements

1. PEC hydrogel samples were transferred into circular holes drilled in rectangular aluminum strips (schematic shown in Fig. 7A). These strips (150 mm × 10 mm × 2 mm) comprise rows of 11 holes of 3 mm diameter

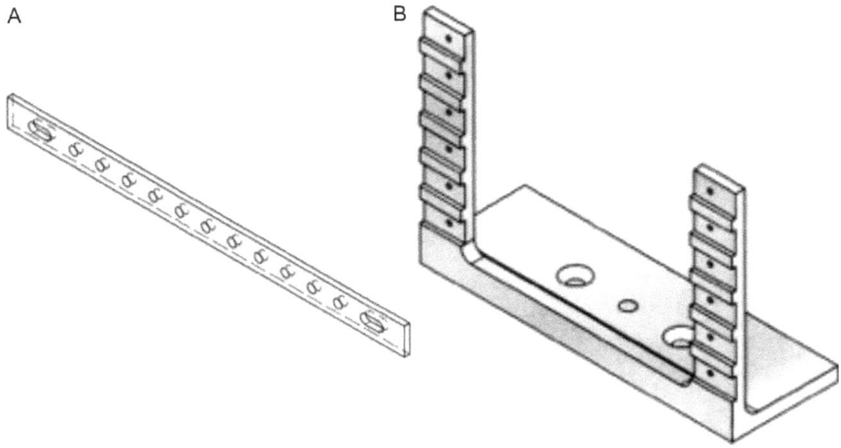

Fig. 7 Schematic depicting (A) the holey metal strips and (B) the strip holder. The metal strips are mounted on the strip holder, which is mounted onto the sample stage.

each, separated by 10 mm, and were mountable onto a strip holder that could accommodate up to six vertically stacked strips (schematic shown in Fig. 7B).

2. 0.15 μL of PEC hydrogel samples were transferred from the Eppendorf tubes into the holes using a positive displacement pipet and sealed from both sides using Kapton tape to minimize water loss.
3. 10 samples and water were loaded onto each strip, and multiple strips were mounted on the strip holder. All samples were prepared at least 12 h in advance of the measurements to facilitate equilibration of the hydrogels.

4.4 Small-angle X-ray scattering (SAXS) measurements

4.4.1 APS beamline 12-ID-B

SAXS measurements were carried out at beamline 12-ID-B at the Advanced Photon Source, Argonne National Laboratory. Scattering measurements were carried out with the distance between the scattering samples and the detector maintained at 3.6 m. In this configuration, measurements with 13 keV X-rays using a Pilatus 2 M detector enabled scattering intensity spectra to be collected for wave vector (q) range of 0.002–0.5 $\mathring{A}^{-1}$. All measurements were conducted at room temperature. In a typical measurement:

1. Strips preloaded with samples were loaded onto the strip holder, and the strip holder was mounted on to the sample stage. The sample stage is equipped with stepper motors that enable its movement vertically and

horizontally, such that the rectangular X-ray beam (300 μm width × 50 μm height) can be pointed at a chosen spot on the samples.

2. SAXS spectra were collected from multiple spots in each hydrogel specimen. Multiple spots were chosen to ensure measurements of representative spectra for spatially non-homogenous hydrogels.
3. At each spot, three successive measurements were carried out, with an exposure time of 0.1 s each. Repeated measurements were conducted to detect any radiation damage—the scattering intensities do not superimpose with each other in case of sample damage during X-ray exposure.
4. The one-dimensional (1-D) scattering intensity (I_{sample}) were calculated from the two-dimensional scattering spectra (Fig. 8A) using the matSAXS package. If the 1-D spectra superposed over each other (indicating minimal radiation damage), they were combined to obtain the averaged SAXS spectra (see Fig. 8B).
5. The averaged scattering spectra from the multiple spots were compared for consistency, with the realization that they wouldn't superpose owing to minor differences in the scattering sample volume at different spots (see Fig. 8C).
6. One of such spectra was chosen as the representative spectra for the sample.
7. Scattering spectra were also measured for Milli-Q water sandwiched between Kapton tape to serve as appropriate background for subsequent data analysis.

4.5 Modeling SAXS spectra of PEC hydrogels

Scattering spectra $I(q)$ from a collection of scatterers can be modeled, in its most general form, as a combination of form factor $P(q)$ and structure factor $S(q)$ as $I(q) \sim P(q)S(q)$ (Glatter & Kratky, 1982; Guinier & Fournet, 1955). The form factor $I(q)$ is the contribution from individual scatterers and can be describe using well-established models to extract scatterer shape and dimensions. The structure factor $S(q)$ is the contribution to $I(q)$ from the spatial correlations among the scatterers arising from the inter-scatterer interactions. It is described as the Fourier transform of the pair correlation function $g(r)$ as

$$S(q) = 1 + 4\pi \int (g(r) - 1) \frac{\sin qr}{qr} r^2 dr \tag{17}$$

with $g(r)$ related to the potential of mean force $U(r)$ as $g(r) = \exp(-U(r)/k_B T)$. Here, k_B and T are the Boltzmann constant and temperature, respectively.

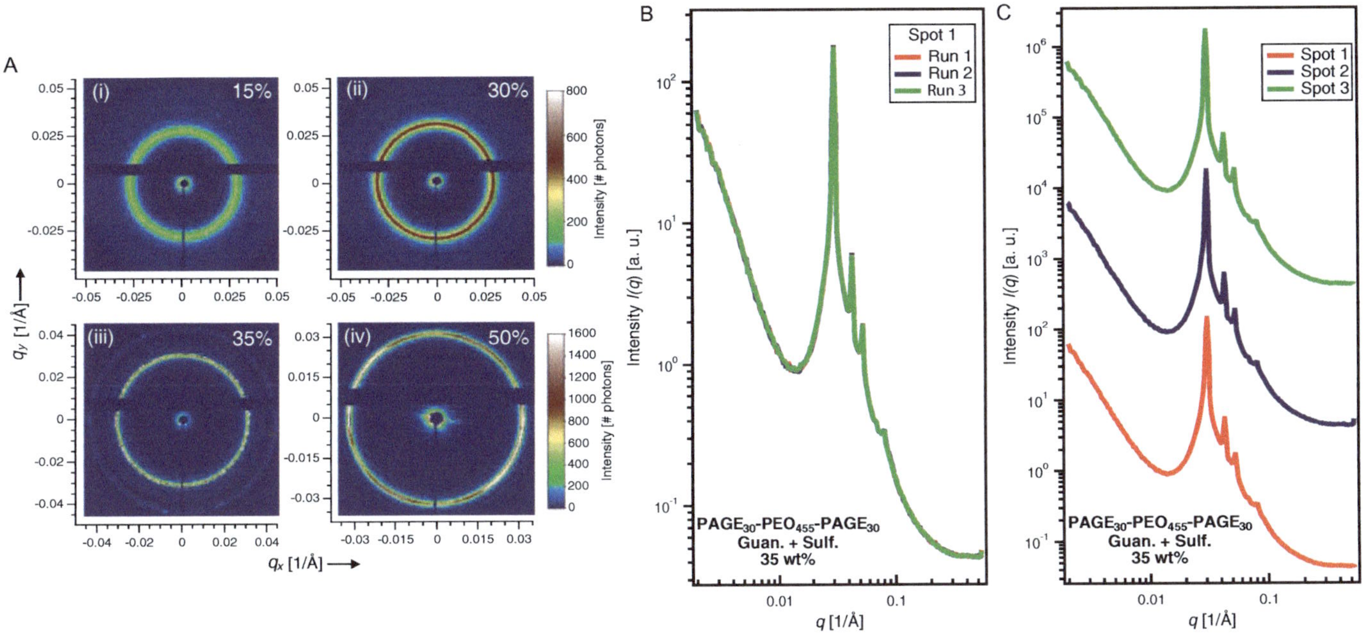

Fig. 8 (A) Two-dimensional (2-D) spectra obtained from SAXS measurements of PEC hydrogels comprising oppositely charged guanidilated and sulfonated $PAGE_{30}$-PEO_{455}-$PAGE_{30}$ block polyelectrolytes with total polymer concentrations $\Phi = $ (i) 15, (ii) 30, (iii) 35, and (iv) 50 wt%. The images are false colored with the corresponding intensity color scales (in arbitrary units) shown at the end of the row. (B and C) One-dimensional (1-D) spectra obtained from azimuthal averaging of the 2-D spectra for PEC hydrogels comprising oppositely charged guanidilated and sulfonated $PAGE_{30}$-PEO_{455}-$PAGE_{30}$ block polyelectrolytes with $\Phi = 35$ wt%. Spectra shown in (B) are obtained from successive experiments at the same spot in the hydrogel. Spectra shown in (C) are the averaged scattering intensities I(q) from different spots in the hydrogel. The averaging is carried out over three runs at each spot. The scattering intensities have arbitrary units and are scaled vertically for clarity. *Panel (A) Adapted with permission from Srivastava, S., Levi, A.E., Goldfeld, D.J., Tirrell, M.V. (2020). Structure, morphology, and rheology of polyelectrolyte complex hydrogels formed by self-assembly of oppositely charged triblock polyelectrolytes.* Macromolecules, 53*(14), 5763–5774. https://doi.org/10.1021/acs.macromol.0c00847. Copyright 2020 American Chemical Society.*

Thus, in weakly correlated (dilute) systems, $S(q){\sim}1$. In strongly correlated (concentrated) yet disordered systems, $S(q)$ has significant contributions and can be described using models with appropriately modeled inter-scatterer interactions. For instance, for disordered spheres with hard sphere interparticle interactions, the structure factor can be described as

$$S(q) = \left[-24\phi\left\{\frac{(1+2\phi)^2}{(1-\phi)^4}\left(\frac{\sin qD - qD\cos qD}{(qD)^3}\right) + \frac{3\phi(2+\phi)^2}{2(1-\phi)^4}\right.\right.$$
$$\left(\frac{(q^2D^2-2)\cos qD - 2qD\sin qD + 2}{(qD)^4}\right) - \frac{\phi(1+2\phi)^2}{2(1-\phi)^4}$$
$$\left.\left.\left(\frac{\left((qD)^4 - 12(qD)^2 + 24\right)\cos qD - \left(4(qD)^3 - 24qD\right)\sin qD - 24}{(qD)^6}\right)\right\}\right]^{-1} \tag{18}$$

with ϕ and D denoting the particle volume fraction and the effective particle diameter, respectively. Furthermore, the location of the principal peak of $S(q)$, q^* is also related to the characteristics inter-scatterer distance d as $d = 2\pi/q^*$. Additionally, in systems comprising ordered assemblies of scatterers, the structure factor contains sharp Bragg reflection peaks. The relative location of the Bragg peaks in such $S(q)$ and consequently $I(q)$ spectra can be correlated with the morphology and the ordering of the scatterers. Table 1 shows the relative position of the secondary, higher q Bragg peaks with respect to the principal peak located at q^*.

In PEC hydrogels, the PEC domains are expected to have a substantial scattering length density contrast as compared to their surroundings owing to a combination of higher polymer density and the presence of higher atomic number sulfur atoms in the PEC domains. Thus, the scattering intensity measured in the SAXS experiments can be considered to be arising

Table 1 Relative positions of the secondary scattering peaks with respect to the principal scattering peak (q^*) for ordered assembly of scatterers.

Body centered cubic spheres	$\sqrt{2}$	$\sqrt{3}$	$\sqrt{4}$	$\sqrt{5}$	$\sqrt{6}$	$\sqrt{7}$
Hexagonally close packed cylinders	$\sqrt{3}$	$\sqrt{4}$	$\sqrt{7}$	$\sqrt{9}$	$\sqrt{12}$	$\sqrt{13}$
Lamellae	2	3	4	5	6	7
Gyroid	$\sqrt{4/3}$	$\sqrt{7/3}$	$\sqrt{8/3}$	$\sqrt{10/3}$	$\sqrt{11/3}$	$\sqrt{12/3}$

primarily from the scattering contributions from the PEC domains. Correspondingly, $I(q)$ can be analyzed and modeled to obtain structural attributes of the PEC domains. For PEC hydrogels with disordered spherical PEC domains, $I(q)$ can be described as (Krogstad, Choi, et al., 2014; Srivastava et al., 2017)

$$I(q) = \int D(R)\left[F(q) + A_{mic}^2(q)(S(q) - 1)\right] dR \tag{19}$$

This form of the model accounts for small but non-negligible polydispersity in size of the PEC domains. The size polydispersity is modeled as a Gaussian distribution with a mean radius R_c and a width of the distribution σ_R as

$$D(R) = \frac{1}{\sqrt{2\pi}\sigma_R} \exp\left(-\frac{(R - R_c)^2}{2\sigma_R}\right) \tag{20}$$

Furthermore, $F(q)$ and $A_{micelle}(q)$ represent the form factor of the core-shell PEC assemblies and the radial scattering length distribution, respectively. For spherical PEC domains, $F(q)$ can be described as

$$\begin{aligned} F(q) = {} & N_{agg}^2 \beta_{core}^2 A_{core}^2(q) + N_{agg} \beta_{corona}^2 P_{chain}(q) \\ & + 2N_{agg}^2 \beta_{core} \beta_{corona} A_{core}(q) A_{corona}(q) \\ & + N_{agg}(N_{agg} - 1) \beta_{corona}^2 A_{corona}^2(q) \end{aligned} \tag{21}$$

Here, N_{agg} represents the aggregation number in the PEC domains, and β_{core} and β_{corona} are the total excess scattering lengths of the PEC domains and the neutral corona, described as $v_{core}(\rho_{core} - \rho_{solv})$ and $v_{corona}(\rho_{corona} - \rho_{solv})$. v_{core}, v_{corona}, ρ_{core}, ρ_{corona} and ρ_{solv} represent the volume of a single chain in the PEC domain and the corona, and the scattering length densities of the PEC domains, the neutral cores and the solvent, respectively. Also,

$$\begin{aligned} A_{core}^2(q) &= \left(\frac{3(\sin qR - qR\cos qR)}{(qR)^3}\right)^2 \exp\left(-q^2 l_{int}^2\right); \\ P_{chain}(q) &= \frac{2\left(\exp\left(-q^2 R_g^2\right) - 1 + q^2 R_g^2\right)}{q^4 R_g^4}; \text{and} \\ A_{corona}(q) &= \frac{4\pi \int \rho_{corona}(r) \frac{\sin qr}{qr} r^2 dr}{4\pi \int \rho_{corona}(r) r^2 dr} \exp\left(-\frac{q^2 l_{int}^2}{2}\right) \end{aligned} \tag{22}$$

Here, R, R_g and l_{int} represent the radius of the PEC domains, the radius of gyration of the neutral polymer chain in the corona and the length of the interfacial region between the PEC domain and the neutral corona. Lastly,

$$A_{micelle}(q) = N_{agg}(\beta_{core}A_{core}(q) + \beta_{corona}A_{corona}(q)) \tag{23}$$

In typical SAXS experiments, the corona is not expected to contribute significantly toward the scattering spectra owing to the near similarity of the scattering length densities of the corona and the solvent, $\rho_{corona} \approx \rho_{solv}$. Also, if $l_{int} \sim 0$, then the expression describing $I(q)$ simplifies to

$$\begin{aligned} I(q) &= N_{agg}^2 \beta_{core}^2 S(q) \int D(R) \left[\left(\frac{3(\sin qR - qR \cos qR)}{(qR)^3} \right)^2 \right] dR \\ &= N_{agg}^2 \beta_{core}^2 S(q) \int D(R) P(q, R) dR \end{aligned} \tag{24}$$

The readers are advised to refer to recent papers from our group (Krogstad, Choi, et al., 2014; Srivastava et al., 2017, 2020) for further details of the model.

4.6 PEC hydrogel SAXS spectra analysis

Analysis of the 1-D scattering data for PEC hydrogels was carried out in two stages. First, the data were organized, the scattering intensity from the samples was obtained by subtracting the background scattering intensity I_{bkg} from I_{sample} as $I(q) = I_{sample} - I_{bkg}$ and peak assignments were carried out in IGOR Pro using the IRENA package (Ilavsky & Jemian, 2009). Representative $I(q)$ spectra from PEC hydrogels composed of oppositely charged guanidilated and sulfonated $PAGE_{30}$-PEO_{455}-$PAGE_{30}$ and $PAGE_{60}$-PEO_{455}-$PAGE_{60}$ block polyelectrolytes with total polymer concentration $\Phi = 5$–50 wt% is shown in Fig. 9. In the samples wherein two spectra were observed consistently, both spectra are presented in the figure. In all the spectra, principal $S(q)$ peaks were identified and the characteristic inter-domain distances as a function of Φ were calculated as $d(\Phi) = 2\pi/q^*$. Further, for PEC hydrogels that exhibited Bragg reflection peaks indicating ordering of PEC domains, the respective locations of the principal and the secondary $S(q)$ peaks were compared to identify PEC domain morphology and ordering. Representative examples are shown in Fig. 9.

For PEC hydrogels with disordered spherical PEC domains, further analysis was carried out to determine the size R_C, size dispersity σ_R and volume fraction ϕ of the PEC domains. $I(q)$ spectra were compared with the

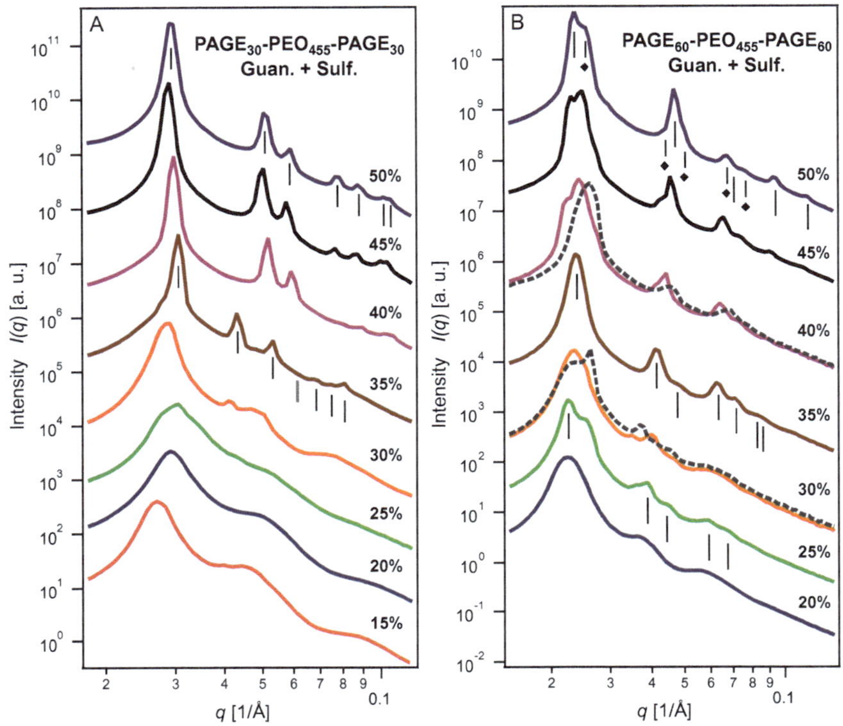

Fig. 9 1-D SAXS spectra for PEC hydrogels comprising oppositely charged guanidilated and sulfonated (A) $PAGE_{30}$-PEO_{455}-$PAGE_{30}$ and (B) $PAGE_{60}$-PEO_{455}-$PAGE_{60}$ triblock polyelectrolytes with varying total polymer concentrations Φ. Bragg reflection peaks, originating from ordered PEC domains, are denoted by vertical bars. Their positions are as follows (A) 35 wt%: q^*, $\sqrt{2}q^*$, $\sqrt{3}q^*$, $\sqrt{5}q^*$, $\sqrt{6}q^*$, and $\sqrt{7}q^*$ (the gray bar at $\sqrt{4}q^*$ indicates an expected but missing peak); (A) 50, (B) 25, 35, and 50 wt%, with ♦: q^*, $\sqrt{3}q^*$, $\sqrt{4}$ q^*, $\sqrt{7}q^*$, and $\sqrt{9}q^*$ (50 wt% in (A) and 35 wt% in (B) also have $\sqrt{12}q^*$ and $\sqrt{13}q^*$); and (b) 50 wt%: q*, $2q^*$, $3q^*$, $4q^*$, and $5q^*$. q^* is the position of the principal scattering peak. For samples where substantially different spectra were observed in different regions of the samples, the second spectra are shown using dashed lines. *Figure and caption are adapted with permission from Srivastava, S., Levi, A.E., Goldfeld, D.J., Tirrell, M.V. (2020). Structure, morphology, and rheology of polyelectrolyte complex hydrogels formed by self-assembly of oppositely charged triblock polyelectrolytes.* Macromolecules, *53(14), 5763–5774. https://doi.org/10.1021/acs.macromol.0c00847. Copyright 2020 American Chemical Society.*

predictions from the model described in Eqs. (18, 20 and 24) to obtain the parameters that provide the best fit of the model to the experimental data. These fits were obtained by minimization of the difference between the experimental data and the model using the in-built *fmincon* function in

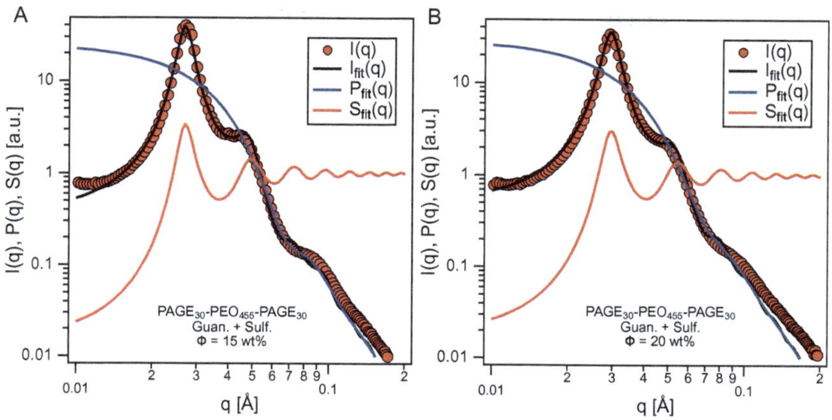

Fig. 10 1-D SAXS spectra for PEC hydrogels comprising oppositely charged guanidilated and sulfonated $PAGE_{30}$-PEO_{455}-$PAGE_{30}$ triblock polyelectrolytes with total polymer concentrations Φ = (a) 15 wt% and (b) 20 wt%. Also shown are the model fits for the form factor *P(q)*, structure factor *S(q)*, and scattering intensity *I(q)*. *Adapted with permission from Srivastava, S., Levi, A.E., Goldfeld, D.J., Tirrell, M.V. (2020). Structure, morphology, and rheology of polyelectrolyte complex hydrogels formed by self-assembly of oppositely charged triblock polyelectrolytes.* Macromolecules, *53(14), 5763–5774. https://doi.org/10.1021/acs.macromol.0c00847. Copyright 2020 American Chemical Society.*

MATLAB. Representative *I*(*q*) spectra for PEC hydrogels composed of oppositely charged guanidilated and sulfonated $PAGE_{30}$-PEO_{455}-$PAGE_{30}$ block polyelectrolytes with total polymer concentration $\Phi = 15$ and 20 wt %, with disordered yet strongly correlated spherical PEC domains are shown in Fig. 10A and B, along with the corresponding fits from the model described by Eqs. (18, 20 and 24). From the fits to the scattering data, the radii of the PEC domains were estimated to be 6.62 ($\pm$1.1) and 6.38 ($\pm$1.3) nm, respectively, for the 15 and 20 wt% PEC hydrogels.

5. Summary

SAXS is a powerful technique that enables nanoscale investigation of polymer conformation and structure associated with phase-separated PECs. In this chapter, we described in detail the sample preparation and scattering analysis routines for (i) macroscale PECs, (ii) PEC micelles, and (iii) PEC hydrogels, thus spanning a range of PEC-based materials. Advances in synthetic chemistry have afforded greater control over charge density, sequence, and chain architecture, thus pushing exploration of the PEC structure-function continuum. However, in addition to the chemical nature of PECs, these electrostatically dominated materials are also influenced by their bound and free counterions, water content, and other environmental factors that govern stability and chain conformation. As more sophisticated PEC materials judiciously combine aspects of these interconnected parameters, SAXS can directly provide in-depth conformational and structural characterization without external chemical labeling or conjugation. This makes SAXS an extremely valuable tool for better understanding the static and dynamic properties associated with LLPS biological condensates (Marin et al., 2020; Schramm et al., 2019).

SAXS can provide insight into far more than simple characterization of size and morphology. In this chapter, we conveyed how inter-chain interactions of polymer chains can be quantified, ion density effects on nanoparticle behavior can be systematically detailed, and careful modeling can elucidate ordered structures in electrostatic-driven hydrogels. Looking forward, SAXS techniques like time-resolved SAXS (Amann et al., 2019; Takahashi, Narayanan, & Sato, 2017; Wu, Ting, Meng, & Tirrell, 2020; Wu, Ting, & Tirrell, 2020) and anomalous SAXS (ASAXS), which is SAXS with varying energy of incident X-rays that can be adjusted near the absorption edge of elements such as counterion constituents of PEC to reveal element-specific form factors, can enhance our understanding of time scales involved in PEC formation and the influence of ion distribution on PEC materials behavior, respectively, vital for designing novel dynamic materials and developing intracellular LLPS physics (Goerigk, Huber, & Schweins, 2007; Huber & Scheler, 2012; Krishnamoorthy et al., 2018; Pabit, Finkelstein, & Pollack, 2009; Sztucki et al., 2012). Lastly, future work on model development for PE solutions and PECs provides an interesting and exciting challenge.

Acknowledgments

The authors thank Dr. Xiaobing Zuo, Dr. Byeongdu Lee, Dr. Soenke Seifert, and Dr. Mrinal Bera at the Advanced Photon Source, Argonne National Laboratory for assisting with SAXS measurements and many great discussions over the years. This research used resources of the Advanced Photon Source, a U.S. Department of Energy (DOE) Office of Science User Facility operated for the DOE Office of Science by Argonne National Laboratory under Contract No. DE-AC02-06CH11357. The authors also thank Dr. Thomas Weiss, Dr. Ivan Rajkovic, and Dr. Tsutomu Matsui at the Stanford National Accelerator Laboratory for their assistance in scattering experiments and insightful discussions throughout the years. Use of the Stanford Synchrotron Radiation Lightsource, SLAC National Accelerator Laboratory, is supported by the U.S. Department of Energy, Office of Science, Office of Basic Energy Sciences under Contract No. DE-AC02-76SF00515. The SSRL Structural Molecular Biology Program is supported by the DOE Office of

Biological and Environmental Research, and by the National Institutes of Health, National Institute of General Medical Sciences (including P41GM103393). The contents of this publication are solely the responsibility of the authors and do not necessarily represent the official views of NIGMS or NIH. A.B.M. acknowledges support from The Welch Foundation (C-2003-20190330). S.S. acknowledges support from the University of California, Los Angeles. J.M.T. acknowledges support from the NIST-CHiMaD Postdoctoral Fellowship, funded under financial assistant award 70NANB19H005 from U.S. Department of Commerce, National Institute of Standards and Technology (NIST) as part of the Center for Hierarchical Materials Design (CHiMaD).

References

Amann, M., Diget, J. S., Lyngsø, J., Pedersen, J. S., Narayanan, T., & Lund, R. (2019). Kinetic pathways for polyelectrolyte coacervate micelle formation revealed by time-resolved synchrotron SAXS. *Macromolecules*, *52*(21), 8227–8237. https://doi.org/10.1021/acs.macromol.9b01072.

Audus, D. J., & Fredrickson, G. H. (2016). Field-based simulations of nanostructured polyelectrolyte gels. In W. Udomkichdecha, A. Mononukul, T. Böllinghaus, & J. Lexow (Eds.), *Materials for energy infrastructure* (pp. 1–9). Singapore: Springer. https://doi.org/10.1007/978-981-287-724-6_1.

Audus, D. J., Gopez, J. D., Krogstad, D. V., Lynd, N. A., Kramer, E. J., Hawker, C. J., et al. (2015). Phase behavior of electrostatically complexed polyelectrolyte gels using an embedded fluctuation model. *Soft Matter*, *11*(6), 1214–1225. https://doi.org/10.1039/C4SM02299H.

Banani, S. F., Lee, H. O., Hyman, A. A., & Rosen, M. K. (2017). Biomolecular condensates: Organizers of cellular biochemistry. *Nature Reviews. Molecular Cell Biology*, *18*, 285–298. https://doi.org/10.1038/nrm.2017.7.

Barrat, J.-L., & Joanny, J.-F. (1996). Theory of polyelectrolyte solutions. *Advances in Chemical Physics*, *94*, 1–66. https://doi.org/10.1002/9780470141533.ch1.

Brangwynne, C. P. (2013). Phase transitions and size scaling of membrane-less organelles. *The Journal of Cell Biology*, *203*(6), 875–881. https://doi.org/10.1083/jcb.201308087.

Brangwynne, C. P., Tompa, P., & Pappu, R. V. (2015). Polymer physics of intracellular phase transitions. *Nature Physics*, *11*(11), 899–904. https://doi.org/10.1038/nphys3532.

Buhler, E., & Boué, F. (2004). Chain persistence length and structure in hyaluronan solutions: Ionic strength dependence for a model semirigid polyelectrolyte. *Macromolecules*, *37*(4), 1600–1610. https://doi.org/10.1021/ma0215520.

Chen, W.-R., Butler, P. D., & Magid, L. J. (2006). Incorporating intermicellar interactions in the fitting of SANS data from cationic wormlike micelles. *Langmuir*, *22*(15), 6539–6548. https://doi.org/10.1021/la0530440.

Chong, P. A., & Forman-Kay, J. D. (2016). Liquid–liquid phase separation in cellular signaling systems. *Current Opinion in Structural Biology*, *41*, 180–186. https://doi.org/10.1016/j.sbi.2016.08.001.

des Cloizeaux, J. (1973). Form factor of an infinite Kratky-Porod chain. *Macromolecules*, *6*(3), 403–407. https://doi.org/10.1021/ma60033a017.

Colby, R. H. (2009). Structure and linear viscoelasticity of flexible polymer solutions: Comparison of polyelectrolyte and neutral polymer solutions. *Rheologica Acta*, *49*, 425–442. https://doi.org/10.1007/s00397-009-0413-5.

Darling, A. L., Liu, Y., Oldfield, C. J., & Uversky, V. N. (2017). Intrinsically disordered proteome of human membrane-less organelles. *Proteomics*, *18*(5–6), 1700193. https://doi.org/10.1002/pmic.201700193.

Dobrynin, A. V., Colby, R. H., & Rubinstein, M. (1995). Scaling theory of polyelectrolyte solutions. *Macromolecules*, *28*(6), 1859–1871. https://doi.org/10.1021/ma00110a021.
Dobrynin, A. V., & Rubinstein, M. (2005). Theory of polyelectrolytes in solutions and at surfaces. *Progress in Polymer Science*, *30*(11), 1049–1118. https://doi.org/10.1016/j.progpolymsci.2005.07.006.
Ermi, B. D., & Amis, E. J. (1998). Domain structures in low ionic strength polyelectrolyte solutions. *Macromolecules*, *31*(21), 7378–7384. https://doi.org/10.1021/ma980579.
Glatter, O., & Kratky, O. (Eds.). (1982). *Small angle X-ray scattering* Academic Press.
Goerigk, G., Huber, K., & Schweins, R. (2007). Probing the extent of the Sr2+ ion condensation to anionic polyacrylate coils: A quantitative anomalous small-angle X-ray scattering study. *The Journal of Chemical Physics*, *127*, 154908. https://doi.org/10.1063/1.2787008.
Guinier, A., & Fournet, G. (1955). *Small angle scattering of X-rays*. John Wiley.
Huber, K., & Scheler, U. (2012). New experiments for the quantification of counterion condensation. *Current Opinion in Colloid & Interface Science*, *17*(2), 64–73. https://doi.org/10.1016/j.cocis.2012.01.005.
Hunt, J. N., Feldman, K. E., Lynd, N. A., Deek, J., Campos, L. M., Spruell, J. M., et al. (2011). Tunable, high modulus hydrogels driven by ionic coacervation. *Advanced Materials*, *23*(20), 2327–2331. https://doi.org/10.1002/adma.201004230.
Ilavsky, J., & Jemian, P. R. (2009). Irena: Tool suite for modeling and analysis of small-angle scattering. *Journal of Applied Crystallography*, *42*, 347–353. https://doi.org/10.1107/S0021889809002222.
Josef, E., & Bianco-Peled, H. (2012). Conformation of a natural polyelectrolyte in semidilute solutions with no added salt. *Soft Matter*, *8*(35), 9156–9165. https://doi.org/10.1039/c2sm25733e.
Keating, C. D. (2012). Aqueous phase separation as a possible route to compartmentalization of biological molecules. *Accounts of Chemical Research*, *45*(12), 2114–2124. https://doi.org/10.1021/ar200294y.
Krishnamoorthy, K., Hoffmann, K., Kewalramani, S., Brodin, J. D., Moreau, L. M., Mirkin, C. A., et al. (2018). Defining the structure of a protein–spherical nucleic acid conjugate and its counterionic cloud. *ACS Central Science*, *4*(3), 378–386. https://doi.org/10.1021/acscentsci.7b00577.
Krogstad, D. V., Choi, S.-H., Lynd, N. A., Audus, D. J., Perry, S. L., Gopez, J. D., et al. (2014). Small angle neutron scattering study of complex coacervate micelles and hydrogels formed from ionic diblock and triblock copolymers. *The Journal of Physical Chemistry. B*, *118*(45), 13011–13018. https://doi.org/10.1021/jp509175a.
Krogstad, D. V., Lynd, N. A., Choi, S.-H., Spruell, J. M., Hawker, C. J., Kramer, E. J., et al. (2013). Effects of polymer and salt concentration on the structure and properties of triblock copolymer coacervate hydrogels. *Macromolecules*, *46*(4), 1512–1518. https://doi.org/10.1021/ma302299r.
Krogstad, D. V., Lynd, N. A., Miyajima, D., Gopez, J., Hawker, C. J., Kramer, E. J., et al. (2014). Structural evolution of polyelectrolyte complex core micelles and ordered-phase bulk materials. *Macromolecules*, *47*(22), 8026–8032. https://doi.org/10.1021/ma5017852.
Lemmers, M., Sprakel, J., Voets, I. K., van der Gucht, J., & Cohen Stuart, M. A. (2010). Multiresponsive reversible gels based on charge-driven assembly. *Angewandte Chemie, International Edition*, *49*(4), 708–711. https://doi.org/10.1002/anie.200905515.
Lemmers, M., Spruijt, E., Beun, L., Fokkink, R., Leermakers, F., Portale, G., et al. (2012). The influence of charge ratio on transient networks of polyelectrolyte complex micelles. *Soft Matter*, *8*(1), 104–117. https://doi.org/10.1039/C1SM06281F.
Lemmers, M., Voets, I. K., Cohen Stuart, M. A., & van der Gucht, J. (2011). Transient network topology of interconnected polyelectrolyte complex micelles. *Soft Matter*, 7(4), 1378–1389. https://doi.org/10.1039/C0SM00767F.

Li, L., Srivastava, S., Andreev, M., Marciel, A. B., de Pablo, J. J., & Tirrell, M. V. (2018). Phase behavior and salt partitioning in polyelectrolyte complex coacervates. *Macromolecules, 51*(8), 2988–2995. https://doi.org/10.1021/acs.macromol.8b00238.

Li, L., Srivastava, S., Meng, S., Ting, J. M., & Tirrell, M. V. (2020). Effects of non-electrostatic intermolecular interactions on the phase behavior of pH-sensitive polyelectrolyte complexes. *Macromolecules, 53*(18), 7835–7844. https://doi.org/10.1021/acs.macromol.0c00999.

Lopez, C. G., Rogers, S. E., Colby, R. H., Graham, P., & Cabral, J. T. (2014). Structure of sodium carboxymethyl cellulose aqueous solutions: A SANS and rheology study. *Journal of Polymer Science Part B: Polymer Physics, 53*(7), 492–501. https://doi.org/10.1002/polb.23657.

Lueckheide, M., Vieregg, J. R., Bologna, A. J., Leon, L., & Tirrell, M. V. (2018). Structure–property relationships of oligonucleotide polyelectrolyte complex micelles. *Nano Letters, 18*(11), 7111–7117. https://doi.org/10.1021/acs.nanolett.8b03132.

Marciel, A. B., Chung, E. J., Brettmann, B. K., & Leon, L. (2017). Bulk and nanoscale polypeptide based polyelectrolyte complexes. *Advances in Colloid and Interface Science, 239*, 187–198. https://doi.org/10.1016/j.cis.2016.06.012.

Marciel, A. B., Srivastava, S., & Tirrell, M. V. (2018). Structure and rheology of polyelectrolyte complex coacervates. *Soft Matter, 14*(13), 2454–2464. https://doi.org/10.1039/C7SM02041D.

Marin, E. W., Hopkins, J. B., & Mittag, T. (2020). Small angle X-ray scattering experiments of monodisperse samples close to the solubility limit. *arXiv:2003.01278 [q-bio.BM]*.

Marras, A. E., Vieregg, J. R., Ting, J. M., Rubien, J. D., & Tirrell, M. V. (2019). Polyelectrolyte complexation of oligonucleotides by charged hydrophobic—Neutral hydrophilic block copolymers. *Polymers, 11*(1), 83. https://doi.org/10.3390/polym11010083.

Marras, A. E., Vieregg, J. R., & Tirrell, M. V. (2020). Assembly and characterization of polyelectrolyte complex micelles. *Journal of Visualized Experiments, 157*, e60894. https://doi.org/10.3791/60894.

Murnen, H. K., Rosales, A. M., Dobrynin, A. V., Zuckermann, R. N., & Segalman, R. A. (2013). Persistence length of polyelectrolytes with precisely located charges. *Soft Matter, 9*(1), 90–98. https://doi.org/10.1039/C2SM26849C.

Muthukumar, M. (2016). Electrostatic correlations in polyelectrolyte solutions. *Polymer Science, Series A, 58*(6), 852–863. https://doi.org/10.1134/S0965545X16060146.

Muthukumar, M. (2017). 50th anniversary perspective: A perspective on polyelectrolyte solutions. *Macromolecules, 50*(24), 9528–9560. https://doi.org/10.1021/acs.macromol.7b01929.

Nakashima, K. K., André, A. A. M., & Spruijt, E. (2020). Enzymatic control over coacervation. *Methods in Enzymology*. https://doi.org/10.1016/bs.mie.2020.06.007.

Ortony, J. H., Choi, S.-H., Spruell, J. M., Hunt, J. N., Lynd, N. A., Krogstad, D. V., et al. (2014). Fluidity and water in nanoscale domains define coacervate hydrogels. *Chemical Science, 5*(1), 58–67. https://doi.org/10.1039/C3SC52368C.

Pabit, S. A., Finkelstein, K. D., & Pollack, L. (2009). Chapter 19—Using anomalous small angle X-ray scattering to probe the ion atmosphere around nucleic acids. *Methods in Enzymology, 469*, 391–410. https://doi.org/10.1016/S0076-6879(09)69019-4.

Patterson, J. P., Robin, M. P., Chassenieux, C., Colombani, O., & O'Reilly, R. K. (2014). The analysis of solution self-assembled polymeric nanomaterials. *Chemical Society Reviews, 43*(8), 2412–2425. https://doi.org/10.1039/C3CS60454C.

Pedersen, J. S., & Schurtenberger, P. (1996). Scattering functions of semiflexible polymers with and without excluded volume effects. *Macromolecules, 29*(23), 7602–7612. https://doi.org/10.1021/ma9607630.

Pergushov, D. V., Müller, A. H. E., & Schacher, F. H. (2012). Micellar interpolyelectrolyte complexes. *Chemical Society Reviews, 41*(21), 6888–6901. https://doi.org/10.1039/c2cs35135h.

Perry, S. L. (2019). Phase separation: Bridging polymer physics and biology. *Current Opinion in Colloid & Interface Science*, *39*, 86–97. https://doi.org/10.1016/j.cocis.2019.01.007.

Perry, S. L., Leon, L., Hoffmann, K. Q., Kade, M. J., Priftis, D., Black, K. A., et al. (2015). Chirality-selected phase behaviour in ionic polypeptide complexes. *Nature Communications*, *6*(1), 6052. https://doi.org/10.1038/ncomms7052.

Prabhu, V. M., Ali, S., Bleuel, M., Mao, Y., & Ma, Y. (2020). Ultra-small angle neutron scattering to study droplet formation in polyelectrolyte complex coacervates. *Methods in Enzymology*. https://doi.org/10.1016/bs.mie.2020.07.001.

Ragnetti, M., & Oberthur, R. C. (1986). Conformation of a chain polyelectrolyte in solution with low molecular weight salt: Small angle neutron scattering measurements. *Colloid & Polymer Science*, *264*(1), 32–45. https://doi.org/10.1007/BF01410306.

Rahalkar, A., Wei, G., Nieuwendaal, R., Prabhu, V. M., Srivastava, S., Levi, A. E., et al. (2018). Effect of temperature on the structure and dynamics of triblock polyelectrolyte gels. *The Journal of Chemical Physics*, *149*(16), 163310. https://doi.org/10.1063/1.5035083.

Schlenoff, J. B., Rmaile, A. H., & Bucur, C. B. (2008). Hydration contributions to association in polyelectrolyte multilayers and complexes: Visualizing hydrophobicity. *Journal of the American Chemical Society*, *130*(41), 13589–13597. https://doi.org/10.1021/ja802054k.

Schramm, A., Bignon, C., Brocca, S., Grandori, R., Santambrogio, C., & Longhi, S. (2019). An arsenal of methods for the experimental characterization of intrinsically disordered proteins—How to choose and combine them? *Archives of Biochemistry and Biophysics*, *676*, 108055. https://doi.org/10.1016/j.abb.2019.07.020.

Sharp, P., & Bloomfield, V. A. (1968). Light scattering from wormlike chains with excluded volume effects. *Biopolymers*, *6*(8), 1201–1211. https://doi.org/10.1002/bip.1968.360060814.

Smolsky, I. L., Liu, P., Niebuhr, M., Ito, K., Weiss, T. M., & Tsuruta, H. (2007). Biological small-angle X-ray scattering facility at the Stanford Synchrotron Radiation Laboratory. *Journal of Applied Crystallography*, *40*(s1), s453–s458. https://doi.org/10.1107/S0021889807009624.

Srivastava, S., Andreev, M., Levi, A. E., Goldfeld, D. J., Mao, J., Heller, W. T., et al. (2017). Gel phase formation in dilute triblock copolyelectrolyte complexes. *Nature Communications*, *8*, 14131. https://doi.org/10.1038/ncomms14131.

Srivastava, S., Levi, A. E., Goldfeld, D. J., & Tirrell, M. V. (2020). Structure, morphology, and rheology of polyelectrolyte complex hydrogels formed by self-assembly of oppositely charged triblock polyelectrolytes. *Macromolecules*, *53*(14), 5763–5774. https://doi.org/10.1021/acs.macromol.0c00847.

Srivastava, S., & Tirrell, M. V. (2016). Polyelectrolyte complexation. *Advances in Chemical Physics*, *161*, 499–544. https://doi.org/10.1002/9781119290971.ch7.

Stetefeld, J., McKenna, S. A., & Patel, T. R. (2016). Dynamic light scattering: A practical guide and applications in biomedical sciences. *Biophysical Reviews*, *8*(4), 409–427. https://doi.org/10.1007/s12551-016-0218-6.

Stewart, R. J., Wang, C. S., & Shao, H. (2011). Complex coacervates as a foundation for synthetic underwater adhesives. *Advances in Colloid and Interface Science*, *167*(1–2), 85–93. https://doi.org/10.1016/j.cis.2010.10.009.

Sztucki, M., Di Cola, E., & Narayanan, T. (2012). Anomalous small-angle X-ray scattering from charged soft matter. *The European Physical Journal Special Topics*, *208*(1), 319–331. https://doi.org/10.1140/epjst/e2012-01627-x.

Takahashi, R., Narayanan, T., & Sato, T. (2017). Growth kinetics of polyelectrolyte complexes formed from oppositely-charged homopolymers studied by time-resolved ultra-small-angle X-ray scattering. *Journal of Physical Chemistry Letters*, *8*(4), 737–741. https://doi.org/10.1021/acs.jpclett.6b02957.

Ting, J. M., Marras, A. E., Mitchell, J. D., Campagna, T. R., & Tirrell, M. V. (2020). Comparing zwitterionic and PEG exteriors of polyelectrolyte complex micelles. *Molecules, 25*(11), 2553. https://doi.org/10.3390/molecules25112553.

Ting, J. M., Wu, H., Herzog-Arbeitman, A., Srivastava, S., & Tirrell, M. V. (2018). Synthesis and assembly of designer styrenic diblock polyelectrolytes. *ACS Macro Letters*, 7(6), 726–733. https://doi.org/10.1021/acsmacrolett.8b00346.

van der Gucht, J., Spruijt, E., Lemmers, M., & Cohen Stuart, M. A. (2011). Polyelectrolyte complexes: Bulk phases and colloidal systems. *Journal of Colloid and Interface Science, 361*(2), 407–422. https://doi.org/10.1016/j.jcis.2011.05.080.

Vernon, R. M., & Forman-Kay, J. D. (2019). First-generation predictors of biological protein phase separation. *Current Opinion in Structural Biology, 58*, 88–96. https://doi.org/10.1016/j.sbi.2019.05.016.

Wu, H., Ting, J. M., Meng, S., & Tirrell, M. V. (2020). In Situ time-resolved small-angle X-ray scattering reveals the formation kinetics of polyelectrolyte complex micelles. *ChemRxiv. Preprint.* https://doi.org/10.26434/chemrxiv.9876395.v2.

Wu, H., Ting, J. M., & Tirrell, M. V. (2020). Mechanism of dissociation kinetics in polyelectrolyte complex micelles. *Macromolecules, 53*(1), 102–111. https://doi.org/10.1021/acs.macromol.9b01814.

Wu, H., Ting, J. M., Weiss, T. M., & Tirrell, M. V. (2019). Interparticle interactions in dilute solutions of polyelectrolyte complex micelles. *ACS Macro Letters, 8*(7), 819–825. https://doi.org/10.1021/acsmacrolett.9b00226.

Wu, H., Ting, J. M., Werba, O., Meng, S., & Tirrell, M. V. (2018). Non-equilibrium phenomena and kinetic pathways in self-assembled polyelectrolyte complexes. *The Journal of Chemical Physics, 149*(16), 163330. https://doi.org/10.1063/1.5039621.

Zhang, F., & Ilavsky, J. (2010). Ultra-small-angle X-ray scattering of polymers. *Journal of Macromolecular Science Polymer Reviews, 50*(1), 59–90. https://doi.org/10.1080/15583720903503486.

Zhou, L., Shi, H., Li, Z., & He, C. (2020). Recent advances in complex coacervation design from macromolecular assemblies and emerging applications. *Macromolecular Rapid Communications, 24*, 2000149. https://doi.org/10.1002/marc.202000149.

CHAPTER NINE

Ultra-small angle neutron scattering to study droplet formation in polyelectrolyte complex coacervates☆

Vivek M. Prabhu[a,*], Samim Ali[a], Markus Bleuel[b,c], Yimin Mao[b,c], and Yuanchi Ma[a]

[a]Materials Science and Engineering Division, Material Measurement Laboratory, National Institute of Standards and Technology, Gaithersburg, MD, United States
[b]Department of Materials Science and Engineering, University of Maryland, College Park, MD, United States
[c]Center for Neutron Research, National Institute of Standards and Technology, Gaithersburg, MD, United States
[*]Corresponding author: e-mail address: vivek.prabhu@nist.gov

Contents

Methods in Enzymology, Volume 646
ISSN 0076-6879
https://doi.org/10.1016/bs.mie.2020.07.001
2021 Published by Elsevier Inc.

Abstract

Associating soft matter such as surfactants, polymers, proteins, and liposomes, may form structures with dimensions not readily accessible by optical methods. Scattering methods can provide detailed information about the mechanism of associative phase separation including nucleation density, size, and shape. Ultra-small angle neutron scattering, a reciprocal space method, provides sensitivity to submicron to micron-scale structures in a non-invasive manner and described in the context of nucleation and growth of dilute droplets formed by a temperature jump into the meta-stable region of polyelectrolyte complex coacervates.

1. Introduction

Biomolecular condensates, such as stress granules, P-granules and nucleoli, are cellular compartments that concentrate proteins and nucleic acids without an encapsulating membrane (Michnick & Bergeron-Sandoval, 2018). These membraneless organelles are hypothesized to be formed via macromolecular phase separation to organize cellular matter on the micron and sub-micron scale (Berry, Brangwynne, & Haataja, 2018; Lin, Brady, Chan, & Ghosh, 2020). The multiphase compartmentalization of oligonucleotides and oligopeptides shows how membraneless structures form by charge complexation pathways of ion-containing polymers (polyelectrolytes) (Mountain & Keating, 2020). Applying phase separation and non-equilibrium routes to form membraneless structures is of active interest and requires a broad range of characterization techniques. These techniques, developed by different disciplines, may offer insight into the study of liquid-liquid and liquid-solid macromolecular phase separation.

Phase separation driven by charge complexation takes place on the nanometer scale with block polyelectrolytes or on the macroscale with linear polyelectrolytes (Srivastava & Tirrell, 2016). The structures and properties formed by synthetic and natural polyelectrolytes are of interest to advanced materials, but also in a broader context to understand molecular mechanisms that take place with biomolecular condensates. Methods of small-angle neutron and X-ray scattering, SANS and SAXS, respectively, are frequently used to characterize macromolecular phases and structure (Higgins & Benoit, 1994).

Often, ultra-small angle neutron scattering (USANS) is combined with SANS to characterize structure from the micron to angstrom length scales (Koga et al., 2008). In the area of polyelectrolyte complex coacervation, a few SANS studies have focused on chain conformation or concentration fluctuations (Fares et al., 2018; Spruijt et al., 2013). However, often the instrument resolution is not enough to observe larger scale structures that are typically present. One must measure at even smaller momentum transfer vectors ($q \equiv |\mathbf{q}|$) accomplished by USANS. An inverse relationship between q and the real-space length scale means that in order to probe larger structures, one measure smaller q. This can be achieved by using radiation of larger wavelength (λ) or measuring the form of the scattering at smaller angles (θ) as will be discussed.

In polyelectrolyte complexes and coacervate liquids or solids the structure in the hydrated state depends upon the sample preparation, polymer structure and chemical composition, and salt concentration. In block and triblock polyelectrolyte complexes, microphase separation occurs, in which the domain size is determined by the characteristic block length. Whereas, with oppositely charged linear polyelectrolytes macrophase separation occurs leading to bulk coacervate phase separated from a dilute phase by a meniscus. Even the formation of micrometer scale structures (Sadman et al., 2019) with homopolymers remains mechanistically poorly understood, however, phase separation via nucleation and growth is speculated. An indication of this mechanism was observed after a temperature quench in a solution of oppositely charged homopolymers leading to liquid-liquid phase separation (Ali & Prabhu, 2019). The methods discussed here refer to bulk viscoelastic liquid rather than stabilizing individual coacervate droplets. Different macromolecular systems may take advantage of higher throughput and microfluidic methods and extrusion techniques as applied to and surfactant-polyelectrolyte mixtures or liposomes (Hann, Niepa, Stebe, & Lee, 2016; Mineart, Kelley, Nagao, & Prabhu, 2017; Wu & Li, 2018; Zook & Vreeland, 2010).

We illustrate the growth of micrometer scale dilute droplets formed in a coacervate characterized by lower critical solution temperature (Ali, Bleuel, & Prabhu, 2019). Upon heating, a phase separated system reaches new equilibrium by forming droplets within the viscoelastic coacervate phase rather than a moving meniscus common in simple liquid (Ali & Prabhu, 2019). USANS is applied to illustrate the kinetics of formation of dilute droplets by examining the scattering below the meniscus that separates the supernatant and dense coacervate.

2. Model polyelectrolyte complex coacervate and liquid-liquid phase separation

2.1 Materials

Sodium poly (styrene sulfonate) (NaPSS) with number average relative molecular mass of 200 kg/mol and poly (diallyldimethylammonium chloride)

(PDADMAC) with number average relative molecular mass of 150 kg/mol were both purchased from Sigma-Aldrich.[a] Their molecular mass and polydispersity were independently characterized by aqueous size exclusion chromatography. The commercial NaPSS and PDADMAC solutions were purified by dialysis and freeze-dried to obtain dry powder. The fraction of hydrated water was determined by thermogravimetric analysis.

2.2 Sample preparation

1. Samples for USANS are made using 99.98 D atom % deuterium oxide (D_2O, Cambridge isotopes) rather than water to provide the contrast for scattering with the polymer
2. Dry NaPSS and PDADMAC were each individually dissolved in deuterium oxide (D_2O) to prepare 1.0 mol/L stock solutions
3. These stock solutions were added to the mixture of predetermined ratio of D_2O and 5.00 mol/L NaCl stock solution and vortex-mixed to give clear NaPSS-NaCl and PDADMAC-NaCl solutions. With designated salt concentration (4.0 mol/L) and polymer concentration (0.3 mol/L)
4. The two solutions with 0.3 mol/L polymer concentration each were then transferred to an Eppendorf tube, and vortex agitated for 30 s. The final mixture contains anionic and cationic polymers at 1:1 stoichiometric ratio of the ion-containing groups
5. The mixture undergoes phase separation at room temperature (22 °C) and subsequently is left undisturbed for 3 h before performing centrifugation with a relative centrifugal force of 10,000*g* m/s^2 for 15 min. This results in polymer-rich complex coacervate at the bottom and polymer-poor supernatant at the top of the Eppendorf tube
6. The tube was sealed using high vacuum grease and parafilm wax
7. The sample was annealed at 60 °C for 24 h and then re-equilibrated at room temperature for 2 days
8. The resulting polyelectrolyte complex solution was stored at 5 °C, which is above the lower critical solution temperature (LCST) of 0 °C, overnight prior to use
9. After the solution had completely separated into two transparent phases with clear meniscus, and there were no visual droplets. The supernatant

[a] Certain equipment, instruments or materials are identified in this paper in order to adequately specify the experimental details. Such identification does not imply recommendation by the National Institute of Standards and Technology, nor does it imply the materials are necessarily the best available for the purpose.

was extracted from the top layer and stored in a clean Eppendorf tube. The dense coacervate phase was transferred directly to the demountable SANS cell with quartz windows, followed by the supernatant. The sample was handled to prevent intermixing but allowed to rest for a minimum of 24h

2.3 Droplets by nucleation and growth

The coacervate exhibits LCST phase behavior (Fig. 1A). Heating, therefore, leads to a stronger driving force for the solution to further phase separate,

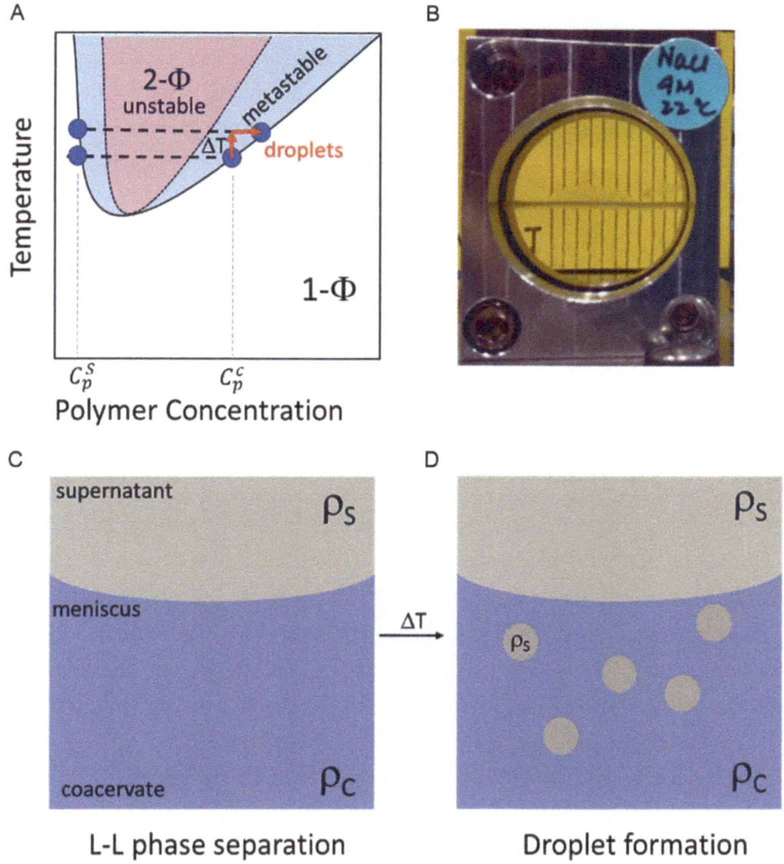

Fig. 1 (A) Schematic of a lower critical solution phase diagram showing two equilibrating phases before and after an increase in temperature (ΔT). A photo (B) of a phase separated sample showing two clear phases separated by a meniscus in a demountable neutron scattering cell of 2.54 cm diameter quartz windows. After a temperature jump (C and D), dilute droplet nucleation and growth occurs in the dense coacervate phase. The neutron optical contrast between the dense phase (ρ_C) and the supernatant (ρ_S) and equilibrating droplets enabled USANS characterization.

resulting in a change in concentration and volume fraction of the coacervate. This new equilibrium state (via tie line) is achieved by nucleation and growth of dilute droplets within the dense coacervate phase (Fig. 1D). These droplets are kinetically trapped because of the viscoelastic nature of the dense phase (Ali & Prabhu, 2018). The size scale and stability of these structure depends upon different physicochemical parameters and is the topic of current research.

1. Sample at room temperature are left unperturbed for a minimum of 24 h, but up to 1 week. A 1 in. x ½ in. piece of cadmium foil can be attached to sample cell to block the incident neutron beam from passing through the top supernatant phase or reflecting from the meniscus. Cadmium foils are readily available at most neutron scattering facility guide halls
2. Instrument temperature block should be set at the desired temperature (45 °C) with USANS beam passing through the coacervate domain immediately below the meniscus
3. Sample is transferred to the heating block and the sample scan are started
4. Data are collected across the multiple configurations with predefined counting time estimated prior to the kinetics experiment for a sample at 45 °C
5. Sample scattering are recorded at 10 min intervals

3. USANS instrument and experimental practice

USANS probes a q-range between $\approx 1 \times 10^{-4}$ nm^{-1} and $\approx 1 \times 10^{-2}$ nm^{-1}, that is realized via a double crystal diffractometer, also known as Bonse-Hart camera. In this design, the first perfect single crystal serves as a monochromator, and the second the analyzer. Sample is placed between the two crystals. The scattered waves are reflected by the analyzer to the detector. Unlike small-angle neutron and X-ray scattering (SAXS/SANS) instruments where the scattering angle is determined by the sample-to-detector distance (SDD). The detector position in the Bonse-Hart design is fixed, and the scattering angle is precisely determined by the rocking angle of the analyzer. Details of the working principles of ultra-small angle scattering instrument (for both X-ray and neutron) has been introduced elsewhere (Glatter & Kratky, 1982). For USANS, undesirable sources of background are reduced by the double crystals (Agamalian, Wignall, & Triolo, 1997; Barker et al., 2005). USANS instruments are only possible at large-scale facilities equipped with a neutron source that are available to general users. Efficient communication between the users and the local facility contact is critical for a successful experiment.

USANS with liquid samples are often loaded into a cell with flat quartz windows similar to optical spectroscopy. The amount of sample needed depends on how strongly it scatters, which in turn is determined by the contrast (discuss in Section 3.1), concentration, etc. Typically, a 1-mm pathlength liquid cell requires $\approx 0.5\,\text{mL}$ sample. No dust freeing step is needed, though it is a crucial step in a light scattering experiment. An empty cell also needs to be measured for background subtraction purposes. Two types of transmission coefficients can be measured, depending on the detector used. In the first case, the counts due to neutron beam as accepted by the main detector are measured with and without the sample. The ratio between the two, denoted as T_{rock}, is used for the purposes of intensity normalization. In practice, a separate detector is used to monitor transmitted intensity. The transmitted intensities are measured when the rocking angle of the analyzer is set at a fixed angle $>2°$. And a different transmission coefficient can be calculated in a similar way, denoted as T_{wide}. The ratio of $T_{\text{rock}}/T_{\text{wide}}$ serves as an empirical measure; the users are advised to pay attention to possible existence of multiple scattering when this value is less than 90%. Again, the users are encouraged to work closely with the local facility contact to set up an optimal procedure. Useful information can also be found in the operational manuals and data reduction documents provided by facility supports (Kline, 2006).

3.1 Neutron scattering length density contrast

The neutron contrast for scattering is provided by the scattering length density (ρ) and may be estimated from the chemical composition. On-line resources are available (NIST SLD Calculator, 2020) to calculate ρ of mixtures using the known coherent atomic scattering lengths, estimates of the density or partial molar specific volumes, and volume fraction of components. Due to the difference in scattering length between hydrogen and deuterium, deuterium oxide is often substituted for water to increase the contrast between polymer and solvent. In the present case, the two-phase region is represented by a polyelectrolyte-rich coacervate and a supernatant almost devoid of polyelectrolytes. Both phases are of high salinity and abundant in deuterium oxide. Upon heating, further phase separation occurs via the formation of polyelectrolyte-deficient droplets in the coacervate phase, because the new equilibrium condition favors denser coacervate with higher polymer concentration (Fig. 1A). The kinetics and the molecular mechanism of such phase separation remain poorly understood. One may assume that the average scattering length density of each the supernatant and coacervate can be estimated by the volume-fraction weighted scattering length density. In the present case, we estimate the average solvent to be 4.0 mol/L of NaCl in D_2O leading to $\rho_S = 5.9 \times 10^{-6}$ Å^{-2}. The scattering length density of the dense coacervate is estimated to be the D_2O solution of 0.15 mol/L NaPSS, 0.15 mol/L PDADMAC and 4.0 mol/L NaCl leading to $\rho_C = 4.7 \times 10^{-6}$ Å^{-2}.

3.2 Particle shape and size

USANS is an elastic scattering method. Thus, general principles for data interpretation is the same as that for small-angle scattering. The reduced data is the scattering intensity as a function of modulus of momentum transfer written as

$$q \equiv |\boldsymbol{k}_s - \boldsymbol{k}_i| = \frac{4\pi}{\lambda} \sin\theta, \tag{1}$$

where $\boldsymbol{k}_s$ and $\boldsymbol{k}_i$ are the wave vectors of the scattered and the incident beam, respectively. θ is half of the scattering angle. The scalar q is used for isotropic system. The task of data analysis is to extract structural information from the $I(q) \sim q$ relationship, known as the scattering function. For particulate system, the low-q scattering intensity can be approximated by

$$I(q) \propto \exp\left(-q^2 R_g^2/3\right) \tag{2}$$

known as the Guinier approximation, where R_g is the radius of gyration of the object. Therefore, in principle, R_g can be obtained by directly manipulating the scattering function. The so-called Guinier plot, which is a $\mathrm{Log}[I(q)] \sim q^2$ plot, can be constructed for this purpose, and R_g is readily derived from the slope, in the range when $qR_g < \approx 1$. Note, however, that USANS data are slit smeared (Glatter & Kratky, 1982; Ilavsky et al., 2009), data need to be desmeared before such manipulation can be carried out. Therefore, it is often preferred to fit the scattering function directly using a smeared model.

For a suspension of particles, the scattering function is attributed to interference of the waves scattered by all subunits of a particle (intra-particle scattering) and that by different particles (inter-particle scattering). If ignoring the latter, the scattering function is determined by density variation inside a particle. Assume a monodisperse system containing homogenous particles, $I(q)$ is written as

$$I(q) = \phi \Delta\rho^2 V^2 P(q) \tag{3}$$

where ϕ is the number fraction of the particles, $\Delta\rho$ is the scattering length density difference between the homogenous particles and the surrounding (Fig. 1D), V is the volume of the particle. $P(q)$ is called the form factor that describes intensity modulation due to intra-particle scattering. It is the squared modulus of the Fourier transform of the particle, $F(q)$, after angular averaging. Form factors of simple geometric particles can be analytically given. Common form factors for a homogenous sphere, hollow sphere, long rod, and thin disk are listed in Table 1.

Fig. 2 shows, for demonstration purposes, form factors of homogenous sphere, rigid rod, and flat disk. Notice that the three particles have the same R_g of 150 nm. The sphere has a radius of 194 nm; the rigid rod is 517 nm long and the cross-sectional radius is 20 nm; the disk has a radius of 212 nm and a thickness of 20 nm. The relationship between the physical dimensions and the R_g of simple geometric particles has been tabulated elsewhere (Glatter & Kratky, 1982). Rough detection ranges of USANS, static light scattering (SLS), and small-angle neutron scattering (SANS), are indicated. Note that in SLS, the wavelength is much longer (e.g., a He-Ne laser produces a 632.8 nm line) as compared to that in SANS and USANS. In practice, it is often advantageous to use these techniques as complementary tools. It is clear, that in the Guinier regime when $qR_g < \approx 1$, all three curves reach a plateau in the logarithmic plot. The full features that differentiate the three particle shapes appears at higher angles. This suggests that when carrying out model fitting, it is of critical importance to have morphology evidence from

Table 1 A list of form factors for some simple geometric particles.

Particle	$P(q) \equiv \lvert F(q)\rvert^2$	Remarks
Sphere	$\left[3\frac{\sin(qR) - qR\cos(qR)}{(qR)^3}\right]^2$	R: sphere radius
Hollow sphere	$\left[\frac{V(R_1)F_S(q,R_1) - V(R_2)F_S(q,R_2)}{V(R_1) - V(R_2)}\right]^2$	R_1, R_2: core and shell radii; $V(R) = 4\pi R^3/3$; $F_S(R)$: $F(q)$ of a sphere
Long rod	$\frac{4\pi}{qL}\left[\frac{J_1(qR)}{qR}\right]^2$	L: rod length; R: cross-sectional radius; $L \gg R$
Circular thin disk	$\frac{2}{(qR)^2}\left[1 - \frac{1}{qR}J_1(2qR)\right]$	R: disk radius; $R \gg$ thickness

J_1 is the first order Bessel function of the first kind.

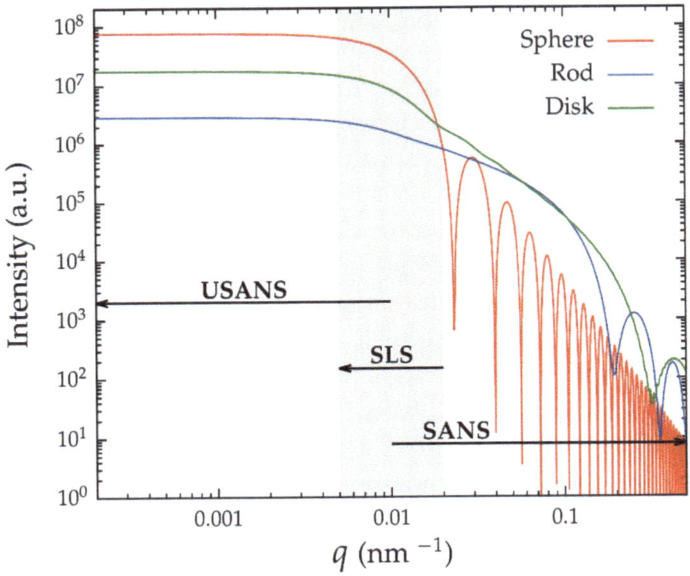

Fig. 2 Form factor of homogenous sphere, rigid rod, and flat disk. The sphere has a radius of 195 nm; length and cross-sectional radius of the rod are 517 and 20 nm, respectively; radius and thickness of the disk are 212 and 20 nm, respectively. All three particles have the same R_g of 150 nm. Rough detection ranges of USANS, SLS, and SANS techniques are indicated, see the text for details. Profiles are generated using SASVIEW package (SASView Version 5.01, 2020).

other techniques such as microscopy. In our case, the coacervate phase was first inspected using optical microscopy as a preliminary check, and spherical droplets were evident, USANS was then used to provide a statistical measure of characteristic length during the phase separation.

3.3 Particle size distribution and polydispersity

The above discussion is for a monodisperse system. Polydispersity may have a complex influence on the scattering function, particularly when a structure factor needs to be considered, and the particle size and shape are coupled with their spatial positions. The structure factor contains average statistical information on how particles interact and spatially distributed. A selection of structure factors developed for different interaction potentials and closures are available to combine with form factors to model the scattering. The treatment of suspensions of uniform spherical particles with a size distribution of $D(R)$ are routinely analyzed. If only the form factor is considered, Eq. (3) can be modified in the following form

$$I(q) = \Delta\rho^2 \int D(R) V^2(R) P(q) \mathrm{d}R \tag{4}$$

Size distributions can be accommodated by weighting of the scattered intensity with different distributions such as Gaussian, Schulz-distribution, or log normal distribution. These procedures are incorporated in the SASView and Igor data analysis packages (Kline, 2006; SASView Version 5.01, 2020) and can be used without any programming expertise.

The time-resolved USANS data taken at various time (12, 24, 36 and 48 min) were fit to the smeared form of (Eq. 4), as shown in Fig. 3. In the present example, monodisperse-sized droplets (with $D(R)=1$) are not consistent with the data, therefore a distribution function must be applied. Here we assume Gaussian distribution in which $<R>$ is the mean radius and σ is the standard deviation. The polydispersity, PD, is defined as the relative standard deviation of the radius, and therefore $PD=\sigma/<R>$. An example of the results are summarized in Table 2 and Fig. 4, from which two main conclusions can be drawn: (1) both the number ($N \sim \Phi/<R>^3$) and the polydispersity of the dilute droplets increase with time, while their average radii remain largely unaffected; and (2) the droplet-coacervate interfaces are smooth, as indicated by the slope of -4 at high-q. Additionally, the goodness of fit statistics, χ^2, allows for direct comparison between the currently used polydisperse sphere model and other from factors, in order to understand alternative trial structures. The uncertainties (error bars) in Table 2 are estimated by one standard deviation of the fits to the data by the non-linear regression model.

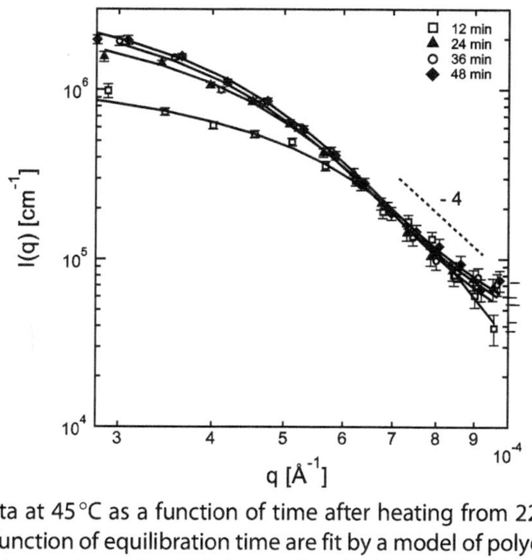

Fig. 3 USANS data at 45 °C as a function of time after heating from 22 °C to 45 °C. The data shown as a function of equilibration time are fit by a model of polydisperse spheres. At the high q region, a slope of -4 is observed and expected for sharp interfaces.

Table 2 Parameters obtained from USANS modeling by polydisperse spheres.

t, min	Φ vol. fraction	<R>, μm	PD = σ/<R>	χ^2
12	0.073 ± 0.001	3.6 ± 0.2	0.14 ± 0.06	0.76
24	0.114 ± 0.002	4.0 ± 0.2	0.29 ± 0.05	0.60
36	0.126 ± 0.003	3.9 ± 0.2	0.35 ± 0.06	1.8
48	0.134 ± 0.002	3.9 ± 0.2	0.34 ± 0.06	0.79

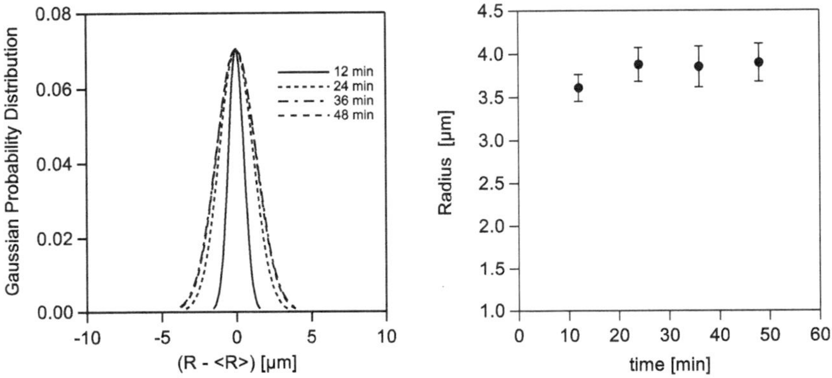

Fig. 4 Gaussian size distribution and average radius of the dilute droplets formed within the dense coacervate phase. The uncertainties (error bars) shown are estimated by one standard deviation of the fits to the data by the non-linear regression model.

4. Related techniques

4.1 Ultra-small-angle X-ray scattering

Historically USAXS instrument appears earlier than USANS (Glatter & Kratky, 1982). Both the instrumentations are based on the Bonse-Hart design. Both in-house and synchrotron-X-ray-based USAXS are available. The latter, owing to high photon flux, enables much faster data collection (Diat et al., 1995; Ilavsky et al., 2009). Compared with USANS, the low-q limit of USAXS is slightly higher, $\approx 0.001\ \text{nm}^{-1}$. USAXS can be combined with small-angle X-ray scattering and even wide-angle scattering (WAXS), which provides an opportunity to conduct simultaneous investigations on hierarchical structures at multiple length scales (Bolze & Gateshki, 2019; Ilavsky et al., 2009). However, when dealing with biological samples, one needs to pay special attention to possible radiation damage caused by X-ray. This issue is often negligible in neutron scattering.

4.2 Very-small angle neutron scattering

At the National Institute for Standards and Technology Center for Neutron Research a newly commissioned very-small angle neutron scattering (VSANS) (NCNR VSANS, 2020) provides the widest range structures possible, in a single instrument. However, USANS remains the highest resolution instrument to measure micron scale structures.

4.3 3D cross correlation static light scattering

One of the major challenges in studying large particles at high concentrations is the presence of multiple scattering. 3D cross correlation static laser light scattering (Block & Scheffold, 2010), experimentally suppresses this effect. An example from the author's laboratory is shown in Fig. 5 from a suspension of very low polydispersity 0.157 μm radius polystyrene latex particles at a concentration of 0.1% by mass. The traditional 2D experiment result is compared to that of the 3D cross-correlation, with the correction to eliminate the multiple scattering effect. As can be seen in the figure, the first

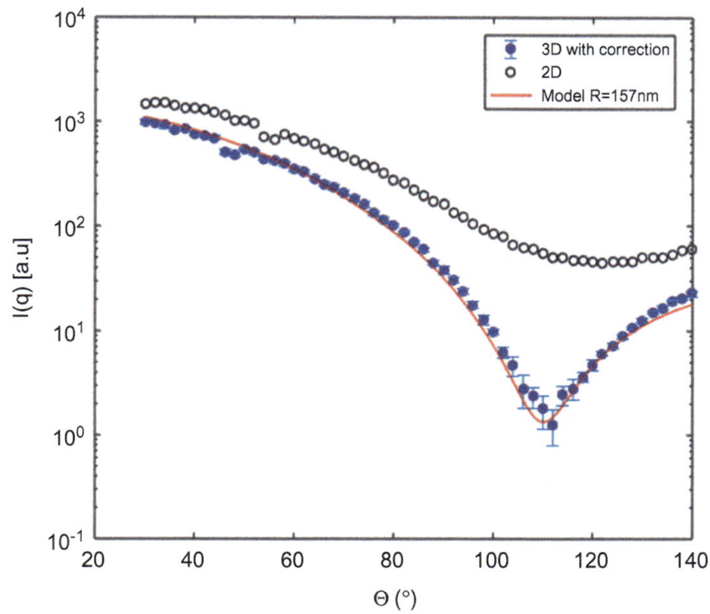

Fig. 5 3D static light scattering on polystyrene latex particle standards with radius of 157 nm (solid symbols). The comparison between the 2D (open symbols) and the 3D data demonstrates the significance of eliminating multiple scattering effect in obtaining the correct form factor of the particles.

scattering minimum is possible only with the 3D measurement, while in 2D, this minimum is significantly smeared. It is worth noting that such particles may be better characterized by USANS, as the data analysis in laser light scattering is complicated by Mie scattering.

5. Safety considerations and standards

As with any measurements involving radiation, an appropriate safety assessment is made before any neutron scattering measurement. This includes providing elemental composition and mass estimates and neutron beam flux intensity on the sample to estimate for activation. After measurements are performed, a radiological survey and clearance of samples exposed to the neutron beam are performed by the trained user and health physics professionals before samples can leave the neutron facility or be shipped. Each neutron facility will have stated procedures and dedicated health physics personnel to guide the user. Organic materials containing C, H, O, N are typically cleared within 24 h.

6. Pros and cons

Pros	Cons
Since neutrons are neutral particles, they do not interact with the electron cloud, such that sample charging or beam damage does not occur as in by electron or X-ray imaging and scattering	Access is available only through user facilities with onsite expertise. Not a real-space measurement method, such as optical, fluorescence, or electron microscopies and therefore requires learning experience to data interpretation and design of experiments
Neutron are transparent to materials such as aluminum, titanium and stainless; therefore, novel sample environments from these materials can be used to study caustic systems or those under pressure	Some elements such as cadmium and boron (present in borosilicate glass) strongly absorb neutrons. Therefore, non-traditional cells and sample elemental composition should be tested for absorption and activation before experimentation
Neutron scattering can be used to study cloudy, high concentration, high light absorption, or low light transmission samples that often invalidates optical approaches	The neutron beam intensity is orders of magnitude lower than synchrotron radiation and therefore require comparatively longer measurement times or larger sample volumes

7. Summary

Ultra-small angle neutron scattering provides high sensitivity to micrometer scale structures. Using established data analysis routines and access to user facilities the size, shape and size distribution may be determined without substantial labeling efforts apart from changes from water to deuterium oxide. Additional contrast variation methods, applicable to neutron scattering, enable further quantification of interactions and intra-particle structure.

Acknowledgments

The authors thank the support of the National Institute of Standards and Technology (NIST) Materials Genome Initiative. Access to USANS was provided by the Center for High Resolution Neutron Scattering, a partnership between NIST and the National Science Foundation under Agreement No. DMR-1508249.

References

Agamalian, M., Wignall, G., & Triolo, R. (1997). Optimization of a Bonse-Hart ultra-small-angle neutron scattering facility by elimination of the rocking-curve wings. *Journal of Applied Crystallography*, *30*, 345–352. https://doi.org/10.1107/S0021889896014343.

Ali, S., Bleuel, M., & Prabhu, V. M. (2019). Lower critical solution temperature in polyelectrolyte complex coacervates. *ACS Macro Letters*, *8*, 289–293. https://doi.org/10.1021/acsmacrolett.8b00952.

Ali, S., & Prabhu, V. M. (2018). Relaxation behavior by time-salt and time-temperature superpositions of polyelectrolyte complexes from coacervate to precipitate. *Gels*, *4*, 11. https://doi.org/10.3390/gels4010011.

Ali, S., & Prabhu, V. M. (2019). Characterization of the ultralow interfacial tension in liquid-liquid phase separated polyelectrolyte complex coacervates by the deformed drop retraction method. *Macromolecules*, *52*, 7495–7502. https://doi.org/10.1021/acs.macromol.9b01491.

Barker, J. G., Glinka, C. J., Moyer, J. J., Kim, M. H., Drews, A. R., & Agamalian, M. (2005). Design and performance of a thermal-neutron double-crystal diffractometer for USANS at NIST. *Journal of Applied Crystallography*, *38*, 1004–1011. https://doi.org/10.1107/S0021889805032103.

Berry, J., Brangwynne, C. P., & Haataja, M. (2018). Physical principles of intracellular organization via active and passive phase transitions. *Reports on Progress in Physics*, *81*, 046601 https://doi.org/10.1088/1361-6633/aaa61e.

Block, I. D., & Scheffold, F. (2010). Modulated 3D cross-correlation light scattering: Improving turbid sample characterization. *The Review of Scientific Instruments*, *81*, 123107. https://doi.org/10.1063/1.3518961.

Bolze, J., & Gateshki, M. (2019). Highly versatile laboratory X-ray scattering instrument enabling (nano-)material structure analysis on multiple length scales by covering a scattering vector range of almost five decades. *The Review of Scientific Instruments*, *90*, 123103 https://doi.org/10.1063/1.5130061.

Diat, O., Bösecke, P., Ferrero, C., Freund, A. K., Lambard, J., & Heintzmann, R. (1995). Ultra-small-angle X-ray scattering with a Bonse-Hart camera on the high brilliance beamline at the ESRF. *Nuclear Instruments and Methods in Physics Research Section A*, *356*, 566–572. https://doi.org/10.1016/0168-9002(94)01253–9.

Fares, H. M., Ghoussoub, Y. E., Delgado, J. D., Fu, J., Urban, V. S., & Schlenoff, J. B. (2018). Scattering neutrons along the polyelectrolyte complex/coacervate continuum. *Macromolecules*, *51*, 4945–4955. https://doi.org/10.1021/acs.macromol.8b00699.

Glatter, O., & Kratky, O. (1982). *Small angle X-ray scattering*. London; New York: Academic Press.

Hann, S. D., Niepa, T. H. R., Stebe, K. J., & Lee, D. (2016). One-step generation of cell-encapsulating compartments via polyelectrolyte complexation in an aqueous two phase system. *ACS Applied Materials & Interfaces*, *8*, 25603–25611. https://doi.org/10.1021/acsami.6b07939.

Higgins, J. S., & Benoit, H. C. (1994). *Polymers and neutron scattering*. New York: Oxford University Press.

Ilavsky, J., Jemian, P. R., Allen, A. J., Zhang, F., Levine, L. E., & Long, G. G. (2009). Ultra-small-angle X-ray scattering at the advanced photon source. *Journal of Applied Crystallography*, *42*, 469–479. https://doi.org/10.1107/S0021889809008802.

Kline, S. (2006). Reduction and analysis of SANS and USANS data using IGOR Pro. *Journal of Applied Crystallography*, *39*, 895–900.

Koga, T., Hashimoto, T., Takenaka, M., Aizawa, K., Amino, N., Nakamura, M., et al. (2008). New insight into hierarchical structures of carbon black dispersed in polymer matrices: A combined small-angle scattering study. *Macromolecules*, *41*, 453–464. https://doi.org/10.1021/ma0718671.

Lin, Y.-H., Brady, J. P., Chan, H. S., & Ghosh, K. (2020). A unified analytical theory of heteropolymers for sequence-specific phase behaviors of polyelectrolytes and polyampholytes. *The Journal of Chemical Physics*, *152*, 045102 https://doi.org/10.1063/1.5139661.

Michnick, S. W., & Bergeron-Sandoval, L.-P. (2018). Why does biopolymer condensation matter? *Nature Reviews. Molecular Cell Biology*, *19*, 613–614. https://doi.org/10.1038/s41580-018-0023-0.

Mineart, K. P., Kelley, E. G., Nagao, M., & Prabhu, V. M. (2017). Processing-structure relationships of poly(ethylene glycol)-modified liposomes. *Soft Matter*, *13*, 5228–5232. https://doi.org/10.1039/c7sm00960g.

Mountain, G. A., & Keating, C. D. (2020). Formation of multiphase complex coacervates and partitioning of biomolecules within them. *Biomacromolecules*, *21*, 630–640. https://doi.org/10.1021/acs.biomac.9b01354.

NCNR VSANS. (2020). [WWW Document]. URL. https://www.nist.gov/ncnr/chrns-vsans-very-small-angle-neutron-scattering.

NIST SLD Calculator. (2020). [WWW Document]. URL. https://www.ncnr.nist.gov/resources/activation/.

Sadman, K., Delgado, D. E., Won, Y., Wang, Q., Gray, K. A., & Shull, K. R. (2019). Versatile and high-throughput polyelectrolyte complex membranes via phase inversion. *ACS Applied Materials & Interfaces*, *11*, 16018–16026. https://doi.org/10.1021/acsami.9b02115.

SASView Version 5.01. (2020). [WWW Document]. URL. http://www.sasview.org/.

Spruijt, E., Leermakers, F. A. M., Fokkink, R., Schweins, R., van Well, A. A., Stuart, M. A. C., et al. (2013). Structure and dynamics of polyelectrolyte complex coacervates studied by scattering of neutrons, X-rays, and light. *Macromolecules*, *46*, 4596–4605. https://doi.org/10.1021/ma400132s.

Srivastava, S., & Tirrell, M. V. (2016). Polyelectrolyte complexation. In S. A. Rice & A. R. Dinner (Eds.), *Vol. 161, Advances in chemical physics* (pp. 499–544). Malden: Wiley-Blackwell.

Wu, C., & Li, Y. (2018). Universal scaling of phase diagrams of polymer solutions. *Macromolecules*, *51*, 5863–5866. https://doi.org/10.1021/acs.macromol.8b01234.

Zook, J. M., & Vreeland, W. N. (2010). Effects of temperature, acyl chain length, and flow-rate ratio on liposome formation and size in a microfluidic hydrodynamic focusing device. *Soft Matter*, *6*, 1352–1360. https://doi.org/10.1039/b923299k.

CHAPTER TEN

Incorporation of proteins into complex coacervates

Whitney C. Blocher McTigue[a] and Sarah L. Perry[a,b,*]
[a]Department of Chemical Engineering, University of Massachusetts Amherst, Amherst, MA, United States
[b]Institute for Applied Life Sciences, University of Massachusetts Amherst, Amherst, MA, United States
*Corresponding author: e-mail address: perrys@engin.umass.edu

Contents

Abstract

Complex coacervates have found a renewed interest in the past few decades in various fields such as food and personal care products, membraneless cellular compartments, the origin of life, and, most notably, as a mode of transport and stabilization of drugs. Here, we describe general methods for characterizing the phase behavior of complex coacervates and quantifying the incorporation of proteins into these phase separated materials.

Methods in Enzymology, Volume 646
ISSN 0076-6879
https://doi.org/10.1016/bs.mie.2020.06.006

1. Introduction

The encapsulation of proteins and other biomacromolecules is an area of tremendous activity, as such materials are finding increasing utility in applications such as drug delivery, environmental remediation, personal care products and biocatalysis. Proteins are generally very sensitive to their environment, and typical methods used for encapsulation can decrease or even destroy the activity of these molecules. Complex coacervation is a method that can be used to encapsulate proteins without using harsh conditions that may denature the protein cargo. This method of sequestration is a viable platform for a variety of different areas such as food science (Schmitt & Turgeon, 2011; Yeo, Bellas, Firestone, Langer, & Kohane, 2005), personal care products (Carvalho, Estevinho, & Santos, 2016; Martins, Barreiro, Coelho, & Rodrigues, 2014), and medicine (Kuo et al., 2014), because of the ability to generate biocompatible formulations and drive high levels of encapsulation without the need for organic solvents (Black et al., 2014; Kishimura, Koide, Osada, Yamasaki, & Kataoka, 2007; Vehlow et al., 2016; Water et al., 2014).

1.1 Complex coacervation and protein incorporation

Complexation occurs when oppositely-charged polyelectrolytes interact under favorable conditions such that the electrostatic attraction and entropic gains can drive phase separation. Coacervation is a purely aqueous strategy that can also be leveraged for the triggerable release protein cargo (Lim, Ping, & Miserez, 2018; Lindhoud, de Vries, Schweins, Cohen Stuart, & Norde, 2009); Lindhoud, Voorhaar, et al., 2009. These materials are versatile and have been shown usable for various delivery techniques such as an injectable protein carrier (Johnson & Wang, 2013, 2014; Nishida, Tamura, & Yui, 2018) and for oral delivery (Bourganis, Karamanidou, Kammona, & Kiparissides, 2017; Chapeau et al., 2017). This method will focus on using complex coacervation as an aqueous protein encapsulation technique, and experimental strategies related to characterizing the concentration of protein present in such formulations.

1.2 Complex coacervate phase behavior

There are many variables that alter the ability of polyelectrolytes to undergo complex coacervation, including the salt concentration, pH, charge density and chemistry of the polymers, etc. Here, we will describe an experimental strategy for characterizing aspects of the phase behavior of complex coacervates, without consideration for other formulation-relevant questions such as the size and temporal stability of a dispersion of coacervate droplets etc. One consequence of this experimental focus is the use of concentration on an ionizable monomer basis, rather than units of mass of polymer per volume that are more typical when considering polymeric materials. Furthermore, this discussion assumes that the coacervate materials in question are able to fully equilibrate (i.e., form liquid droplets, rather than kinetically-trapped gels or solid complexes).

Complex coacervation involves the interaction of oppositely-charged polyelectrolytes. Thus, while physical and chemical aspects of the polymer such as length, charge density and/or degree of ionization, as well as hydrophobicity can modulate the phase behavior, we can typically rely on general intuition regarding electrostatic effects and charge neutrality. For complex coacervation, the ability to phase separate will be maximized when an equal number of oppositely-charged groups are present. While this condition may be straightforward to predict with some simple polymer systems, factors such as pH-dependent degree of ionization can play a role. These effects are highlighted in Fig. 1A and B, which show schematic depictions of a two-dimensional phase diagram as a function of the concentration of polycation and polyanion, with the concentration of salt present in the system defining the contours. The height (i.e., salt stability) of the two-phase region is maximized along the line of equimolar charge (black line). For systems where all of the potentially ionizable monomers are charged, this results in a symmetric phase diagram (Fig. 1A). However, if the pH of the solution has been changed such that only half of the ionizable groups on one of the polymers are charged (e.g., the apparent pKa of the polycation), then the phase diagram becomes asymmetric, as in Fig. 1B.

These same trends can be observed through a "stoichiometry" experiment, which typically uses turbidity to determine the polymer ratio that gives maximum coacervate yield (Chollakup, Beck, Dirnberger, Tirrell, & Eisenbach, 2013; Chollakup, Smitthipong, Eisenbach, & Tirrell, 2010; Cummings & Obermeyer, 2018; Obermeyer, Mills, Dong, Flores, & Olsen, 2016; Perry, Li, Priftis, Leon, & Tirrell, 2014; Priftis, Megley, Laugel, & Tirrell, 2013; Priftis & Tirrell, 2012). A stoichiometry experiment varies the relative amount of polycation to polyanion while keeping the total polymer concentration constant (corresponding to the gray lines in Fig. 1A and B). For systems where the polycation and polyanion can be considered as

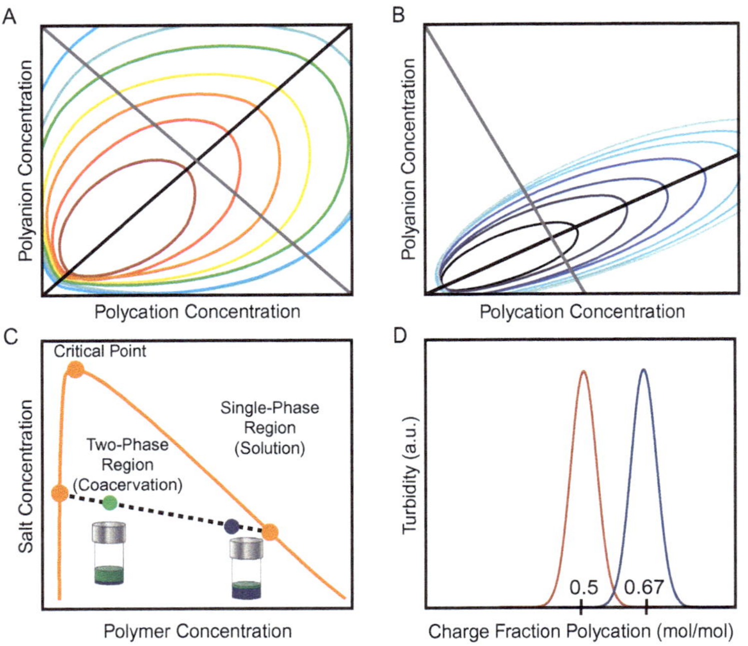

Fig. 1 A schematic contour plot of the three-dimensional phase envelope for complex coacervation as a function of relative polycation concentration, polyanion concentration, and salt concentration (shown as contours) for (A) a system where the relative degree of ionization of the polycation and polyanion is the same, and (B) a system where the relative degree of ionization for the polycation is half that of the polyanion. The black line in both plots indicates the salt-polymer phase behavior for a mixture of equal numbers of ionized cationic and anionic species, shown as the binodal curve in (C). The gray line traces out the effect of changing the charge stoichiometry of the system. (C) A schematic illustration of a typical salt vs. polymer concentration phase diagram defined by the black lines in (A and B). Coacervation occurs in the two-phase region beneath the binodal curve. A sample prepared within this two-phase region will phase separate into a polymer-dense coacervate phase and a polymer-poor supernatant phase, connected by a tie-line. While identical coacervate and supernatant phases will be formed from any sample prepared along such a tie-line, the relative position on the line dictates the fraction of the resulting sample that will be coacervate vs. supernatant, as per the lever rule. A sample prepared at relatively high polymer concentration (blue dot) will produce a much larger volume of coacervate than one prepared at lower polymer concentrations (green). The relative volumes of coacervate (blue) and supernatant (green) are indicated in the depicted vials. (D) Schematic depiction of the results of a stoichiometry experiment, which tests the effect of changing the relative amounts of polycation and polyanion at constant total polymer concentration and constant solution conditions. The turbidity signal maps out conditions where phase separation occurs. For the system where the degree of ionization of the polymeric species is the same, a maximum in turbidity is observed at a charge fraction of 0.5 (red, corresponding to the gray line in (A)). This result shifts to a cationic charge fraction of 0.67 if the degree of ionization of the polycation is half that of the polyanion (blue, corresponding to the gray line in (B)).

fully ionized (or have equal levels of ionization), the peak in the turbidity would be expected at a mole fraction of 0.5 with respect to one of the polymer species, or a 1:1 equimolar ratio (Fig. 1D, red curve). However, if one of the polymer species is only half charged, this will result in a shift in the observed signal. For instance, in the example shown in Fig. 1B where only half of the monomers on the polycation are ionized, a turbidity peak would be observed at a charge fraction of 0.67, corresponding to the condition where two cationic monomers are needed for every one anionic monomer, or a 1:2 ratio (Fig. 1D, blue curve). This condition of charge neutrality identifies the point where the maximum number of polymer chains will be incorporated into the coacervate phase; complexation will still occur at off stoichiometric conditions, but with a decreased level of coacervation. From a theoretical perspective, the composition of the bulk coacervate phase should be the same, with the difference being purely one of yield. However, experimentally, the preparation of off-stoichiometry dispersions of coacervate droplets can result in the recruitment of excess polymer to the surface of the droplet, imparting colloidal stability (Perry et al., 2014; Priftis & Tirrell, 2012).

For a given ratio of polycation-to-polyanion, the phase diagram for coacervation is then shown as a one-dimensional binodal curve, typically as a function of salt and total polymer concentration (Fig. 1C). This binodal curve represents the slice through the larger two-dimensional phase space at constant polymer composition (i.e., the black line). Samples prepared at a composition within this two-phase region will phase separate into two liquid phases, the polymer-rich coacervate phase and the polymer-poor supernatant phase. The composition of the resulting coacervate and supernatant phases is defined by the tie-line that connects these two points. One interesting (and potentially unintuitive) result of this phase behavior is that increasing the amount of polymer present in the initial sample mixture will not result in a commensurate increase in the polymer concentration in the resulting coacervate phase. Instead, samples prepared at different points along the tie-line will result in samples with different quantities of the same coacervate phase, as defined by the lever rule. Samples prepared at higher polymer concentrations will result in a larger coacervate volume, and vice versa.

There are numerous strategies for weakening or completely overcoming the interactions that cause complex coacervation. For example, changes in the solution pH or the addition of an excess of one of the polymers can shift the solution conditions outside of the two-phase window (Comert & Dubin, 2017; Kaibara, Okazaki, Bohidar, & Dubin, 2000). As suggested by Fig. 1A–C, ionic strength is also an important variable for controlling coacervation (Lindhoud, de Vries, et al., 2009; Lindhoud, Voorhaar, et al., 2009; Perry et al., 2014). The addition of salt can facilitate screening of the electrostatic interactions and reduces the entropic gains associated with complexation (Perry et al., 2014; Yan et al., 2013). However, it is important to define the conditions for which this salt dissolution is being defined. As shown in Fig. 1C, the *critical point* is the highest salt concentration for which phase separation can be observed. However, most experiments do not operate near the critical condition and are therefore interested in determining the concentration of salt above which phase separation is no longer observed for a given sample condition. This salt concentration is typically referred to as the salt resistance (Li et al., 2018; Madinya, Chang, Perry, & Sing, 2020), and is dependent upon the choice of polymers, the coacervate composition, and the identity of the salt used. As was discussed regarding the effects of polymer concentration on coacervate composition, the salt resistance for any sample prepared along a given tie-line will be the same. However because tie-lines for coacervation tend to be non-horizontal (i.e., there is preferential partitioning of salt out of the coacervate phase), it is possible to change the salt resistance by changing polymer concentration, although this usually requires a significant change so as to move off of one tie-line and onto another (Chang et al., 2017). In the context of encapsulation studies, the salt resistance is an important parameter because it determines the concentration of salt required to dismantle the coacervate.

This introduction is intended to provide a foundational understanding of coacervate phase behavior to facilitate the use of coacervation for protein encapsulation. Thus far, our discussion has focused on complex coacervates formed from two species, a polycation and a polyanion. While the addition of protein does not alter these design rules, the chemical complexity of proteins can make interpretation of experimental data more challenging. For instance, most proteins carry a mixture of positive and negative charges. While electrostatic intuition would dictate that we consider only the net charge of a protein, there are examples where clustering of charges has allowed for complex coacervation to occur "on the wrong side of the isoelectric point," such that it would *appear* that complexation is occurring between two species of the same charge (Comert, Malanowski, Azarikia, & Dubin, 2016; Cooper, Dubin, Kayitmazer, & Turksen, 2005; Xu, Liu, Faisal, Si, & Guo, 2017; Xu, Mazzawi, Chen, Sun, & Dubin, 2011). It is also common to explore coacervation between a protein and two oppositely-charged polymers, particularly for cases where the protein of interest is only weakly charged (Blocher McTigue & Perry, 2019; Lindhoud & Claessens, 2016; Lindhoud, de Vries, et al., 2009; Lindhoud, Voorhaar, et al., 2009). In these cases it is necessary to consider the net charge of all three species, and to carefully balance the ionic strength of the system, as changes in the salt concentration will disfavor the incorporation of the more weakly charged protein in favor of stronger electrostatic interactions between the two more strongly charged polymers.

In the following sections we will discuss experimental strategies to characterize the coacervation phase behavior of a system of two oppositely-charged polymers containing a protein cargo, as well as the incorporation of protein into the coacervate phase. These methods are intended to be general, and can be adapted to fit specific situations, such as the complexation of a protein with only a single, oppositely-charged polymer.

2. Materials, equipment, and reagents

The materials required for these experiments include two oppositely-charged polyelectrolytes, a protein of interest, salt, buffer (if desired), and acid/base for pH adjustments. We recommend that all solutions of polymer, protein, buffer (if desired) and salt be adjusted to the same pH. Coomassie brilliant blue G-250 dye can be purchased alone or as part of a Bradford assay kit.

Samples will be prepared via pipetting in microcentrifuge tubes and transferred into well plates (96- or 384-well plates are common). We recommend the use of a vortex mixer during sample preparation, an optical microscope for sample visualization, and a plate reader with UV/vis spectrophotometry capabilities. It is also possible to perform samples in cuvettes using a UV/vis spectrophotometer. Turbidity measurements are typically performed at a wavelength of light in the middle of the visible spectrum, and away from the absorbance peak of any of the materials (~562 nm is common). For quantifying the concentration of protein present, the Coomassie dye used in the Bradford assay is analyzed at a wavelength of 595 nm, and the experiment can be performed in a well plate, cuvette, or other small volume UV/vis setup (e.g., NanoDrop). The use of absorbance at 280 nm requires the use of either UV-compatible cuvettes or a NanoDrop-type setup.

3. Safety considerations

These techniques are safe to perform in a standard laboratory setting with the use of appropriate personal protection equipment such as safety glasses, lab coats, and gloves. It is recommended that experiments involving Coomassie dye be performed inside a chemical fume hood. For specific materials, refer to the safety data sheets.

To keep solutions sterile and dust free, keep the lids on all reagents, removing the lid only when pipetting, placing the cap back each time, though the lid may remain loose. Similarly, we recommend keeping all microcentrifuge tubes closed except when adding solution or transferring samples. The presence of dust can alter turbidity results. Care should also be taken to avoid cross-contamination of samples from pipette tips.

4. Protocols

4.1 Characterizing coacervate phase behavior

While the ultimate goal of an experiment might be the encapsulation of a target protein, we recommend first characterizing the phase behavior of your coacervate system in terms of charge stoichiometry and salt resistance. These information will help in the planning of experiments related to protein encapsulation and will facilitate the interpretation of the resulting protein encapsulation data. These experiments are typically performed at relatively low concentrations of protein and polymer to limit reagent requirements. All experiments can be scaled up in terms of volumes and/or concentrations, though it is important to ensure that all samples are fully mixed and equilibrated.

4.2 Polymer-polymer stoichiometry experiments

Stoichiometry experiments examine coacervate formation as a function of the ratio of polycation to polyanion at constant polymer concentration. Thus, while polymer stock solutions can be prepared at any concentration, we recommend the use of ionizable monomer concentration on a molar basis. The use of monomer concentration circumvents issues with polymer polydispersity, and allows for the easy analysis of results in terms of the stoichiometry of electrostatic interactions. A stock solution concentration of 10 mM monomer is generally sufficient for turbidity experiments.

A typical stoichiometry experiment will span the range of possible charge fractions, which we will express in terms of the mole fraction of ionizable monomers of the polycation present in our sample, to observe both a peak in the data, and clear baselines. The data in Figs. 3 and 4 and the experimental recipes listed in Tables 1 and 2 span the range of 0.1–0.9. These data points can be equally spaced for initial experiments. However, once the location of the turbidity peak is known (or if its location is estimated based on the charge state of the polymers), it is useful to sample the concentration space around the peak more closely. Additionally, it is important to ensure that the final polymer concentration in the prepared samples results in a sufficiently high level of turbidity. This signal should be distinguishable

Table 1 Sample preparation for complexation between poly(L-lysine) (K_{50}) and poly(D,L-glutamate) (E_{50}), degree of polymerization $N=50$, pH 7.0.

	Charge fraction $K_{50}(+)$	Volume 10 mM $E_{50}(-)$ (μL)	Volume 10 mM $K_{50}(+)$ (μL)	Volume water added (μL)
1	0.100	10.8	1.2	108.0
2	0.200	9.6	2.4	108.0
3	0.300	8.4	3.6	108.0
4	0.400	7.2	4.8	108.0
5	0.425	6.9	5.1	108.0
6	0.450	6.6	5.4	108.0
7	0.475	6.3	5.7	108.0
8	0.500	6.0	6.0	108.0
9	0.525	5.7	6.3	108.0
10	0.550	5.4	6.6	108.0
11	0.575	5.1	6.9	108.0
12	0.600	4.8	7.2	108.0
13	0.700	3.6	8.4	108.0
14	0.800	2.4	9.6	108.0
15	0.900	1.2	10.8	108.0
Blank	–	0.0	0.0	120.0

Components were added from right to left as outlined in the protocol. The final monomer concentration is 1 mM.

Table 2 Sample preparation for a salt curve for between poly(L-lysine) (K_{50}) and poly(D,L-glutamate) (E_{50}), degree of polymerization $N = 50$, in 10 mM HEPES buffer pH 7.0.

	Charge fraction $K_{50}(+)$	Volume 10 mM $E_{50}(-)$ (μL)	Volume 10 mM $K_{50}(+)$ (μL)	Volume 2 M NaCl added (μL)	Volume water added (μL)
1	0.500	6.0	6.0	0.0	108.0
2	0.500	6.0	6.0	1.5	106.5
3	0.500	6.0	6.0	3.0	105.0
4	0.500	6.0	6.0	4.5	103.5
5	0.500	6.0	6.0	6.0	102.0
6	0.500	6.0	6.0	9.0	99.0
7	0.500	6.0	6.0	12.0	96.0
8	0.500	6.0	6.0	18.0	90.0
9	0.500	6.0	6.0	21.0	87.0
10	0.500	6.0	6.0	24.0	84.0
11	0.500	6.0	6.0	27.0	81.0
12	0.500	6.0	6.0	30.0	78.0
13	0.500	6.0	6.0	36.0	72.0
14	0.500	6.0	6.0	42.0	66.0
15	0.500	6.0	6.0	48.0	60.0
16	0.500	6.0	6.0	54.0	54.0
Blank	–	0.0	0.0	0.0	120.0

Components were added from right to left as outlined in the protocol. The final monomer concentration is 1 mM.

from that of a blank solution at the same salt/buffer concentration, but in the absence of polymer. In our experience, samples should be prepared at a final concentration of at least 1 mM (with respect to the total number of monomers), though this threshold concentration is a function of the path length through the sample.

To facilitate the preparation of fully equilibrated coacervate samples, we recommend that any salt, buffer, and excess water are combined first, followed by the addition of one of the polymers. It is then important to ensure that the sample is well mixed (e.g., vortexing for 5–10 s) before the second polymer is added, and that the sample is mixed again after the addition of the second polymer. It is also possible to prepare the polymer stock solutions at a specified concentration of salt/buffer, thereby eliminating the need to add these components separately, though this approach is less flexible in terms of adjusting experimental parameters. Once samples have been prepared, they should be mixed well and transferred to a well plate or cuvette for turbidity analysis. An example turbidity result for a stoichiometry experiment between poly(L-lysine) and poly(D,L-glutamate), degree of polymerization $N = 50$, pH 7.0 is shown in Fig. 2A. The coacervate samples, as formed, should be a dispersion of droplets that gives the sample a cloudy and possibly opalescent appearance. At this point, the time between sample preparation and analysis is an important consideration, as the coacervate droplets can coalesce and settle over time and thus give variable turbidity readings. Generally, the turbidity signal should not be sensitive to differences of a few minutes, though this time scale can vary significantly based on the identity of the polymer system and the solution conditions.

While all experimental results should be replicated in order to ensure reproducibility (i.e., "technical" replicates), we also recommend that samples be prepared in such a way that each sample can be split and analyzed separately (a repeat measurement of the sample, similar to a "biological" replicate). The use of three replicates/repeats will allow for the statistical analysis of the resulting data. In terms of sample preparation, the total volume of sample prepared for a repeated measurement should be $>3\times$ the volume

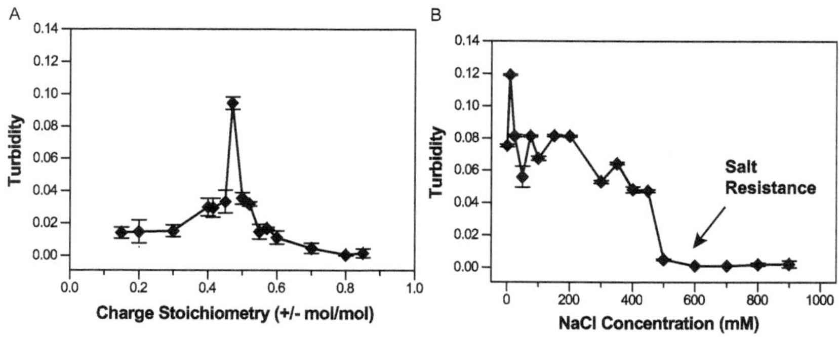

Fig. 2 Turbidity data from (A) stoichiometry experiments involving poly(L-lysine) and poly(D,L-glutamate) with a degree of polymerization $N = 50$, pH 7.0 with no added buffer and (B) salt resistance experiments as a function of increasing NaCl concentration. The final monomer concentration was 1 mM. Lines connecting the data points are a guide for the eye.

needed for the three individual samples. For example, 35 μL of sample is needed for a turbidity measurement using a 384-well plate. However, to ensure that sufficient sample volume is available for pipetting, a total sample volume of 120 μL might be prepared.

After turbidimetry, samples should be inspected visually using an optical microscope. Standard brightfield microscopy with a 40 × objective is typically sufficient for this procedure, though more advanced techniques such as phase contrast or differential interference contrast (DIC) can enhance the ability to visualize samples. For samples prepared at low polymer concentration, it may be difficult to distinguish very small droplets. The size of these coacervate droplets can be increased either by increasing the polymer concentration present in the sample, or by allowing the sample more time for the droplets to coalesce. The main goals in visualizing coacervate samples are to confirm the liquid vs. solid nature of the resulting materials, and provide secondary confirmation of trends (e.g., the presence or absence of coacervates) suggested via turbidity. Coacervate droplets should appear as circular/spherical structures either floating in solution or adhered onto a surface, whereas solid precipitation typically appears as fractal aggregates (Fig. 3).

An important consideration in the preparation of coacervates is whether they are fully equilibrated. This question can be answered by testing whether or not the results of an experiment are sensitive to the order of polymer addition.

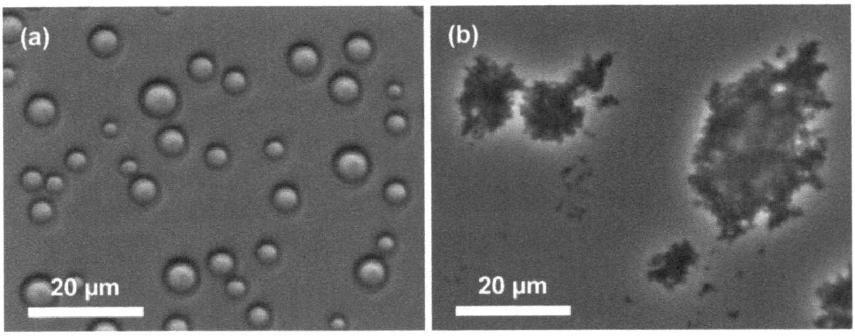

Fig. 3 Optical micrographs of (A) liquid complex coacervate droplets of poly(L-lysine) and poly(D,L-glutamate), and (B) fractal solid precipitates resulting from the interaction of poly(L-lysine) and poly(L-glutamate). All samples were prepared using polymers with degree of polymerization $N=100$ at a total monomer concentration of 6 mM in 100 mM NaCl, pH 7.0.

4.3 Polymer-polymer salt resistance experiments

The goal of salt resistance experiments is to identify the concentration of salt (for a given total polymer concentration and stoichiometric ratio) above which phase separation is no longer observed. This provides information on the location of the binodal curve, and informs the design of experimental procedures where dissolution of the coacervate phase is needed. The magnitude of the salt resistance is dependent upon the length and identity of the polymers, the choice of salt, the solution pH, etc. Generally speaking, increases in polymer length, hydrophobicity, and charge density will result in higher values of the salt resistance at the same polymer concentration.

In designing a salt resistance experiment for a new polymer, it may be necessary to perform a screening-level experiment to identify the general range of salt concentrations over which the experiment should be performed. Generally, the salt resistance can be identified as a clear decrease in the turbidity signal with increasing salt concentration. However, it is useful to combine turbidimetry measurements with direct visualization of the samples via optical microscopy to confirm results. An example of salt resistance data for the system of poly(L-lysine) and poly(D,L-glutamate) degree of polymerization $N=50$, pH 7.0 is shown in Fig. 2B.

A general step-by-step protocol for these polymer-only experiments is given below.

4.4 Experimental protocol for coacervate samples (no protein)

1. Set up microcentrifuge tubes for each sample and a blank, labeling appropriately
2. Pipette the appropriate amount of water into each tube (as needed)
3. Pipette the appropriate amount of buffer solution into each tube (as needed)
4. Pipette the appropriate amount of salt solution into each tube (as needed)
5. Pipette the appropriate amount of the first polyelectrolyte solution into each tube
6. Vortex each tube for 5–10 s
7. Pipette the appropriate amount of the second polyelectrolyte solution into each tube, vortexing for 5–10 s immediately after each addition
8. Transfer aliquots of each sample to the well plate or cuvette for turbidity analysis
9. Inspect each sample via optical microscopy

4.5 Polymer-polymer-protein stoichiometry experiments

The design of a stoichiometry experiment for a system that combines protein in the presence of a polycation and polyanion varies only slightly in design and intent. While more extensive experiments that vary the relative amounts of each polymer and the quantity of protein can be performed to map out the complete phase behavior of this more complex system, simpler experiments that consider the effect of polymer charge stoichiometry in the presence of a constant level of protein can also be performed (see Table 3 for an example recipe). Here, we will discuss the design of these

Table 3 Sample preparation for complexation between poly(L-lysine) (K_{50}) and poly(D,L-glutamate) (E_{50}), degree of polymerization $N=50$, in 10 mM HEPES buffer pH 7.0 with bovine serum albumin (BSA).

	Charge fraction $K_{50}(+)$	Volume 10 mM $E_{50}(-)$ (μL)	Volume 10 mM $K_{50}(+)$ (μL)	Volume 2 mg/mL BSA (−) added (μL)	Volume 0.5 M HEPES added (μL)	Volume water added (μL)
1	0.100	151.2	16.8	6.0	4.80	61.2
2	0.200	134.4	33.6	6.0	4.80	61.2
3	0.300	117.6	50.4	6.0	4.80	61.2
4	0.400	100.8	67.2	6.0	4.80	61.2
5	0.425	96.6	71.4	6.0	4.80	61.2
6	0.450	92.4	75.6	6.0	4.80	61.2
7	0.475	88.2	79.8	6.0	4.80	61.2
8	0.500	84.0	84.0	6.0	4.80	61.2
9	0.525	79.8	88.2	6.0	4.80	61.2
10	0.550	75.6	92.4	6.0	4.80	61.2
11	0.575	71.4	96.6	6.0	4.80	61.2
12	0.600	67.2	100.8	6.0	4.80	61.2
13	0.700	50.4	117.6	6.0	4.80	61.2
14	0.800	33.6	134.4	6.0	4.80	61.2
15	0.900	16.8	151.2	6.0	4.80	61.2
Blank	0.000	0.0	0.0	0.0	4.80	235.2

Components were added from right to left as outlined in the protocol. The final monomer concentration is 7 mM and the final protein concentration is 50 μg/mL.

simpler experiments and how the addition of a protein as a third charged macromolecule can alter the phase behavior of the system.

Our initial stoichiometry experiments with just the two-polymer system served to identify the composition corresponding to maximum coacervate yield, corresponding to a charge-neutral mixture of the two polymers. However, the addition of a charged protein would be expected to shift this optimal condition. As can be seen in Fig. 4A, the addition of negatively-charged bovine serum albumin (BSA) to a mixture of poly(L-lysine) and poly(D,L-glutamate) degree of polymerization $N=50$, in 10 mM HEPES, pH 7.0 results in a shift in the resulting turbidity signal to higher mole fractions of the polycation. A shift in the opposite direction, to "net negative" conditions would be expected for positively-charged proteins (Blocher McTigue & Perry, 2019; Lindhoud & Claessens, 2016; Lindhoud, Norde, & Cohen Stuart, 2009).

As for the two-polymer system, it is important to determine whether the order of mixing has any effect on the resulting coacervates.

4.6 Polymer-polymer-protein salt resistance experiments

Salt resistance experiments can also be performed for samples including proteins. However, most proteins would be expected to have a lower charge content and charge density than the associated polymers. As such, it is a

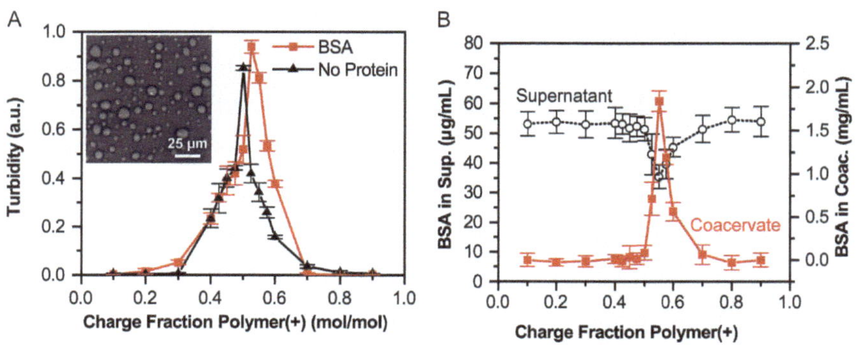

Fig. 4 (A) Turbidity data from stoichiometry experiments involving poly(L-lysine) and poly(D,L-glutamate) with a degree of polymerization $N=50$ (black) and for the same system with the anionic protein bovine serum albumin (BSA) (red) in 10 mM HEPES, pH 7.0. The inset optical micrograph shows the formation of coacervate droplets with BSA. (B) The corresponding concentration of protein in the supernatant (black) and coacervate (red) phases, as determined using a Bradford assay. *Blocher McTigue, W. C., Perry, S. L. (2019). Design rules for encapsulating proteins into complex coacervates.* Soft Matter, 15, *3089–3103. Reproduced by permission of The Royal Society of Chemistry.*

reasonable assumption that the salt resistance of the three-macromolecule system should be lower than that of the polymer-only system.

A general step-by-step protocol for these polymer-polymer-protein experiments is given below.

4.7 Experimental protocol for coacervate samples (with protein)

1. Set up microcentrifuge tubes for each sample and a blank, labeling appropriately
2. Pipette the appropriate amount of water into each tube (as needed)
3. Pipette the appropriate amount of buffer solution into each tube (as needed)
4. Pipette the appropriate amount of salt solution into each tube (as needed)
5. Pipette the appropriate amount of the first polyelectrolyte solution into each tube
6. Vortex each tube for 5–10 s
7. Pipette the appropriate amount of the protein solution into each tube, vortexing for 5–10 s immediately after each addition
8. Pipette the appropriate amount of the second polyelectrolyte solution into each tube, vortexing for 5–10 s immediately after each addition
9. Transfer aliquots of each sample to the well plate or cuvette for turbidity analysis
10. Inspect each sample via optical microscopy

4.8 Quantifying protein incorporation into complex coacervates

While turbidity and optical microscopy can be used to determine whether complexation has occurred or not, these measurements do not provide information on the incorporation of protein into the coacervate phase. Instead, separate measures of the protein concentration in the coacervate and supernatant phases must be made, along with a determination of the volume of each phase.

Once the concentrations for both the coacervate and supernatant phases are known, several other parameters may be determined. The first is the encapsulation efficiency (*EE*), which is the percentage of cargo (by mass) sequestered, in this case, by complex coacervates. Measuring the supernatant volume and multiplying by the concentration of protein in the supernatant phase gives the mass of cargo in the supernatant. Subtracting this mass from the total mass of protein added to the system during sample preparation allows for calculation of the mass of protein in the coacervate phase. These masses can then be used to calculate the encapsulation efficiency:

$$EE = \frac{m_{coac.\ protein}}{m_{total\ protein}} x100\% \tag{1}$$

A partition coefficient, on the other hand, is typically defined as the ratio of the concentration of protein in the dense coacervate phase over the concentration of the cargo in the dilute phase, written as:

$$K = \frac{\left[cargo_{coac.\ protein}\right]}{\left[cargo_{sup.\ protein}\right]} \tag{2}$$

Finally, loading or loading capacity is the amount of protein in the coacervate compared to the total mass of the coacervate, i.e., it describes what fraction of the coacervate phase is comprised of the cargo:

$$LC = \frac{m_{coac.\ protien}}{m_{coac.\ total}} x100\% \tag{3}$$

Below, we will detail strategies for assaying the protein concentration in coacervate samples using a colorimetric Bradford assay, and via direct measurement of protein absorbance.

4.9 Bradford assay

The protein quantification experiments described here are intended to be run in parallel with a stoichiometry-type experiment, as described above, with half of the total sample volume used for turbidity experiments, and the other half used for protein quantification. As in the case of turbidity experiments, it is important to measure both repeat and replicate samples in order to allow for statistical analysis of the results.

The Coomassie brilliant blue G-250 dye used in the Bradford assay is typically described as interacting with basic amino acids in hydrophobic pockets, but primarily responds to arginine residues, as well as histidine, tryptophan, tyrosine, and phenylalanine to a lesser extent (Bio-Rad Laboratories, n.d.; Olson & Markwell, 2007; Stoscheck, 1990). Typically, users buy a kit that may come with protein standards (e.g., bovine serum albumin, BSA) to aid in the creation of a calibration curve. However, individual proteins interact differently with the Coomassie dye, based on their amino acid sequence. Therefore, we recommend creating calibration curves directly with the protein of interest. Furthermore, all calibration curves must be prepared at the solution conditions (i.e., pH, salt concentration, etc.) expected in the final sample. One consequence of this is that separate calibration curves are required for supernatant samples, which can be measured directly, and coacervate samples, which must be performed at a higher salt concentration in order to dismantle the coacervate (Fig. 5).

In addition to considering the specific interactions of the Coomassie dye with the protein at different solution conditions, it is critical to determine the potential background signal that might result from the dye interacting with the polymers used in the coacervate. While the interaction of the dye with poly(L-lysine) and poly(D,L-glutamate) has been shown to be minimal (Blocher McTigue & Perry, 2019), much stronger interactions occur when more hydrophobic polymers such as poly(styrene sulfonate) (PSS), poly(diallyldimethylammonium chloride) (PDADMAC), and methacrylate-based polyelectrolytes. The interaction between polymer and dye can be tested by mixing a 1:1 volume ratio of the dye with a solution of polymer at the concentration intended for use in experiments. Qualitatively, a strong interaction between the dye and the polymers can be observed visually as a color change from brackish-brown to blue. Quantitatively, it is important to consider whether the level of background signal resulting from the polymers in the coacervate will swamp the potential absorbance signal from the protein by itself. This comparison can be done via an absorbance measurement at 595 nm. While it is possible to perform a background subtraction to account for signal associated with the polyelectrolytes, this correction can become complicated if the concentration of each of the polymers varies across different samples.

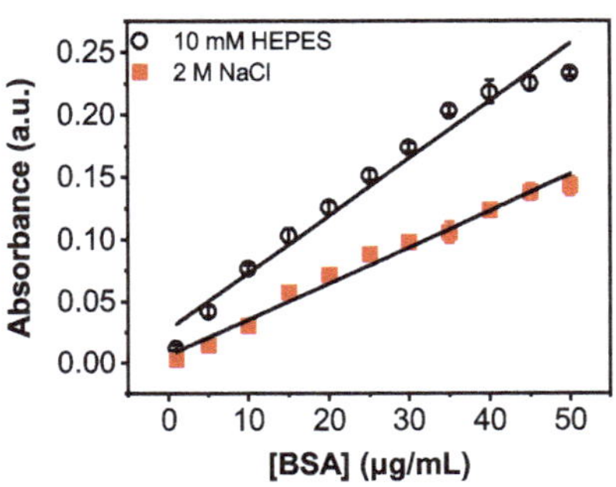

Fig. 5 Standard curves for BSA for a Bradford assay in 10 mM HEPES pH 7.0 (black open circles) with a 1:1 dye to sample ratio and 2 M NaCl (red squares) with a 1:1 dye to sample ratio corresponding to the supernatant and coacervate phases. The black lines represent linear fits.

The linear range of the Bradford assay is a function of the specific interaction of the protein with the Coomassie dye, and the volumetric scale (and thus the absorbance path length) at which the experiment is performed. Different scale protocols have been developed for the Bradford assay depending on the desired sample volumes and the range of concentrations. The main difference in the protocols is the ratio of Coomassie dye solution to protein sample. For example, the "standard" Bradford protocol uses a 50:1 ratio of dye to sample, and has a typical range of ~125–1000 μg/mL. In contrast, the "micro assay" uses a 1:1 ratio of dye to sample and has a linear range of approximately 1–10 μg/mL (Bio-Rad Laboratories, n.d.). These ratios, as well as the sample path length can be tuned to optimize a protocol for a specific protein target and/or solution condition. For more information on the Bradford assay, we encourage readers to refer to Bio-Rad Laboratories (n.d.) and Olson and Markwell (2007).

For the example data shown in Fig. 4 and Table 3, we assayed the uptake of bovine serum albumin (BSA) into coacervates formed from poly(L-lysine) and poly(D,L-glutamate) with a degree of polymerization $N=50$, in 10 mM HEPES, pH 7.0. The experiment tested the incorporation of 50 μg/mL BSA as a function of the charge stoichiometry of the two polymers. A "micro assay" style Bradford assay was used to quantify the protein concentration in both the coacervate and the supernatant phases. Samples were prepared at a total volume of 240 μL, 105 μL of which was used for turbidity measurements (3 repeat samples of 35 μL each), and 115 μL of which was used for protein quantification. After separating the coacervate and supernatant, protocols were developed separately for the two phases.

Having measured the total volume of the supernatant for each sample to facilitate future mass balance calculations, 115 μL of supernatant was then aliquoted into a clean microcentrifuge tube and combined with an equal volume (115 μL) of 1 × Coomassie dye. Three repeat aliquots of 35 μL each were then pipetted into a 384-well plate for absorbance measurements. The absorbance data were converted to concentration values using a standard curve with samples prepared in 10 mM HEPES, pH 7.0.

For the coacervate samples, visual inspection after centrifugation had determined that the volume of coacervate was very small (~0–2.2 μL). We then added 70 μL of 2.0 M NaCl to disassemble the coacervate. This sample was then mixed with an equal volume (70 μL) of 1 × Coomassie dye. For these samples, we chose to neglect the volume of our coacervate in subsequent concentration calculations as an error of ~0–2.2 μL represents a ~3% volume error. Three repeat aliquots of 35 μL each were then pipetted into a 384-well plate for absorbance measurements, as above. The absorbance data for coacervate samples were converted to concentration using a standard curve with samples prepared at a concentration of 2 M NaCl.

The data in Fig. 4A show a strong peak in the turbidity at a mole fraction of polycation of approximately 0.525. This peak in the turbidity data corresponds directly with a strong increase in the concentration of protein present in the coacervate phase (to nearly 2.0 mg/mL, or 2000 μg/mL), and a corresponding decrease in the protein from the supernatant (Fig. 4B). In contrast, for samples at the extremes of the stoichiometric range where no phase separation occurred, we recovered the expected value of 50 μg/mL BSA that was input into the system.

4.10 Experimental protocol for protein quantification using the Bradford assay

1. Perform steps 1–10 as described above for "Experimental protocol for coacervate samples (with protein)." Samples should be prepared at a volume scale 2 × the quantity needed for a turbidity experiment
2. Centrifuge samples to phase separate the coacervate and supernatant phases for 20 min at 14,000 rpm (18,800 × *g*) at 15 °C
3. Use a pipette to transfer the supernatant into a new microcentrifuge tube carefully, noting the volume

   In many instances, such as the samples described in Table 3, it is difficult to directly obtain accurate measurements of coacervate volumes that can be on the order of ~1 μL. To circumvent this issue, we can estimate this volume by subtracting the measured value of the supernatant volume from the measured total sample volume after formulation, but before centrifugation.
4. Pipette a sufficient volume of a concentrated stock solution of salt (e.g., 2 M NaCl) to cause the dissolution of the coacervate phase and create a large enough volume sample to run multiple aliquots for the Bradford assay. The necessary concentration of salt can be determined by a salt resistance experiment, as described above
5. Pipette the necessary volume of Coomassie dye to the supernatant and coacervate samples
6. Vortex each tube for 5–10 s
7. Transfer aliquots of the supernatant and coacervate samples into to the well plate or cuvette, taking care to avoid and/or remove bubbles

8. Measure the absorbance of the samples and a blank at 595 nm
9. Convert the measured absorbance values to protein concentration using the appropriate standard curve

4.11 Utilizing absorbance at 280 nm

While the Bradford assay is performed using a colorimetric readout, measuring absorbance at 280 nm (A280) requires special consideration. Many standard well plates and disposable cuvettes may not be suited for this method as many plastics absorb in the UV. Specialized plastic cuvettes, quartz cuvettes, and small volume absorbance setups such as the NanoDrop or Take3 should be used for these measurements.

The linear range of concentrations measurable by absorbance at 280 nm generally extends from ~20 to ~3000 μg/mL (Simonian, 2002). However, the specific measurable range will be a function of the extinction coefficient for a given protein, in tandem with potential background interference from other components of the solution, as well as the path length of the sample. For example, the standard range for the small volume NanoDrop system is between 0.1 and 400 mg/mL for BSA, which has an extinction coefficient of 43,824 M^{-1} cm^{-1}, assuming a molecular weight of 66.4 kDa (Thermo Scientific, 2010). However, this range can be heavily influenced by the presence of other molecules present in solution, such as salt, buffer, surfactants, etc. Table 3 lists the maximum allowable concentration for a range of common chemicals related to these absorbance measurements.

In the specific context of complex coacervate samples, it is important to test whether or not the polyelectrolytes used to form the coacervate absorb at 280 nm. While it is possible to perform a background subtraction to account for signal associated with the polyelectrolytes, this correction can become complicated if the concentration of each of the polymers varies across different samples.

In each case, there is a known path length to use for l in Beer's equation:

$$A = c\varepsilon l \tag{4}$$

The extinction coefficient can be found using programs such as ProtParam or looking into the literature (Gasteiger et al., 2005). If there is no known extinction coefficient and no sequence, a standard assumption is that an absorbance reading of 1.0 is equal to 1 mg/mL based on a 0.1% or 1 mg/mL protein concentration producing an absorbance at 280 nm of 1.0 with a path length of 10 mm or 1 cm (Stoscheck, 1990).

In general, the molar extinction coefficient (in units of M^{-1} cm^{-1}) is approximated as:

$$\varepsilon = 5500W + 1490Y + 125C \tag{5}$$

where W is the number of tryptophans, Y is the number of tyrosines, and C is the number of cysteines in the protein sequence. Each number before the amino acids is the molar absorptivity at 280 nm for that residue. It is noteworthy that extinction coefficients are frequently reported using a range of different units, thus knowing the conversions is sometimes necessary for analysis. To convert to a percent extinction coefficient, which has units of $(g/100\ mL)^{-1}\ cm^{-1}$, the formula is:

$$\varepsilon_{1\%} = \frac{10\varepsilon}{MW} \tag{6}$$

where MW is the molecular weight of the protein. Another common form is the 0.1% extinction coefficient and has units of $(mg/mL)^{-1}\ cm^{-1}$. Converting to the 0.1% extinction coefficient follows:

$$\varepsilon_{0.1\%} = \frac{\varepsilon_{\%}}{10} = \frac{\varepsilon}{MW} \tag{7}$$

The use of a program to calculate protein concentration using absorbance at 280 nm may have correction calculations, appropriate reading of the literature and user guides will help determine such corrections. If it is suspected that the protein sample has nucleic contamination, Eq. 8 may be used for an approximate determination: (Olson & Markwell, 2007; Stoscheck, 1990):

$$[protein](mg/mL) = (1.55\ x\ A_{280}) - (0.76\ x\ A_{260}) \tag{8}$$

However, a potentially simpler method for protein quantification at 280 nm is to create a standard curve of the protein. This is useful for systems that employ various buffer conditions that may affect results and does not require a NanoDrop program or use of an extinction coefficient. Standard curves are made by preparing known sample concentrations at desired conditions and running them at 280 nm, subtracting out a blank. It is good for scientists, however, to understand where the math comes from if a NanoDrop or similar system is used. Note that many of these systems have additional blank subtraction and correction procedures built in.

4.11.1 Experimental protocol for protein quantification using A280

1. Perform steps 1–4 as described above for the Bradford protocol
2. Pipette an aliquot of known volume for absorbance measurements
   a. A Take3 or NanoDrop system allow for the use of ~2 µL with a known path length
   b. A UV-compatible cuvette can be used with a larger sample volume
3. Measure the absorbance of the samples and a blank at 280 nm
4. Convert the measured absorbance values to protein concentration using the appropriate standard curve
   a. Take3/NanoDrop have built in programs for 280 nm and are recommended for specific moieties, such as BSA and IgG, with appropriate extinction coefficients
   b. A standard curve may be used to determine protein concentration if extinction coefficient is unknown or the extinction coefficient can be calculated via code or from another program such as ProtParam (Gasteiger et al., 2005).

4.12 Standard curves

For all of the methods described, it is necessary to create a standard curve to determine the concentration of protein from the measured absorbance signal. While BSA is commonly used as a protein standard to generate a "universal" set of standard curves, the colorimetric signal from the Coomassie dye and/or the absorbance at 280 nm can vary widely from protein to protein, and as a function of solution conditions. Thus, we advise the creation of a standard curve for each protein at every set of experimental conditions used. Table 4 outlines limits for several common buffers, salts, etc., for the Bradford assay and 280 nm (Stoscheck, 1990). Generally speaking, measurements at 280 nm are sensitive to molecules that have double bonds between carbons or carbon and oxygen (Stoscheck, 1990), while the Bradford assay does not tolerate high concentrations of detergents (Olson & Markwell, 2007).

Based on the various limits of detection for the different methods and the anticipated concentrations of proteins, standard curves should be prepared at the relevant solution conditions for the supernatant and coacervate samples. It is recommended that standard curves have at least five points and completely span the range of interest. Example standard curves corresponding to the supernatant and coacervate samples from Fig. 4 are shown in Fig. 5.

Table 4 Concentration limits for protein assays using Coomassie dye and absorbance at 280 nm.

Concentration limits for protein assays[a]

Substance[b]		Dye[c]	280 nm[d]
Acids and bases			
	HCl	0.1 M	>1 M
	NaOH	0.1 M	>1 M
	PCA		10%
	TCA		10%
Buffers			
	Acetate	0.6 M	0.1 M
	Ammonium sulfate	1 M	>50%
	Citrate	50 mM	5%
	Glycine	0.1 M	1 M
	HEPES	100 mM	
	Phosphate	2 M	1 M
	Tris	2 M	0.5 M
Detergents			
	Brij 35		1%
	CHAPS		10%
	Deoxycholate	0.25%	0.30%
	Digitonin		10%
	Lubrol PX		10%
	Octylglucoside		10%
	SDS	0.10%	0.10%
	Triton X-100	0.10%	0.02%
	Triton X-100(R)		>10%
	Tween 20		0.30%
Reductants			
	Dithiothreitol	1 M	3 mM
	2-Mercaptoethanol	1 M	10 mM

Table 4 Concentration limits for protein assays using Coomassie dye and absorbance at 280 nm.—cont'd

Concentration limits for protein assays[a]

Substance[b]		Dye[c]	280 nm[d]
Miscellaneous			
	DNA/RNA	0.25 mg	1 μg
	DMSO		20%
	EDTA	0.1 M	30 mM
	Glycerol	100%	40%
	KCl	1 M	100 mM
	NaCl	5 M	>1 M
	Sucrose	1 M	2 M
	Urea	6 M	>1 M

[a]This table is a guide. Test buffer mixtures as described in the text. Values preceded by (<) or (>) symbols indicate that that tolerable limit for the chemical is unknown but is, respectively, less than or greater than the amount shown. Blank spaces indicate data were unavailable.
[b]PCA, Perchloric acid; TCA, trichloroacetic acid; HEPES, *N*-2-hydroxyehtylpiperazine-*N*′-2-ethanesulfonic acid; CHAPS, 3-[(3-cholamidopropyl)dimethylammino]propanesulfonic acid; SDS, sodium dodecyl sulfate; (R), reduced; DMSO, dimethyl sulfoxide; EDTA ehtylenediamine-tetraacetic acid.
[c]Values indicate the concentration of the chemical in a 25 μL sample.
[d]Values indicate the final concentration of the chemical which does not produce an absorbance of 0.5 compared to an equivalent water blank.
Values adapted from Olson, B. J. S. C., Markwell, J. (2007). *Assays for determination of protein concentration* (pp. Unit3.4–3.4.29). Hoboken, NJ: John Wiley & Sons, Inc., Vol. Chapter 3.

The best practice is to make new curves for every new cargo as they vary from protein to protein, as seen comparing the curves from Fig. 5 and Fig. 6A. Additionally, varying the ratio of dye to sample can change the linear range of a protein. In the case of hen egg white lysozyme (HEWL), a ratio of two parts dye to one part sample is used to extend the range to 50 μg/mL (Fig. 6A). If the same ratio was used as for BSA in Fig. 5, the result is no longer linear (Fig. 6B). The more dye added, the longer the linear range, this also allows for smaller sample volumes.

Table 5 shows an example recipe for a standard curve is outlined in Figs. 5 and 6 to create a curve that would place the target concentration in the middle of the range at pH 7.0 in 10 mM HEPES. The goal is to create a linear curve over the desired range of concentrations using the Bradford. A curve for both the supernatant and coacervate phases is required.

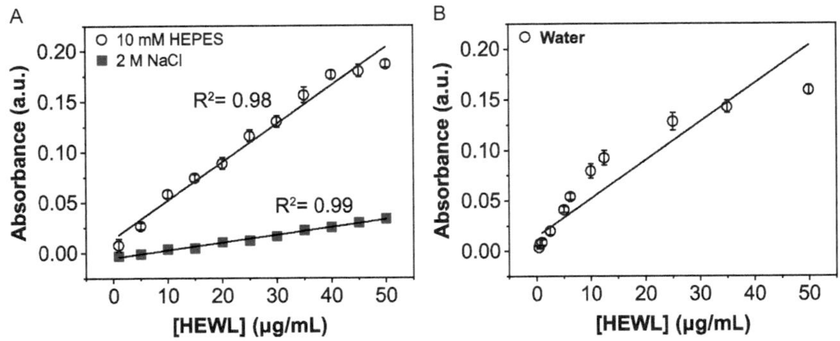

Fig. 6 (A) Standard curves for HEWL for a Bradford assay in 10 mM HEPES pH 7.0 (black open circles) with a 2:1 dye to sample ratio and 2 M NaCl (blue squares) with a 1:1 dye to sample ratio corresponding to the supernatant and coacervate phases. (B) Standard curve for HEWL for a Bradford assay in water with a 1:1 dye to sample ratio, which shows a shorter linear range. The black lines represent linear fits to the data.

Table 5 Standard curve example table for bovine serum albumin (BSA) or hen egg white lysozyme (HEWL) for the supernatant phase.

Point	(Protein) (μg/mL)	0.2 mg/mL protein (μL)	0.5 M HEPES (μL)	Water (μL)
1	1	0.6	2.4	117.0
2	5	3.0	2.4	114.6
3	10	6.0	2.4	111.6
4	15	9.0	2.4	108.6
5	20	12.0	2.4	105.6
6	25	15.0	2.4	102.6
7	30	18.0	2.4	99.6
8	35	21.0	2.4	96.6
9	40	24.0	2.4	93.6
10	45	27.0	2.4	90.6
11	50	30.0	2.4	87.6
12	55	33.0	2.4	84.6
13	60	36.0	2.4	81.6
14	65	39.0	2.4	78.6
15	70	42.0	2.4	75.6
Blank	0	0	2.4	117.6

Total solution volume is 120 μL and can be used for a 96 (120 μL/well) or 384-well plate (35 μL/well × 3).

The supernatant will have the conditions set by sample preparation, while the coacervate phase will necessitate whatever conditions are used to dismantle the dense phase, i.e., 2M NaCl. To convert Table 5 from samples related to the supernatant phase to those from the coacervate phase, replace the total volume of buffer (HEPES) and water with an equal amount of salt.

5. Analysis and statistics

To facilitate statistical analysis, repeat samples and replicate experiments should be performed at least three times. A standard *t*-test or ANOVA may be used for comparison between assay results.

Moving beyond raw data to calculated values such as the concentration of protein in the coacervate phase, the encapsulation efficiency, partitioning, etc., require propagation of error. Furthermore, propagation of error should be considered when performing baseline subtraction. To obtain an equation for the propagated error use:

$$\delta R = \sqrt{\left(\frac{\partial R}{\partial x}\delta x\right)^2 + \left(\frac{\partial R}{\partial y}\delta y\right)^2 + \ldots} \tag{9}$$

where R is the parameter whose error is being calculated and is a function of x, y, etc. The partial derivative of R for a specific independent variable is multiplied by the error of that variable and the whole quantity squared. This is done for each variable for which R is dependent and summed together. Finally, the square root of the summation is taken and that is the error for R. For example, the propagated error associated with performing a baseline subtraction between averaged values of the sample and baseline absorbance is:

$$\delta A_{avg,b} = \sqrt{\left(\frac{\partial}{\partial b}\left(A_{avg} - b\right)\delta b\right)^2 + \left(\frac{\partial}{\partial A_{avg}}\left(A_{avg} - b\right)\delta A_{avg}\right)^2} \tag{10}$$

This type of calculation should be performed for each mathematical manipulation performed during data analysis. When using a standard curve, there is an associated error with the fit, which should be included using error propagation. This error can be determined by using Eq. 9.

6. Summary

Complex coacervates are a novel materials platform for the encapsulation and delivery of a wide range of materials, including proteins. We have discussed strategies for understanding the phase behavior associated with the coacervation of polymeric systems in the absence and presence of protein, along with methods for measuring the concentration of protein in coacervate samples. Here we have shown that it is possible to use established methods and adapt them to coacervate systems for protein analysis.

Acknowledgments

W.C.B.M. was supported by a Fellowship from the Soft Materials for Life Sciences National Research Traineeship Program #NRT-1545399. We would also like to thank PPG for funding support of W.C.B.M. through a grant to the Department of Chemical Engineering at the University of Massachusetts Amherst.

References

Bio-Rad Laboratories. n.d. Quick Start™ Bradford protein assay; pp 1–36.

Black, K. A., Priftis, D., Perry, S. L., Yip, J., Byun, W. Y., & Tirrell, M. (2014). Protein encapsulation via polypeptide complex coacervation. *ACS Macro Letters*, *3*(10), 1088–1091. https://doi.org/10.1021/mz500529v.

Blocher McTigue, W. C., & Perry, S. L. (2019). Design rules for encapsulating proteins into complex coacervates. *Soft Matter*, *15*, 3089–3103. https://doi.org/10.1039/C9SM00372J.

Bourganis, V., Karamanidou, T., Kammona, O., & Kiparissides, C. (2017). Polyelectrolyte complexes as prospective carriers for the oral delivery of protein therapeutics. *European Journal of Pharmaceutics and Biopharmaceutics*, *111*, 44–60. https://doi.org/10.1016/j.ejpb.2016.11.005.

Carvalho, I. T., Estevinho, B. N., & Santos, L. (2016). Application of microencapsulated essential oils in cosmetic and personal healthcare products—A review. *International Journal of Cosmetic Science*, *38*(2), 109–119. https://doi.org/10.1111/ics.12232.

Chang, L.-W., Lytle, T. K., Radhakrishna, M., Madinya, J. J., Vélez, J., Sing, C. E., et al. (2017). Sequence and entropy-based control of complex coacervates. *Nature Communications*, *8*, 1273. https://doi.org/10.1038/s41467-017-01249-1.

Chapeau, A.-L., Bertrand, N., Briard-Bion, V., Hamon, P., Poncelet, D., & Bouhallab, S. (2017). Coacervates of whey proteins to protect and improve the oral delivery of a bioactive molecule. *Journal of Functional Foods*, *38*, 197–204. https://doi.org/10.1016/j.jff.2017.09.009.

Chollakup, R., Beck, J. B., Dirnberger, K., Tirrell, M., & Eisenbach, C. D. (2013). Polyelectrolyte molecular weight and salt effects on the phase behavior and coacervation of aqueous solutions of poly(acrylic acid) sodium salt and poly(allylamine) hydrochloride. *Macromolecules*, *46*(6), 2376–2390. https://doi.org/10.1021/ma202172q.

Chollakup, R., Smitthipong, W., Eisenbach, C. D., & Tirrell, M. (2010). Phase behavior and coacervation of aqueous poly(acrylic acid)−poly(allylamine) solutions. *Macromolecules*, *43*(5), 2518–2528. https://doi.org/10.1021/ma902144k.

Comert, F., & Dubin, P. L. (2017). Liquid-liquid and liquid-solid phase separation in protein-polyelectrolyte systems. *Advances in Colloid and Interface Science*, *239*, 213–217. https://doi.org/10.1016/j.cis.2016.08.005.

Comert, F., Malanowski, A. J., Azarikia, F., & Dubin, P. L. (2016). Coacervation and precipitation in polysaccharide-protein systems. *Soft Matter*, *12*, 4154–4161. https://doi.org/10.1039/C6SM00044D.

Cooper, C. L., Dubin, P. L., Kayitmazer, A. B., & Turksen, S. (2005). Polyelectrolyte–protein complexes. *Current Opinion in Colloid & Interface Science*, *10*(1–2), 52–78. https://doi.org/10.1016/j.cocis.2005.05.007.

Cummings, C. S., & Obermeyer, A. C. (2018). Phase separation behavior of supercharged proteins and polyelectrolytes. *Biochemistry*, *57*(3), 314–323. https://doi.org/10.1021/acs.biochem.7b00990.

Gasteiger, E., Hoogland, C., Gattiker, A., Duvaud, S., Wilkins, M. R., Appel, R. D., et al. (2005). In J. M. Walker (Ed.), *The Proteomics Protocols Handbook* (pp. 571–607). Humana Press.

Johnson, N. R., & Wang, Y. (2013). Controlled delivery of heparin-binding EGF-like growth factor yields fast and comprehensive wound healing. *Journal of Controlled Release*, *166*(2), 124–129. https://doi.org/10.1016/j.jconrel.2012.11.004.

Johnson, N. R., & Wang, Y. (2014). Coacervate delivery systems for proteins and small molecule drugs. *Expert Opinion on Drug Delivery*, *11*(12), 1829–1832. https://doi.org/10.1517/17425247.2014.941355.

Kaibara, K., Okazaki, T., Bohidar, H. B., & Dubin, P. L. (2000). pH-induced coacervation in complexes of bovine serum albumin and cationic polyelectrolytes. *Biomacromolecules*, *1*(1), 100–107. https://doi.org/10.1021/bm990006k.

Kishimura, A., Koide, A., Osada, K., Yamasaki, Y., & Kataoka, K. (2007). Encapsulation of myoglobin in PEGylated polyion complex vesicles made from a pair of oppositely charged block ionomers: A physiologically available oxygen carrier. *Angewandte Chemie, International Edition*, *46*(32), 6085–6088. https://doi.org/10.1002/anie.200701776.

Kuo, C.-H., Leon, L., Chung, E. J., Huang, R.-T., Sontag, T. J., Reardon, C. A., et al. (2014). Inhibition of atherosclerosis-promoting microRNAs via targeted polyelectrolyte complex micelles. *Journal of Materials Chemistry B*, *2*, 8142–8153. https://doi.org/10.1039/C4TB00977K.

Li, L., Srivastava, S., Andreev, M., Marciel, A. B., de Pablo, J. J., & Tirrell, M. V. (2018). Phase behavior and salt partitioning in polyelectrolyte complex coacervates. *Macromolecules*, *51*(8), 2988–2995. https://doi.org/10.1021/acs.macromol.8b00238.

Lim, Z. W., Ping, Y., & Miserez, A. (2018). Glucose-responsive peptide coacervates with high encapsulation efficiency for controlled release of insulin. *Bioconjugate Chemistry*, *29*(7), 2176–2180. https://doi.org/10.1021/acs.bioconjchem.8b00369.

Lindhoud, S., & Claessens, M. M. A. E. (2016). Accumulation of small protein molecules in a macroscopic complex coacervate. *Soft Matter*, *12*(2), 408–413. https://doi.org/10.1039/C5SM02386F.

Lindhoud, S., de Vries, R., Schweins, R., Cohen Stuart, M. A., & Norde, W. (2009). Salt-induced release of lipase from polyelectrolyte complex micelles. *Soft Matter*, *5*(1), 242–250. https://doi.org/10.1039/B811640G.

Lindhoud, S., Norde, W., & Cohen Stuart, M. A. (2009). Reversibility and relaxation behavior of polyelectrolyte complex micelle formation. *The Journal of Physical Chemistry B*, *113*(16), 5431–5439. https://doi.org/10.1021/jp809489f.

Lindhoud, S., Voorhaar, L., de Vries, R., Schweins, R., Cohen Stuart, M. A., & Norde, W. (2009). Salt-induced disintegration of lysozyme-containing polyelectrolyte complex micelles. *Langmuir*, *25*(19), 11425–11430. https://doi.org/10.1021/la901591p.

Madinya, J. J., Chang, L.-W., Perry, S. L., & Sing, C. E. (2020). Sequence-dependent self-coacervation in high charge-density polyampholytes. *Molecular Systems Design and Engineering*, *5*, 632–644. https://doi.org/10.1039/C9ME00074G.

Martins, I. M., Barreiro, M. F., Coelho, M., & Rodrigues, A. E. (2014). Microencapsulation of essential oils with biodegradable polymeric carriers for cosmetic applications. *Chemical Engineering Journal*, *245*, 191–200. (C). https://doi.org/10.1016/j.cej.2014.02.024.

Nishida, K., Tamura, A., & Yui, N. (2018). pH-responsive coacervate droplets formed from acid-labile methylated polyrotaxanes as an injectable protein carrier. *Biomacromolecules*, *19*(6), 2238–2247. https://doi.org/10.1021/acs.biomac.8b00301.

Obermeyer, A. C., Mills, C. E., Dong, X.-H., Flores, R. J., & Olsen, B. D. (2016). Complex coacervation of supercharged proteins with polyelectrolytes. *Soft Matter*, *12*, 3570–3581. https://doi.org/10.1039/C6SM00002A.

Olson, B. J. S. C., & Markwell, J. (2007). *Assays for determination of protein concentration*. Hoboken, NJ: John Wiley & Sons, Inc., (pp. Unit3.4–3.4.29). Vol. Chapter 3.

Perry, S., Li, Y., Priftis, D., Leon, L., & Tirrell, M. (2014). The effect of salt on the complex coacervation of vinyl polyelectrolytes. *Polymers*, *6*(6), 1756–1772. https://doi.org/10.3390/polym6061756.

Priftis, D., Megley, K., Laugel, N., & Tirrell, M. (2013). Complex coacervation of poly(ethylene-imine)/polypeptide aqueous solutions: Thermodynamic and rheological characterization. *Journal of Colloid and Interface Science*, *398*, 39–50. https://doi.org/10.1016/j.jcis.2013.01.055.

Priftis, D., & Tirrell, M. (2012). Phase behaviour and complex coacervation of aqueous polypeptide solutions. *Soft Matter*, *8*(36), 9396–9405. https://doi.org/10.1039/C2SM25604E.

Schmitt, C., & Turgeon, S. L. (2011). Protein/polysaccharide complexes and coacervates in food systems. *Advances in Colloid and Interface Science*, *167*(1–2), 63–70. https://doi.org/10.1016/j.cis.2010.10.001.

Simonian, M. H. (2002). Spectrophotometric determination of protein concentration. *Current Protocols in Cell Biology*, *15*(1), A.3B.1–A.3B.7. https://doi.org/10.1002/0471143030.cba03bs15.

Stoscheck, C. M. (1990). Quantitation of protein. *Methods in Enzymology*, *182*, 50–68.

Thermo Scientific. (2010). *Protein A280*. Thermo Scientific, pp. 1–32.

Vehlow, D., Schmidt, R., Gebert, A., Siebert, M., Lips, K., & Müller, M. (2016). Polyelectrolyte complex based interfacial drug delivery system with controlled loading and improved release performance for bone therapeutics. *Nanomaterials*, *6*(3), 53. 21. https://doi.org/10.3390/nano6030053.

Water, J. J., Schack, M. M., Velazquez-Campoy, A., Maltesen, M. J., van de Weert, M., & Jorgensen, L. (2014). Complex coacervates of hyaluronic acid and lysozyme: Effect on protein structure and physical stability. *European Journal of Pharmaceutics and Biopharmaceutics*, *88*(2), 325–331. https://doi.org/10.1016/j.ejpb.2014.09.001.

Xu, Y., Liu, M., Faisal, M., Si, Y., & Guo, Y. (2017). Selective protein complexation and coacervation by polyelectrolytes. *Advances in Colloid and Interface Science*, *239*, 158–167. https://doi.org/10.1016/j.cis.2016.06.004.

Xu, Y., Mazzawi, M., Chen, K., Sun, L., & Dubin, P. L. (2011). Protein purification by polyelectrolyte coacervation: Influence of protein charge anisotropy on selectivity. *Biomacromolecules*, *12*(5), 1512–1522. https://doi.org/10.1021/bm101465y.

Yan, Y., Kizilay, E., Seeman, D., Flanagan, S., Dubin, P. L., Bovetto, L., et al. (2013). Heteroprotein complex coacervation: Bovine β-lactoglobulin and lactoferrin. *Langmuir*, *29*(50), 15614–15623. https://doi.org/10.1021/la4027464.

Yeo, Y., Bellas, E., Firestone, W., Langer, R., & Kohane, D. S. (2005). Complex coacervates for thermally sensitive controlled release of flavor compounds. *Journal of Agricultural and Food Chemistry*, *53*(19), 7518–7525. https://doi.org/10.1021/jf0507947.

CHAPTER ELEVEN

Measuring the activity and structure of functional RNAs inside compartments formed by liquid-liquid phase separation

Raghav R. Poudyal[a,c], McCauley O. Meyer[b,c], and Philip C. Bevilacqua[a,b,c,*]

[a]Department of Chemistry, Pennsylvania State University, University Park, PA, United States
[b]Department of Biochemistry, Microbiology, and Molecular Biology, Pennsylvania State University, University Park, PA, United States
[c]Center for RNA Molecular Biology, Pennsylvania State University, University Park, PA, United States
*Corresponding author: e-mail address: pcb5@psu.edu

Contents

Abstract

Liquid-liquid phase separation (LLPS) has been known to drive formation of biomolecular compartments, which can encapsulate RNA and proteins among other cosolutes. Such compartments, which lack a lipid membrane, have been implicated in origins of life scenarios as they can easily uptake and concentrate biomolecules, similar to intracellular condensates. Indeed, chemical interactions that drive LLPS *in vitro* have also been shown to lead to similar sub-cellular compartments *in vivo*. Here we describe methods to prepare compartments formed by complex coacervates, which are driven by LLPS of oppositely-charged polyions, and to probe the structures and functions of RNAs in them. These methods can be adapted to study RNA biochemistry in compartments formed by diverse artificial and biological macromolecules.

Methods in Enzymology, Volume 646
ISSN 0076-6879
https://doi.org/10.1016/bs.mie.2020.06.010

1. Rationale

Molecular compartmentalization through liquid-liquid phase separation (LLPS) has been hypothesized to play important roles both in extant biology and on primordial Earth (Poudyal, Pir Cakmak, Keating, & Bevilacqua, 2018). Compartmentalization of biomolecules within phase-separated condensates can lead to dramatic increases in the concentrations of magnesium ions, RNA and DNA, peptides and proteins, and in nucleotides and other small molecules (Frankel, Bevilacqua, & Keating, 2016; Koga, Williams, Perriman, & Mann, 2011). Such increases in the concentrations of functional molecules within the condensates can promote molecular interactions and chemical reactions that are otherwise not accessible in dilute solutions. This includes enzyme-mediated reactions in which collisions between the enzyme and the substrate are enhanced. This has led to a growing desire to understand how compartmentalization of biomolecules by LLPS affects a wide range of chemistries as well as the structure and activity of biomolecules, especially functional RNA molecules.

Complex coacervates, which are formed by LLPS of oppositely-charged polymers, have been investigated as primordial protocells for a number of years, as they form spontaneously and also encapsulate biomolecules (Jia, Hentrich, & Szostak, 2014; Koga et al., 2011; Oparin, 1938). Membraneless organelles in extant biology often share features of complex coacervates such as highly charged regions in intrinsically disordered proteins that bind RNAs (Elbaum-Garfinkle et al., 2015; Putnam, Cassani, Smith, & Seydoux, 2019), making the methods described herein of broad interest. Furthermore, complex biomolecular reactions such as transcription and translation have been demonstrated to occur efficiently inside of certain types of coacervate droplets (Sokolova et al., 2013). However, dissecting the structure-function relationships of biomolecules upon encapsulation becomes difficult with such complex systems as they require many functional macromolecules and metabolites for assay-readout (Nott, Craggs, & Baldwin, 2016), although recent advances have been made in probing the structure of RNAs in LLPS *in vitro* using SHAPE-MaP (Langdon & Gladfelter, 2018). RNA enzymes (ribozymes) and other functional RNAs, on the other hand, are much simpler. They require only the enzyme and the substrate oligonucleotides, along with metal ions such as magnesium ions, and so are better suited for studying biomolecular functions and structure inside compartments formed by LLPS. As such, we have studied the effects of compartmentalization and complex coacervate composition on several different RNA and DNA enzymes that cleave specific oligonucleotide substrates (Poudyal, Guth-Metzler, et al., 2019; Poudyal, Keating, & Bevilacqua, 2019). Here we provide a guide to carry out such experiments inside of complex coacervates.

As encapsulation of biomolecules inside complex coacervates is primarily driven by charge-charge interactions, diverse functional nucleic acids can be concentrated by these compartments. Here, we describe a general framework both for evaluating activity of RNA enzymes and for probing structures of functional RNA molecules inside complex coacervates. At low concentrations of enzyme strand ($[E] < K_D$ between the enzyme E and the substrate S under single turnover conditions), reduced binding of the substrate can significantly lessen the activity of the ribozyme. To offset this effect, encapsulation of oligonucleotides by complex coacervates can dramatically increase the local concentration of the enzyme strand (up to 1000-fold), which leads to activation of the ribozyme (Poudyal, Guth-Metzler, et al., 2019) (Fig. 1).

Our lab has shown that specific complex coacervates can strongly encapsulate RNAs of varying length and structures, with enhanced encapsulation of longer RNAs (Frankel et al., 2016). As such, ribozyme reactions and RNA structure probing can be performed by simply assembling the complex coacervates and then adding the RNA of interest. Complex coacervates can be formed by a variety of polycation and polyanion pairs, allowing exploration of how polymeric properties such as charge density, H-bonding, metal-chelation, *etc.*, affect RNA structure and function (Poudyal et al., 2018). Furthermore, because complex coacervates concentrate nucleic acid molecules primarily on the basis of charge-charge interactions and not the identity of the backbone, these compartments can also be used to probe activity of other functional nucleic acids such as DNA enzymes.

1.1 General equipment

Micropipettors and tips
0.2 and 1.5 mL polypropylene tubes
Thermocycler
Glass plates for casting polyacrylamide gels purchased locally
Apparatus for gel electrophoresis prepared in-house by machine shop

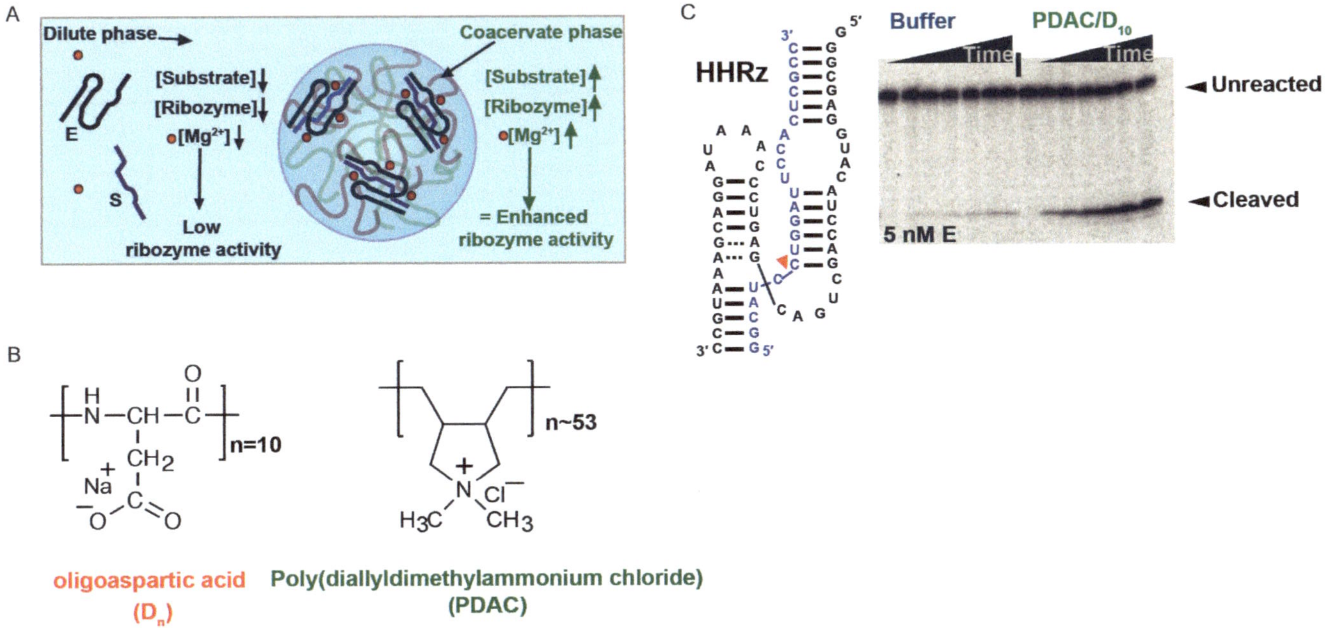

Fig. 1 Overview of ribozyme encapsulation and folding assays in complex coacervates. (A) Schematic showing ribozyme activation in complex coacervates. In dilute phase, the ribozyme (black) and substrate (blue) do not appreciably interact with one another when present at low concentrations. Complex coacervates (shown as droplet) form by the association of polyanions (red) and polycations (green) and encapsulate the ribozyme and substrate RNAs and magnesium ions leading to productive ribozyme-substrate interactions and subsequent activation of catalysis. (B) Examples of polyanions and polycations that interact to form complex coacervates. (C) Gel image showing activation of the hammerhead ribozyme (black) and substrate (blue), with cleavage site indicated by red triangle. Substrate cleavage is poor in the absence of PDAC/D_{10} coacervates (buffer), but enhanced in their presence. *Adapted from Poudyal, R. R., Guth-Metzler, R. M., Veenis, A. J., Frankel, E. A., Keating, C. D., & Bevilacqua, P. C. (2019). Template-directed RNA polymerization and enhanced ribozyme catalysis inside membraneless compartments formed by coacervates.* Nature Communications, 10, *490. doi:10.1038/s41467-019-08353-4.*

Hand-held UV light, shortwave (254 nm)
Fluor-coated thin layer chromatography (TLC) plate for UV-shadowing
NanoDrop or other small volume UV–vis spectrophotometer
Liquid scintillation counter
Gel dryer
Phosphor storage cassettes and screens in a variety of sizes
Amersham Typhoon biomolecular imager
pH probe
Confocal microscope

1.2 General materials

RNase-free water, confirmed by long-term incubation with end-labeled RNAs
DNA templates for RNAs and T7 promoter strand
10 × Transcription buffer: 250 mM $MgCl_2$, 10 mM spermidine, 400 mM Tris–HCl (pH 8.0 at 25 °C), 50 mM dithiothreitol (DTT), and 0.01% TritonX-100
Ribonucleoside 5′ triphosphate (rNTP) mix (10 ×): 100 mM each of ATP, CTP, GTP, and UTP in deionized water, neutralized to pH 7
Bacteriophage T7 RNA polymerase, either commercial or prepared in-house
10% Ammonium persulfate (APS) (should be prepared fresh or can be stored at 4 °C for up to 3 months)
Tetramethylethylenediamine (TEMED)
20% denaturing polyacrylamide solution: 20% polyacrylamide (29:1 polyacrylamide:bis-acrylamide) in 1 × TBE (90 mM Tris, 90 mM borate, and 10 mM EDTA pH 8.0 at 20 °C), 8.3 M urea
Silicon spacers (1.0–1.5 mm thick)
Denaturing gel diluent: 8.3 M urea in 1 × TBE
1 × Gel running buffer (TBE)
Vacuum grease
2 × Gel-loading buffer: 95% formamide 50 mM EDTA, 0.05% (w/v) bromophenol blue sodium salt, 0.05% (w/v) xylene cyanol
Crush/soak buffer: 300 mM sodium acetate (pH 5.2 at 25 °C) and 1 mM EDTA (pH 8.0 at 25 °C) or 10 mM Tris pH 8.0, 1 mM EDTA, and 250 mM sodium chloride
100% ethanol chilled to −20 °C

Phosphatase enzyme (rSAP—recombinant Shrimp Alkaline Phosphatase—supplied by New England Biolabs M0371, 1000 U/mL)
10 × Cutsmart Buffer (Supplied by New England Biolabs)
T4 Polynucleotide Kinase (Supplied by New England Biolabs M0201)
10 × T4 Polynucleotide Kinase Buffer (Supplied by New England Biolabs)
[γ^{32}-P] ATP: 6000 Ci/mmol total ATP (Supplied by Perkin Elmer)
Stratagene Autorad markers
X-ray film BioMax (Supplied by Kodak)
Developer Solution (Supplied by Kodak)
Fixer Solution (Supplied by Kodak)
10 × Ribozyme Reaction Buffer 250 mM Tris pH 8.0, 10 mM $MgCl_2$, 50 mM NaCl
Polycation: Polydiallyldimethylammonium chloride 8.5 kDa (Polysciences Cat No: 24828–100) or oligolysine (10-mer, 30-mer, 50-mer and 100-mer from Alamanda Polymers)
Polyanion: Oligoaspartic acid (10-mer, 30-mer, 50-mer and 100-mer from Alamanda Polymers)
Stopping solution–8.3 mM polyacrylic acid 1.8 kDa (to displace RNA from interactions with polycation), 166 mM EDTA in 70% formamide
Drawing paper to lift gels off of plates and prepare for drying (14 × 17 in)
Microscope cover slips (22 × 22 mm, VWR)
Fluorescent dye-labeled RNA (IDT)
Sodium periodate, made fresh (50 mM)
Sodium acetate (0.5 M, pH 5.2)
Hydrazide-functionalized fluorophore
Anhydrous dimethylsulfoxide (DMSO)
Amicon Ultra-0.5 centrifugal filter unit (10 kDa)

2. Preparation of reagents

2.1 Preparation of RNAs by *in vitro* transcription

To prepare RNAs by *in vitro* transcription, order DNA templates (IDT) for desired RNAs as reverse complements along with binding site for T7 promoter sequence. Schematic of DNA templates is shown in Fig. 2. The T7 promoter sequence, which is depicted as top strand base-paired to the RNA template in the figure below, is GCCTAATACGACTCACTATA<u>G</u>, where the underlined G also corresponds to the first nucleotide of the desired transcript.

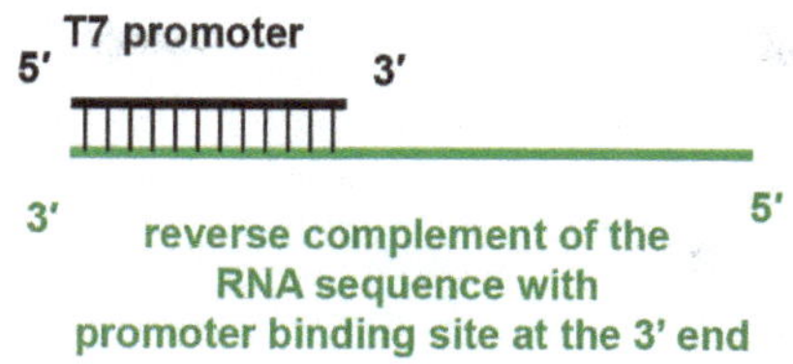

Fig. 2 Design of DNA oligonucleotides for *in vitro* transcription. Note that the T7 promoter, on the top, is annealed to the reverse complement of the desired RNA sequence, on the bottom.

2.1.1 Procedure for generating RNA from in vitro *transcription*

A. Reagents for transcription

2 μL, 10 μM bottom strand DNA
2.5 μL, 10 μM top strand DNA
20 μL of 10 × transcription buffer
6 μL of each of 100 mM rNTPs (24 μL total)
6 μL of T7 polymerase (purified in-house)
145 μL of water (total volume 200 μL)

Note: The amount of T7 polymerase needed can vary depending on the preparation. We typically perform a titration from 1% to 10% enzyme per transcription reaction and compare yields to find the optimal amount after the enzyme purification. In some cases, too much enzyme can inhibit the reaction. If T7 polymerase is purchased commercially, follow manufacturer's protocol.

B. Steps for setting up transcription reaction

1. Add water, 10 × transcription buffer, and both top and bottom strand DNA oligonucleotides in a 1.5 mL polypropylene tube.
2. To anneal the T7 promoter, heat the mixture at 65 °C for 30 s and let it cool at room temperature for 3 min.
3. Add the four rNTP stock solutions.
4. Add T7 RNA polymerase, which should be pre-mixed by flicking the tube.
5. Mix well by pipetting with the pipette set to at least 50 μL. Watch the tube carefully for the mixing lines and mix until they disappear. Quick spin to collect droplets.
6. Incubate the reaction at 37 °C for 3 h.

C. Steps for purification of RNA

1. To purify the full-length RNA away from the aborted transcription products, use denaturing polyacrylamide gel electrophoresis.

For separating RNAs from transcription reactions, we typically use 6.5 × 9.5 × 0.125 in. glass plates with 0.04 in. silicon spacers (1 mm). For the above gel size, we first make 40 mL of 10% acrylamide by mixing equal volumes of a urea 20% acrylamide stock and a denaturing gel diluent solution. We cast the polyacrylamide gel by adding 40 μL of TEMED and 400 μL of 10% ammonium persulfate to the 40 mL of 10% acrylamide. Take the time to mix the solution well after each addition to avoid local quick polymerization. Pour and let the gel polymerize for 10–20 min and save the leftover acrylamide—this can be used to approximate when the gel in glass plates has polymerized and should be allowed to polymerize since it is much less toxic than the unpolymerized stocks solutions.

2. Setup the electrophoresis equipment while the gel is polymerizing. Pre-run the gel for 30–60 min at 25 W, 1000 V, and 300 mA. Periodically check the electrophoresis apparatus for any leakage of buffer. Apply vacuum grease if needed, being sure to turn off the power supply before touching the apparatus to avoid electrical shock.
3. Once the polyacrylamide gel has pre-run for the appropriate length of time, add an equal volume of the 2× gel loading buffer to the transcription reaction. Heat the resulting solution at 90 °C for 1–2 min to denature secondary structures of RNA. If the secondary structures are not denatured, RNAs with strong secondary structure and/or multiple conformations can result in gels with large smears or several bands for the same RNA.
4. Load the transcription reactions containing the gel loading buffer on the gel and run the electrophoresis system for 1.5–2 h, until the xylene cyanol band (top band) has migrated about 1/2–1/3 way down the gel (Sambrook, Fritsch, & Maniatis, 1989).
5. Peel off one of the glass plates and cover the gel with plastic wrap. Then, putting your hand over the plastic wrap, flip the gel over and carefully peel off the other glass plate. Cover this side of the gel with plastic wrap as well.
6. To isolate the RNA using UV-shadowing, place the plastic-wrapped gel on top of a fluorescent TLC plate and hold a 254 nm handheld UV light over the gel in a dark room. Be sure to protect your face and eyes by wearing a face shield. As nucleotides absorb UV, the location of RNA will appear as a dark band in the background of the fluorescing TLC plate. Outline the RNA band using a marker on the plastic.

Note: Try to minimize the UV exposure time as it can damage the RNA. This is one of the main reasons for outlining the RNA on the plastic itself with a marker.

7. Excise the appropriate RNA band from the gel using a clean razor and further dice the gel slice into numerous small (~0.5–1 mm) cubes. Transfer the gel cubes to a 1.5 mL polypropylene tube and add 0.55 mL of Crush/soak buffer. Incubate at 4 °C on a rotating tumbler overnight.
8. To precipitate the RNA, centrifuge the tube containing the RNA gel slices at 5000 × *g* for 30 s. Transfer the supernatant (~500 μL) to a fresh tube. Add 2 volumes (1 mL) of 100% ethanol pre-chilled at −20 °C. Mix the tube by vortexing and inverting and store at −80 °C for 2 h (or on powdered dry ice for 30 min).
9. Pellet the RNA by centrifuging in a table-top centrifuge at 14,000 × *g* for 30 min. Use a refrigerated centrifuge (or one in a cold room) to aid precipitation and to minimize temperature-induced RNA degradation. Decant the supernatant and dry the tube using a speedvac or open air until all the solution has evaporated.
10. Resuspend the RNA in 30 μL RNase-free water and measure its concentration using a small volume UV–vis spectrophotometer.

Notes: RNAs are significantly less stable than DNA due to the potential for nucleophilic attack from the 2′ OH. As such special precautions must be taken during synthesis, handling, and storage of RNA molecules.

(i) Always use RNase-free water for all procedures
(ii) Always perform experiments wearing gloves to prevent RNases from hands contaminating tubes and pipettes
(iii) Do not breath directly into samples
(iv) For long-term storage, we suggest storing RNA as a precipitate in ethanol at −80 °C

2.1.2 Steps for 5′-end labeling of RNA using [γ^{32}-P] ATP

These procedures must be performed in area designated for the use of radiation with proper PPE and shielding.

2.1.2.1 Radiation safety

We recommend using double gloves and a disposable lab coat whenever working with radiation. ^{32}P is a high-energy beta emitter so keep all radioactive samples behind clear plexiglass shields to limit exposure. Never work with radiation alone. Be sure to refer to and follow the health and safety guidelines from your institution for more comprehensive instructions on working with radioactive materials.

1. To remove the triphosphates on the 5′-end of the freshly transcribed RNA, use a phosphatase enzyme (Note that RNA and DNA made by solid-phase synthesis does not have a 5′-triphosphate and so is ready to directly use for radiolabeling in the next step). To a 0.2 mL polypropylene tube, add about 10 pmol of transcribed RNA, 10 × Cutsmart Buffer, 1 U of recombinant shrimp alkaline phosphatase (NEB), and water to a total volume of 10 μL. Incubate this reaction at 37 °C for 30 min. Inactivate the phosphatase by heating the reaction at 65 °C for 5 min (Note: It is imperative to inactivate the phosphatase in order to successfully label the RNA, as any active phosphatase remaining will remove the 5′-monophosphate added by the kinase in the next step).
2. Start 5′-end labeling of RNA by adding 6–7 μL of dephosphorylated RNA to the reaction, 1 μL of 10 × T4 Polynucleotide Kinase Buffer, 1 μL T4 Polynucleotide Kinase enzyme, and finally 1–2 μL of [γ^{32}-P] ATP: 6000 Ci/mmol to a total volume of 10 μL. Incubate the reaction at 37 °C for 30 min. After 30 min, add 10 μL of the 2 × gel loading buffer.
3. While the kinase reaction is incubating for 30 min, make a 10% denaturing polyacrylamide gel and run electrophoresis as described in C1–C4. Once the gel has run for desired time (generally 1–1.5 h works well), peel one of the gel plates off and wrap the gel with plastic on the open side. Place the gel into a phosphor screen cassette, and proceed to a dark room for exposing the gel.
4. Place Stratagene autorad markers on the benchtop to expose them to light allowing them to luminesce. Place Stratagene autorad markers on the periphery of the gel away from the expected RNA band.
5. Turn off the lights and place an X-ray film on top of the gel with the autorad markers. Close the lid of the cassette firmly such that the X-ray film does not move during exposure. Let the X-ray film expose for 2–5 min, after which place the X-ray film in the tray containing developer solution for 30 s to 1 min. Next, briefly place the X-ray film in water to wash off excess developer solution. Move the washed X-ray film to the tray containing fixer solution and soak for 30 s. Wash the film with water and let it dry in the air for ~15 min.

Note: The X-ray film may need to be exposed for longer time if the source radioactive nucleotide is old (more than two half-lives) OR if labeling efficiency is low.

6. To isolate the RNA from the gel, place the X-ray film on a lightbox and turn the light on. Set the gel with the plastic side up onto the X-ray film. Align autorad makers on the gel to the spots on the film. Using a marker, outline the location of the RNA band. Follow steps 7–9 outlined in Section 2.1.1.C to obtain the labeled RNA, then resuspend the RNA in 30 μL of water.

Note: Check the counts on the gel and also in the tube with crush and soak buffer to make sure that an accurate cut has been made, meaning that radioactive RNA has been transferred from the gel. If counts in the tube are <100,000 cpm based on Geiger counter reading from 1 cm away, re-expose the gel to verify that the 5′-end-labeled RNA has been isolated and not accidentally left behind on the gel.

7. To measure the concentration of 5′-end-labeled RNA, add 1 μL of the labeled RNA to approximately 10 mL of scintillation fluid and measure the counts per minute using a liquid scintillation counter (LSC). Prior to estimating concentration of the RNA, calculated the specific activity for the source NTP by using the following equation:

$$\mathrm{SA(calculated)} = \frac{D_F}{\frac{1}{SA\ (calibrated)} - \frac{1-D_F}{SA\ (theoretical)}}.$$

where SA(calibrated) = specific activity for the nucleotide (usually 6000 ci/mmol), SA theoretical = 9120, and decay factor (based on the half-life of ^{32}P) since the calibration date.

Note: Unless there is a dedicated UV–vis spectrophotometer for radiation work, DO NOT attempt to measure absorbance for ^{32}P-labeled RNA by bringing the radiation outside in the general lab area. In most cases, the concentration of the RNAs are not accurately detectable by table top UV–vis after the above procedure anyhow.

3. RNA structure and function inside complex coacervates

Background: Determination of RNA function and RNA structure within complex coacervates involves generating the complex coacervates and then adding the RNAs of interest. Once the reactions have been incubated for the desired amount of time, competing polyanions are added to displace the RNA from the polycations within the complex coacervates, after which gel electrophoresis can be carried out to understand the outcome of the experiment. We have developed and applied these methods to study both ribozyme activity and tRNA structure inside diverse complex coacervates.

3.1 Steps for measuring ribozyme activity inside complex coacervates

1. Before generation of complex coacervates, renature the 50 nM ribozyme (10 ×) and trace radiolabeled substrate (~1 nM) (10 ×) separately in 25 mM Tris pH 8.0 and 5 mM NaCl by heating the tube at 80 °C for 3 min. These tubes can be left equilibrating at room temperature until initiation of the reaction. Volumes needed of the 10 × solutions typically vary between 10 and 50 μL depending on the volume/number of experiments.
2. To generate complex coacervates, first add 10 μL of water, 2 μL 10 × ribozyme reaction buffer, and 2 μL each of polycation and polyanion. Order of addition is very important as complex coacervates can form as soon as the polycation and polyanions are added. Final concentration of total positive and negative charge from polyions should be held at 10 mM. For ease of experimental design, stock concentrations of polyanion and polycations are set at 100 mM total charge (10 ×). For example, stock concentration of PDAC 8.5 kDa with 53 monomers is made at 1.88 mM and oligoaspartic acid with 10 monomers is made at 10 mM. Once both the polyions are added, the solution turns visibly turbid due to liquid-liquid phase separation, indicating formation of complex coacervates.
3. Next, add 2 μL of the renatured 10 × ribozyme solution, followed by addition of 2 μL 10 × substrate to the coacervate solutions (total volume of 20 μL) to initiate the reaction. Mix well by pipetting with a pipette set to at least 10 μL after each of the components is added.
4. To stop the reaction, add 40 μL of "stopping solution." Since the stopping solution contains polyacrylic acid, which itself is a polyanion, it will displace the substrate and the ribozyme from the complex coacervates.
5. Make 20% polyacrylamide gel by following steps 1–3 outlined in Section 2.1.1.C. The running time for the gel can vary depending on the size of the substrate and the cleaved product. We typically run the gel for about 45 min for the hammerhead ribozyme system.
6. Once the electrophoresis is complete, peel one of the glass plates off and cover the gel with drawing paper. Flip the gel over and peel off the other

glass plate. Now cover the top of the gel with the plastic wrap. To dry the gel, place the plastic side up on a gel dryer and place the dryer cover on top of it. Dry the gel at 70 °C for 30 min with vacuum turned on. Once done, turn off the heat and let it cool to room temperature for 30 min. Note that if the vacuum seal is broken while the gel is still hot, cracks can appear on the gel and completely ruin gel analysis. If you are in a hurry, you can add a small amount of powdered dry ice to the sealed top of the gel to aid cooling.

7. After drying, place the gel on a phosphor cassette and cover with phosphor screen. Store the cassette at room temperature overnight. Image the phosphor screen using Amersham Typhoon Biomolecular Imager according to manufacturer's instructions.

Note 1: It is important to take note of the orientation of the gel on the phosphor screen for further analysis, as the full-length band of the substrate should migrate slower and be the upper band and the cleaved band is the lower band. The xylene cyanol band (light-blue) should be on top and the bromophenol blue band (dark blue) should be on the bottom.

Note 2: The above protocol is useful for performing a single time-point experiment. To perform a kinetic experiment, the reaction must be aliquoted into different tubes right after initiation so that stopping solution can be added at desired timepoints. Working quickly is important because over time, complex coacervates can settle and stick to the plastic tubes resulting in decrease of signal.

3.1.1 Data analysis

The images of gels are obtained in either .gel file or .tiff file formats, which can be opened in different bioimaging programs such as ImageJ or ImageQuant. To quantify ribozyme activity, measure the signal of the uncleaved substrate (shorter migration) and cleaved substrate (longer migration). Correct these measurements from background in the image by measuring intensity at three arbitrary positions away from the radioactive bands, average these three measurements, and subtract the average value from all other data points. Then calculate fraction cleaved by dividing signal from the cleaved fragment by the sum of signals from both cleaved and uncleaved fragments.

3.1.2 Optimization of complex coacervate composition

We have found that optimization of complex coacervate composition can significantly improve ribozyme reactions. It is therefore necessary to try different polycation/polyanion pairs to generate complex coacervates. In our previous studies, we have found that very strong ribozyme-polycation interactions can inhibit ribozyme reactions potentially by misfolding the ribozyme, and that addition of excess polyanions can rescue this inhibition (Poudyal, Keating, et al., 2019). A systematic approach to identifying optimal complex coacervates is therefore needed. First, test ribozyme activity in different pairs of complex coacervates generated at 1:1 (−/+) total charges. For example, generate complex coacervates using different lengths of oligolysine and oligoaspartic acid. For each polyion pair, excess polyanions can then be added such that total charge ratio (−/+) is 1:1, 2:1, 3:1, *etc.*

3.1.3 Measuring RNA partitioning using fluorescence microscopy

Partitioning of RNA and other molecules can be measured using confocal fluorescence microscopy, where fluorescence intensity inside coacervate droplets can be compared against a standard curve to calculate the concentration of a desired fluorescently–tagged molecule inside complex coacervates.

1. To measure RNA concentrations inside complex coacervates, first make standards of fluorescently-labeled RNA. We typically prepare 10 μL solutions containing 1 × ribozyme reaction buffer and 0–200 μM RNA. Once these standards are made, store them in the dark until complex coacervates are assembled. These standards will be used to calculate the concentration of RNA inside the coacervate droplets from fluorescent images.
2. Generate 20 μL of complex coacervate solutions, as described in step 2 of Section 3.1. Instead of adding 2 μL each of 10 × ribozyme and substrate in step 3 of Section 3.1, add 0.5 μL of 10 μM fluorescently-labeled RNA and 3.5 μL of water such that the final concentration is 250 nM.
3. Place 10 μL of the complex coacervate sample on a glass coverslip and obtain fluorescent images of complex coacervates using an appropriate confocal microscope. We typically use 63 × objective to observe complex coacervates. For fluorescence microscopy, we routinely use Alexa-488 labeled RNA, for which we excite the complex coacervates using a 488 nm laser to collect the images. Make sure that images are not saturated by controlling laser power and gain values for the photomultiplier tube. We typically use settings where the pixel intensity inside the complex coacervate is approximately 100. The settings used to obtain these images must be recorded as they must be same when measuring standards.

4. Once the fluorescent images of the complex coacervates have been collected, pipette 5 μL of the standards on the glass coverslips and measure the fluorescence of the solution. Starting with the highest concentration standard makes image collection easy, as the field-of-view appears as designated false color for the emission window.
5. Obtain pixel intensity values for 0–200 μM RNA using ImageJ or other available software. Perform a background subtraction (0 μM) for all concentrations, and generate a standard curve.
6. To determine the concentration of RNA inside the complex coacervates, measure the fluorescence intensity within a droplet and use the standard curve to determine the concentration of RNA (RNA_{IN}). Also determine the concentration of RNA outside the droplet in the dilute phase (RNA_{OUT}).
7. Determine partitioning coefficient (K) by using the equation:

$$\mathrm{K} = RNA_{IN}/RNA_{OUT}$$

Note: Adjustments in the concentration of the standards may be necessary such that the pixel intensity for the complex coacervates is in the linear region of the standard curve.

3.1.4 Related technique for fluorescently labeling RNA

For experiments with long RNAs, purchasing fluorescently-labeled RNAs directly from the vendors can be costly. Chemical methods to fluorescently label RNAs on the 5′- and the 3′-ends are available. Outlined here are steps to generate RNAs with fluorescent tags on the 3′-end, which take advantage of the unique cis-diol on this terminus of the RNA.

1. Add 5 μL of freshly prepared sodium periodate (50 mM), 10 μL of 0.5 M sodium acetate (pH 5.2), and 25 μL of water to 10 μL of 50 μM RNA. Incubate the reaction at room temperature for 90 min. This opens the ring of the 3′-terminal ribose.
2. Add 300 μL of 300 mM sodium acetate (pH 5.2), followed by addition of 800 μL of 100% ethanol. Place the tube in dry ice for 2 h, and centrifuge at 14000 × *g* to precipitate the RNA. Follow with vacuum centrifugation to dry the pellet.
3. Resuspend the pellet in 2 μL of 10 mM hydrazide dye derivative fluorophore (dissolved in anhydrous DMSO), 2 μL of 0.5 M NaOAc pH 5.2, and 16 μL of water. Mix thoroughly pipetting up and down

with pipette volume set at 20 μL. Incubate the reaction in the dark at 4 °C overnight to make a conjugate with the dye.

4. Add 25 μL of water, and repeat step 2 to precipitate the RNA and remove excess unincorporated dye.
5. To further purify the RNA and remove all free dye, resuspend it in 400 μL of water. Place the solution in an Amicon Ultra-0.5 centrifugal filter unit (10 kDa) and centrifuge at 14000 × *g* for 5 min. Discard the flow-through and add 400 μL of water, followed by centrifugation again. Repeat these steps until 2.4 mL of water has been added. After the final run, either pipette out the retained volume in the filter or follow manufacturer's protocol to obtain the RNA solution.
6. Measure absorbance at 260 nm and at the desired excitation wavelength for the fluorophore. Calculate fraction of the RNA 3′-end-labeled with the fluorophore by comparing concentration of total RNA (Absorbance at 260 nm) with concentration of the fluorescently-labeled RNA (Absorbance at excitation maxima).

3.1.5 Notes

1. After the conjugation reaction between the oxidized RNA and the fluorophore, sodium cyanoborohydride can be used to further reduce the hydrazone bond to a more stable secondary amine. (Please follow necessary safety precautions while handling sodium cyanoborohydride as it is a strong reductant.)
2. Gel extraction can also be used instead of centrifugal filtration method to purify the RNA. This method ensures complete separation between unincorporated and RNA-tagged fluorophore. However, if using UV-shadowing, the procedure must be performed with minimal irradiation to prevent photo-bleaching of the fluorophore.
3. All procedures must be performed in the dark as much as possible to prevent photo-bleaching of the fluorophore.

3.2 Determination of RNA structure in complex coacervates

Background: Ribozyme cleavage assays report only on the active fold of the RNA. To report on the potential misfolded/unfolded states, one can use in-line probing in an effort to attain nucleotide-level resolution of RNA folding interactions. In-line probing (ILP) works by taking advantage of base-catalyzed hydrolysis of RNA leading to strand cleavage, preferentially in flexible, single-stranded regions. Base-catalyzed hydrolysis of RNAs occurs primarily *via* hydroxide ions coordinated to Mg^{2+} ions (Soukup & Breaker, 1999) (Fig. 3). Water coordinated to Mg^{2+} has its pK_a downwardly shifted, greatly increasing the proportion of hydroxide ions near neutral pH. These hydroxide ions can deprotonate the 2′ hydroxyl of the ribose sugar at any position along the RNA. This leads to efficient nucleophilic attack on the nearby phosphate backbone subsequently resulting in strand cleavage. After being subjected to ILP, a 5′ radiolabeled RNA can be fractionated on a sequencing denaturing polyacrylamide gel and the resulting pattern of high and low reactivity can be used to deduce the structure of the RNA.

This technique was originally developed under dilute buffer *in vitro* conditions, but it can also be used with great effect inside of LLPSs such as complex coacervates (Cakmak, Choi, Meyer, Bevilacqua, & Keating, 2020). First, an RNA is subjected to ILP under dilute buffer conditions, in the

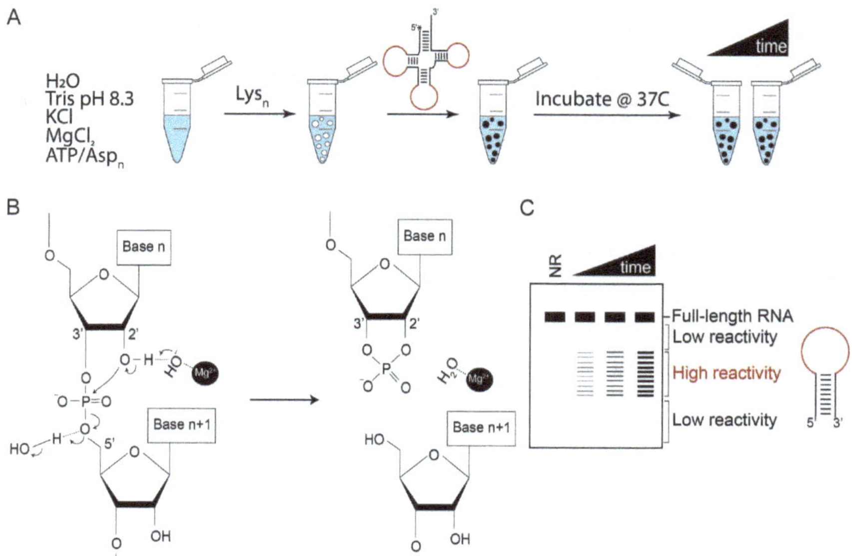

Fig. 3 In-line probing inside of coacervates. (A) General procedure for coacervate formation, and subsequent ILP experiment. (B) Magnesium-bound hydroxide ions deprotonate the 2′ hydroxyl of single-stranded/flexible bases, which performs a nucleophilic attack on the phosphate backbone causing strand cleavage. (C) 5′-end-labeled RNAs are fractionated on denaturing polyacrylamide gels and a pattern of reactivity corresponding to structure can be elucidated. *Adapted from Cakmak, F. P., Choi, S., Meyer, M. O., Bevilacqua, P. C., & Keating, C. D. (2020). Prebiotically-relevant low polyion multivalency can improve functionality of membraneless compartments.* bioRxiv. *doi:10.1101/2020.02.23.961920.*

absence of complex coacervates. This helps to determine the native fold of RNAs under standard, dilute buffer conditions. Then, the cleavage pattern derived from this control experiment can be compared to subsequent experiments in coacervates or other systems formed by LLPS. This is important to do because upon compartmentalization, RNA structures can be modulated by interactions with the surrounding compartment.

3.2.1 *Steps for measuring RNA structure* via *in-line probing (ILP) inside complex coacervates*

1. Dilute the RNA of interest with water to 10 × the assay concentration in 10 μL total (generally 500 pM of 5′ ^{32}P RNA is used).
2. Renature 10 × RNA by incubating at 90 °C for 2 min, and cool at room temperature for at least 5 min. This allows the RNA to fold into the native structure at the given condition.
3. While 10 × RNA is cooling, prepare coacervates. To prepare coacervates, add 15 μL of water, 2.5 μL of 10 × ILP buffer (100 mM Tris, pH 8.3, 150 mM KCl, 5 mM MgCl2), 2.5 μL of 10 × polyanion and 2.5 μL of 10 × polycation to a total volume of 22.5 μL in a 0.2 mL tube. Once both the polyions are present, the solution turns visibly turbid due to liquid-liquid phase separation.
4. To initiate the reaction, add 2.5 μL of 10 × RNA to the prepared coacervate solution from step 3, and mix well by pipetting with a pipette set to at least 15 μL.
5. Immediately after adding the RNA, take out a 5 μL sample of the reaction and place it in a new 0.2 mL tube labeled "0."
6. To stop the reaction, add 15 μL of "stopping solution." Since the stopping solution contains polyacrylic acid, it will displace the RNA from interactions with the coacervates.
7. Take 5 μL timepoints at 24 and 48 h, repeating step 6 to stop the reactions (see note 2 in Section 3.1 for aliquoting reactions to take timepoints).
8. Pour a 10% sequencing polyacrylamide gel following steps 1–2 under Section 2.1.1.C.
9. Pre-run the gel for ~30–45 min until the glass is warm to the touch.
10. Once the gel is warm, incubate the timepoints at 90 °C for 3 min to denature secondary structures.
11. Load the gel and run it for 1–1.5 h at 60 W. (Time will vary here depending on how long the RNA is that needs to be resolved. It may be necessary to load the same gel twice, in different sets of lanes, or run two different gels to get the necessary resolution of the entire

RNA if the RNA is >90 nucleotides long. We suggest starting with 1–1.5 h, but time should be optimized for the RNA of interest)

12. After the gel has run for the desired distance, refer to steps 5 and 6 in Section 3.1 to disassemble the gel, dry it, and expose a phosphor screen to the gel.
13. (Optional) to quantify the intensities of each band on the gel, semi-automated footprinting analysis (SAFA) can be used (Laederach et al., 2008). Additionally, quantified intensities on a per nucleotide basis can be used to provide pseudo-free energy values (Deigan, Li, Mathews, & Weeks, 2009).

3.2.2 Measuring RNA partitioning by liquid scintillation counting

Rationale: While fluorescence partitioning experiments are excellent in that they allow for partitioning coefficients to be calculated, it can be expensive and/or time consuming to buy or make fluorescently-labeled RNA. In addition, there is always the concern that the dye can influence partitioning, especially for shorter RNAs. As many of the other experiments we describe require radiolabeled RNA, it is much simpler to perform RNA partitioning experiments with the radiolabeled RNA than it would be to fluorescently label/buy another RNA.

1. Follow steps 1–4 in Section 3.2.1 to prepare the RNA and coacervate solution for the partitioning experiment.
2. Once the RNA is in the coacervate solution and well mixed by pipetting with a pipette set to 15 μL, immediately take 1 μL of the solution and add it to approximately 10 mL of scintillation fluid and measure the counts per minute using a liquid scintillation counter (LSC). This will be used to compare to the total counts per minute in 1 μL of solution.
3. To determine the amount of RNA remaining in the dilute phase, prepare another identical sample following steps 1–4 in Section 3.2.1, and centrifuge at 14000 × *g* for 1 min.
4. Immediately after centrifugation, take 1 μL of the solution and add it to 10 mL of scintillation fluid and measure the counts per minute (cpm) using a liquid scintillation counter. Be careful not to disturb the coacervates at the bottom of the tube as this will cause the amount of RNA in the dilute phase to be artificially high.
5. To calculate the % RNA in the coacervate phase, divide the cpm from post-centrifugation sample by cpm from pre-centrifugation sample. Then subtract that value from 1 and multiply by 100.

Notes: To have confidence in how much RNA is partitioned, at least three independent partitioning experiments should be performed and averaged.

4. Conclusion

The methods described herein can be utilized to probe the activity of different RNA enzymes, as well as the structures of diverse RNA molecules including those in intracellular condensates. Since oligonucleotide encapsulation in complex coacervates is mainly driven by charge-charge interactions, activity of DNAzymes can also be studied. By applying the methods herein, different complex coacervate compositions and intracellular condensates can be explored for nucleic acid structure and functions.

Acknowledgements

This work was supported by a grant (290363) from the Simons Foundation to R.R.P. and by NASA grant (80NSSC17K0034).

References

Cakmak, F. P., Choi, S., Meyer, M. O., Bevilacqua, P. C., & Keating, C. D. (2020). Prebiotically-relevant low polyion multivalency can improve functionality of membraneless compartments. *bioRxiv*. https://doi.org/10.1101/2020.02.23.961920.

Deigan, K. E., Li, T. W., Mathews, D. H., & Weeks, K. M. (2009). Accurate SHAPE-directed RNA structure determination. *Proceedings of the National Academy of Sciences of the United States of America*, *106*, 97–102. https://doi.org/10.1073/pnas.0806929106.

Elbaum-Garfinkle, S., Kim, Y., Szczepaniak, K., Chen, C. C., Eckmann, C. R., Myong, S., et al. (2015). The disordered P granule protein LAF-1 drives phase separation into droplets with tunable viscosity and dynamics. *Proceedings of the National Academy of Sciences of the United States of America*, *112*, 7189–7194. https://doi.org/10.1073/pnas.1504822112.

Frankel, E. A., Bevilacqua, P. C., & Keating, C. D. (2016). Polyamine/nucleotide coacervates provide strong compartmentalization of Mg(2)(+), nucleotides, and RNA. *Langmuir*, *32*, 2041–2049. https://doi.org/10.1021/acs.langmuir.5b04462.

Jia, T. Z., Hentrich, C., & Szostak, J. W. (2014). Rapid RNA exchange in aqueous two-phase system and coacervate droplets. *Origins of Life and Evolution of Biospheres*, *44*, 1–12. https://doi.org/10.1007/s11084-014-9355-8.

Koga, S., Williams, D. S., Perriman, A. W., & Mann, S. (2011). Peptide-nucleotide microdroplets as a step towards a membrane-free protocell model. *Nature Chemistry*, *3*, 720–724. https://doi.org/10.1038/nchem.1110.

Laederach, A., Das, R., Vicens, Q., Pearlman, S. M., Brenowitz, M., Herschlag, D., et al. (2008). Semiautomated and rapid quantification of nucleic acid footprinting and structure mapping experiments. *Nature Protocols*, *3*, 1395–1401. https://doi.org/10.1038/nprot.2008.134.

Langdon, E. M., & Gladfelter, A. S. (2018). Probing RNA structure in liquid-liquid phase separation using SHAPE-MaP. *Methods in Enzymology*, *611*, 67–79. https://doi.org/10.1016/bs.mie.2018.09.039.

Nott, T. J., Craggs, T. D., & Baldwin, A. J. (2016). Membraneless organelles can melt nucleic acid duplexes and act as biomolecular filters. *Nature Chemistry*, *8*, 569–575. https://doi.org/10.1038/nchem.2519.

Oparin, A. I. (1938). *Origin of life*. New York: Macmillan.

Poudyal, R. R., Guth-Metzler, R. M., Veenis, A. J., Frankel, E. A., Keating, C. D., & Bevilacqua, P. C. (2019). Template-directed RNA polymerization and enhanced ribozyme catalysis inside membraneless compartments formed by coacervates. *Nature Communications, 10*, 490. https://doi.org/10.1038/s41467-019-08353-4.

Poudyal, R. R., Keating, C. D., & Bevilacqua, P. C. (2019). Polyanion-assisted ribozyme catalysis inside complex coacervates. *ACS Chemical Biology, 14*, 1243–1248. https://doi.org/10.1021/acschembio.9b00205.

Poudyal, R. R., Pir Cakmak, F., Keating, C. D., & Bevilacqua, P. C. (2018). Physical principles and extant biology reveal roles for RNA-containing membraneless compartments in origins of life chemistry. *Biochemistry, 57*, 2509–2519. https://doi.org/10.1021/acs.biochem.8b00081.

Putnam, A., Cassani, M., Smith, J., & Seydoux, G. (2019). A gel phase promotes condensation of liquid P granules in Caenorhabditis elegans embryos. *Nature Structural & Molecular Biology, 26*, 220–226. https://doi.org/10.1038/s41594-019-0193-2.

Sambrook, J., Fritsch, E. F., & Maniatis, T. (1989). *Molecular cloning: A laboratory manual*. Cold Spring Harbor Laboratory Press.

Sokolova, E., Sprujit, E., Hansen, M. M. K., Dubuc, E., Groen, J., Chokkalingam, V., et al. (2013). Enhanced transcription rates in membrane-free protocells formed by coacervation of cell lysate. *Proceedings of the National Academy of Sciences of the United States of America, 110*, 11693–11697.

Soukup, G. A., & Breaker, R. R. (1999). Relationship between internucleotide linkage geometry and the stability of RNA. *RNA, 5*, 1308–1325. https://doi.org/10.1017/s1355838299990891.

CHAPTER TWELVE

Reversible photocontrol of DNA coacervation

Suzanne Lafon[a,b] **and Nicolas Martin**[a,*]
[a]Université de Bordeaux, CNRS, Centre de Recherche Paul Pascal, UMR5031, Pessac, France
[b]Ecole Normale Supérieure d'Ulm, Paris, France
*Corresponding author: e-mail address: nicolas.martin@crpp.cnrs.fr

Contents

Abstract

Coacervate micro-droplets produced by liquid-liquid phase separation are increasingly used to emulate the dynamical organization of membraneless organelles found in living cells. Designing synthetic coacervates able to be formed and disassembled with improved spatiotemporal control is still challenging. In this chapter, we describe the design of photoswitchable coacervate droplets produced by phase separation of short double stranded DNA in the presence of an azobenzene cation. The droplets can be reversibly dissolved with light, which provides a new approach for the spatiotemporal regulation of coacervation. Significantly, the dynamics of light-actuated droplet formation and dissolution correlates with the capture and release of guest solutes. The reported system can find applications for the dynamic photocontrol of biomolecule compartmentalization, paving the way to the light-activated regulation of signaling pathways in artificial membraneless organelles.

Methods in Enzymology, Volume 646
ISSN 0076-6879
https://doi.org/10.1016/bs.mie.2020.06.013

1. Introduction

Membraneless organelles (MLOs) formed by liquid-liquid phase separation of proteins and RNA play a key role in the organization of intracellular contents and reactions (Brangwynne, Tompa, & Pappu, 2015; Hyman, Weber, & Jülicher, 2014). These biomolecular condensates include nucleoli that synthesize ribosome subunits, centrosomes that nucleate microtubules, and stress granules that segregate RNA under stress conditions. Crucial to the functions of MLOs is their highly dynamic nature that allows them to transiently compartmentalize biomolecules in cells (Wheeler & Hyman, 2018). In particular, the formation and dissolution, localization, size and number of MLOs rely on intracellular concentration profiles of phase separating species, which are dynamically regulated by biochemical reactions (e.g., DNA translation and transcription reactions, and post-translational modifications) (Berry, Brangwynne, & Haataja, 2018; Hondele et al., 2019; Nott et al., 2015; Rai, Chen, Selbach, & Pelkmans, 2018; Shin & Brangwynne, 2017).

Recent efforts to design synthetic coacervate droplets able to recapitulate key features of membraneless organelles in vitro have been made. Coacervate micro-droplets are typically produced by associative liquid-liquid phase separation between oppositely charged polyions or surfactants. These liquid-like droplets provide simple models to rationalize the underlying physicochemical rules of biological phase separation (Perry, 2019), reproduce fundamental functions of biomolecular condensates (Martin, Li, & Mann, 2016; Nakashima, Vibhute, & Spruijt, 2019), and demonstrate primitive forms of compartmentalization ("protocells") (Crowe & Keating, 2018; Martin, 2019). Dynamic aspects are also increasingly considered when designing synthetic coacervates as models of membraneless organelles and protocells (Martin, 2019). Owing to the weak molecular interactions involved, coacervate droplets dynamically respond to environmental cues; for instance, they are readily formed and dissolved by changes in temperature or pH (Douliez et al., 2017; Love et al., 2020; Martin et al., 2018). Biochemical reactions have also recently been exploited to control the formation and dissolution of coacervate droplets, including phosphorylation/dephosphorylation reactions (Aumiller & Keating, 2016; Nakashima, Baaji, & Spruijt, 2018) and polynucleotide synthesis (Deshpande et al., 2019). Improved spatiotemporal control over the formation and dissolution of synthetic coacervates would be highly beneficial to emulate the dynamic regulation of biomolecular condensates: new tools for the fast and local actuation of liquid-liquid phase separation need thus to be developed.

Within this context, optogenetic approaches have recently emerged to activate the formation of membraneless organelles with light *in cellulo* (Shin et al., 2017). Yet, experimental models of synthetic coacervates capable of being dynamically controlled with light in vitro are required. Light is indeed a bio-orthogonal tool that offers great spatiotemporal capabilities and spectral selectivity. Light is also one of the primary energy sources on Earth, making it an interesting tool to control membrane-free compartmentalization in the context of origin-of-life studies. Overall, the design of photoswitchable coacervates would open new perspectives for the spatiotemporal regulation of biomolecule compartmentalization.

2. Photoswitchable coacervation: The azoTAB/*ds*DNA system

We recently reported an approach for the reversible photocontrol of DNA coacervation in vitro using molecular azobenzene photoswitches (Martin et al., 2019). In this system, liquid-liquid phase separation of short double stranded DNA (*ds*DNA) is induced by an azobenzene-based cation, azobenzene trimethylammonium bromide (azoTAB). The overall photoswitchable process is summarized in Fig. 1A.

Azobenzene moieties are well known to undergo *trans-cis* photoisomerization under light irradiation. The resulting changes in geometry and dipolar moment at the molecular level have been exploited to control the structure and function of biomolecules. For instance, azoTAB cations have been used to control the compaction of single *ds*DNA chains with light, as first reported by Le Ny and Lee (2006), followed by studies from Diguet, Mani, Geoffroy, Sollogoub, and Baigl (2010) and Sollogoub et al. (2008), who further demonstrated light-mediated regulation of cell-free gene expression via the reversible photocontrol of *ds*DNA condensation (Estévez-Torres et al., 2009). In these studies, long *ds*DNA chains were used (e.g., bacteriophage DNA and plasmids, >1000 base pairs) at relatively low concentrations (typically μM nucleobase concentrations). Observations pointed to a contribution of both electrostatic and hydrophobic forces in the *ds*DNA compaction process.

In comparison, photoswitchable coacervates are produced with relatively short *ds*DNA chains (<200 base pairs) and at higher polynucleotide concentration (~mM nucleobase concentration) (Martin et al., 2019).

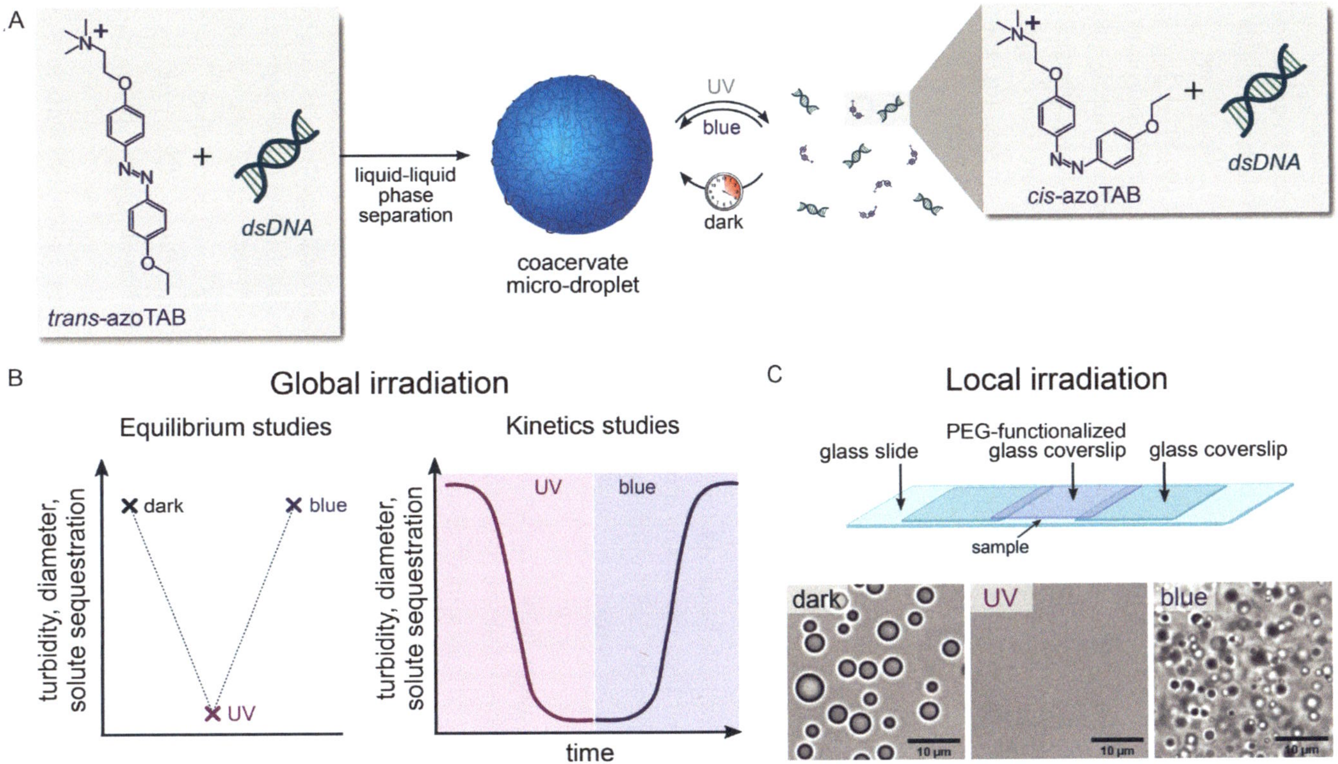

Fig. 1 (A) Scheme of reversible photocontrol of *ds*DNA coacervation with azoTAB. ON/OFF phase separation can be triggered in response to UV and blue light. Temporal programming of coacervate droplets formation is achieved in the dark. (B) Photoswitchable phase separation and the associated sequestration/release of solutes can be characterized under global irradiation of a droplets suspension at equilibrium (photostationary state) or via kinetics studies. (C) These processes can also be monitored under local irradiation by optical microscopy. Droplets are here imaged with an oil immersion objective on an inverted microscope using a PEG-functionalized capillary chamber.

Our observation that coacervation between *ds*DNA and *trans*-azoTAB occurs close to equimolar charge ratio indicates that electrostatic interactions play an important role in the phase separation process. Additional forces involved in the process include hydrophobic interactions between azobenzene moieties (e.g., π-π stacking) and partial intercalation of the planar *trans*-azoTAB between *ds*DNA base pairs (Martin et al., 2019). In comparison, *cis*-azoTAB has a lower affinity for *ds*DNA because it is less hydrophobic and can no longer develop π-π stacking due to its bent geometry: its unbinding from *ds*DNA eventually leads to the dissolution of coacervate micro-droplets. Overall, UV light converts the coacervate-active *trans*-azoTAB into the coacervate-dormant *cis* isomer, resulting in the dissolution of coacervate droplets. Reversibly, *cis-trans* isomerization under blue light favors the re-assembly of coacervate micro-droplets (Fig. 1A).

We have further demonstrated that the dynamics of droplet dissolution and reformation correlates with the release and capture of guest solutes: solutes sequestered within droplets are released under UV light due to coacervate disassembly, and are recaptured in the droplets that reform under blue light. Such a photoswitchable release and capture of solutes can find applications to dynamically control the compartmentalization of biomolecules, opening new perspectives for the regulation of artificial membraneless organelle systems.

This chapter describes the methods for the formulation and reversible photocontrol of azoTAB/*ds*DNA coacervate micro-droplets, including the protocols for the synthesis of the cationic azoTAB photoswitch.

3. Synthesis and photophysical characterization of azoTAB

AzoTAB is produced by a three-step synthesis derived from the method originally described by Hayashita, Kurosawa, Miyata, Tanaka, and Igawa (1994), and further adapted by Le Ny and Lee (2006), Diguet et al. (2010), Estévez-Torres et al. (2009) and Sollogoub et al. (2008)). We detail below the synthesis procedure (summarized in Fig. 2A).

3.1 Protocol for the synthesis of azoTAB

3.1.1 Equipment

1. Round-bottomed flasks (two-necked, 250 mL; single-necked, 150 mL) equipped with septa
2. Magnetic stirring bar

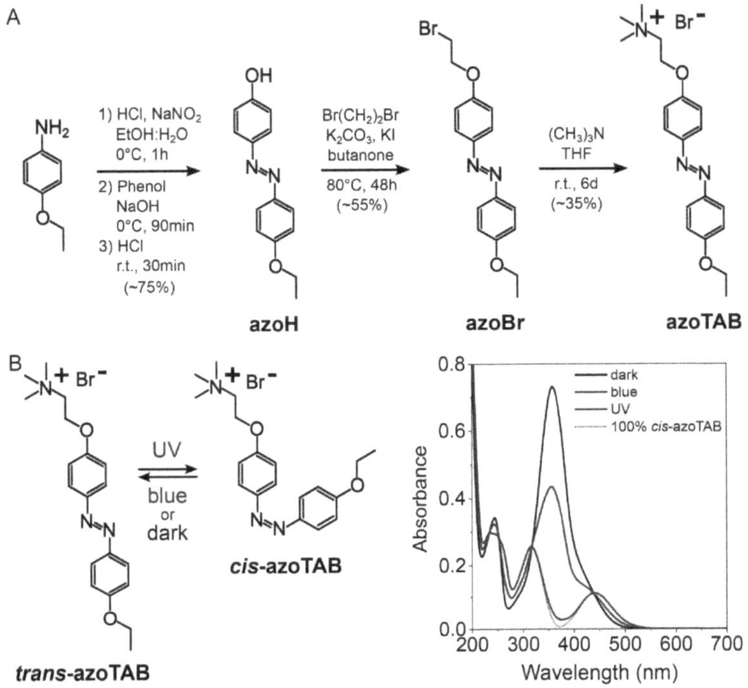

Fig. 2 (A) General overview of the synthesis procedure of azoTAB. (B) Photophysical characterization of azoTAB isomerization. UV/vis spectra of a 30 μM azoTAB solution equilibrated in the dark (black line, corresponding to 100% *trans*-azoTAB), and exposed to UV (purple line) or blue (blue line) light are reported, together with the calculated spectrum for 100% *cis*-azoTAB (light purple line).

3. Thermometer
4. Stirring hotplate
5. Aluminum foil
6. pH test strips
7. Analytical balance
8. Ice bath; oil bath
9. Vacuum filtration set-up: filter, Büchner funnel, conic seal, Büchner flask, vacuum pump
10. Separation funnel (50 mL)
11. Rotary evaporator
12. NMR spectrometer

3.1.2 Chemicals

Commercially available reagents were used as received. The ^{1}H and ^{13}C NMR spectra were recorded at room temperature on a 400 MHz

ECZ400 spectrometer (Jeol); chemical shifts are reported in ppm with protonated solvent as internal reference (^1H, $CHCl_3$ in $CDCl_3$ 7.26 ppm, CHD_2SOCD_3 in CD_3SOCD_3 2.49 ppm; ^{13}C, $^{13}CDCl_3$ in $CDCl_3$ 77.0, $^{13}CD_3SOCD_3$ in CD_3SOCD_3 39.7 ppm); coupling constants (J) are given in Hz.

1. *Para*-ethoxyaniline (Sigma-Aldrich)
2. Hydrochloric acid 37% (Sigma-Aldrich)
3. Sodium hydroxide (Sigma-Aldrich)
4. Sodium nitrite (Sigma-Aldrich)
5. Phenol (Sigma-Aldrich)
6. 1,2-Dibromoethane (Sigma-Aldrich)
7. Potassium carbonate (Sigma-Aldrich)
8. Potassium iodide (Sigma-Aldrich)
9. 33% Trimethylamine solution in ethanol (Sigma-Aldrich)
10. Magnesium sulfate ($MgSO_4$) (Sigma-Aldrich)
11. Ethanol (Sigma-Aldrich)
12. Methyl ethyl ketone (Sigma-Aldrich)
13. Dichloromethane (Sigma-Aldrich)
14. Anhydrous tetrahydrofuran (Sigma-Aldrich)
15. DMSO-d_6 and $CDCl_3$ for NMR experiments (Sigma-Aldrich)

3.1.3 Protocol

All synthetic steps are performed in a fume hood.

3.1.3.1 Synthesis of 4-ethoxy-4′-hydroxy-azobenzene (azoH)

1.1 Working at 0 °C in an ice bath, add concentrated HCl (17 mL, 37%, dropwise addition) and ice (80 g) to a 1:1 *v*/v ethanol:water solution (160 mL) containing *para*-ethoxyaniline (10.3 mL, 80 mmol, 1 equiv.) and sodium nitrite (5.5 g, 80 mmol, 1 equiv.) in a two-necked round-bottomed flask (250 mL) under magnetic stirring. The temperature of the mixture can be checked with a thermometer.

1.2 Stir the mixture for 1 h while maintaining the temperature below 5 °C with an ice bath.

1.3 Carefully add a solution of phenol (7.5 g, 80 mmol, 1 equiv.) and NaOH (6.4 g, 160 mmol, 2 equiv.) dissolved in cold water (42 mL, ~5 °C) to the mixture and stir for 90 min while maintaining the temperature below 5 °C with an ice bath.

1.4 Adjust the pH of the solution to 1 with concentrated HCl (37%) using pH test strips, and leave the mixture to warm to room temperature for 30 min. A brown precipitate should form.

1.5 Recover the precipitate with vacuum filtration, wash thoroughly with water and then dry in vacuum overnight. The typical yield of the reaction is ca. 75%.

1.6 Analyze the dark brown powder obtained (**azoH**) by NMR to yield: ^{1}H NMR (ppm, 400 MHz, $CDCl_3$): $\delta = 7.86$ (d, 2H; Ar-H, $J = 8$ Hz), 7.82 (d, 2H; Ar-H, $J = 8$ Hz), 6.99 (d, 2H; Ar-H, $J = 8$ Hz), 6.94 (d, 2H; Ar-H, $J = 8$ Hz), 4.12 (q, 2H; CH_2, $J = 6$ Hz), 1.46 ppm (t, 3H; CH_3, $J = 4$ Hz); ^{13}C NMR (ppm, 400 MHz, $CDCl_3$): $\delta = 161.0$ (Ar-C), 157.9 (Ar-C), 147.1 (Ar-C), 146.8 (Ar-C), 124.6 (Ar-C), 124.4 (Ar-C), 115.8 (Ar-C), 114.7 (Ar-C), 63.8 (CH_2O), 14.8 ppm (CH_3). These data are in line with the reported values for the same compound (Diguet et al., 2010).

3.1.3.2 Synthesis of 4-ethoxy-(4′-(2-bromoethyloxy)phenyl) azobenzene (azoBr)

2.1 Mix **azoH** (2.4 g, 10 mmol, 1 equiv.), 1,2-dibromoethane (5.6 g, 3 equiv.), potassium carbonate (2.07 g, 1.5 equiv.) and potassium iodide (0.083 g, 0.05 equiv.) in 50 mL of methyl ethyl ketone (butanone) in a single-necked round-bottomed flask (150 mL) equipped with a condenser, and reflux (~80 °C) for 48 h in the dark (use aluminum foil around the flask) under magnetic stirring.

2.2 Filter the reaction mixture hot to remove any solid impurities and salt, and wash the residue with methyl ethyl ketone.

2.3 Collect the filtrate and remove the solvent under reduced pressure by rotary evaporation.

2.4 Dissolve the obtained solid in dichloromethane (20 mL), wash twice with a NaOH solution (1 M, 2×8 mL), then twice with pure water (2×8 mL), dry over $MgSO_4$ and concentrate under reduced pressure to yield an orange powder.

2.5 Analyze the orange powder obtained by ^{1}H NMR. The product may be pure enough to be used without further purification for the next step. Alternatively, the crude product can be recrystallized with hot filtration from ethanol and dried under vacuum overnight. The typical yield of the reaction is ca. 55%.

2.6 Analyze the orange powder obtained (**azoBr**) by NMR to yield: ^{1}H NMR (ppm, 400 MHz, CDC13): $\delta = 7.92$ (q, 4H; Ar-H, $J = 8.2$ Hz), 7.00 (dd, 4H; Ar-H, $J = 8.6$ Hz), 4.37 (t, 2H; CH_2O, $J = 8$ Hz), 4.12 (q, 2H; CH_2O, $J = 6$ Hz), 3.67 (t, 2H; CH_2Br, $J = 7$ Hz), 1.46 ppm (t, 3H; CH_3, $J = 6$ Hz); ^{13}C NMR (ppm, 400 MHz, $CDCl_3$):

$\delta = 161.5$ (Ar-C), 160.2 (Ar-C), 146.8 (Ar-C), 146.3 (Ar-C), 124.8 (Ar-C), 124.6 (Ar-C), 114.9 (Ar-C), 114.8 (Ar-C), 68.0 (CH_2O), 63.9 (CH_2O), 28.8 (CH_2Br), 14.8 ppm (CH_3).

3.1.3.3 Synthesis of azobenzene trimethylammonium bromide (azoTAB)

3.1 Working at room temperature, dissolve 1 g of **azoBr** (2.9 mmol, 1 equiv.) in 80 mL of dry THF in a single-necked round-bottomed flask (150 mL), then add a 33% solution of trimethylamine in ethanol (4.2 mL, 11.5 mmol, 4 equiv.) under magnetic stirring.

3.2 Stir the mixture for 6 days in the dark (use aluminum foil around the flask) at room temperature. A precipitate should form over time. Since trimethylamine is highly volatile, another aliquot of the 33% solution of trimethylamine in ethanol (4.2 mL, 11.5 mmol, 4 equiv.) should be added after 3 days.

3.3 Recover the resulting orange precipitate with vacuum filtration, wash thoroughly with THF, and dry under vacuum overnight.

3.4 Analyze the orange powder obtained by ^1H NMR. The product may be pure enough to be used without further purification. Alternatively, the crude product can be recrystallized from ethanol and dried under vacuum overnight. The typical yield of the reaction is ca. 35%.

3.5 Analyze the orange powder obtained (**azoTAB**) by NMR to yield: ^1H NMR (ppm, 400 MHz, DMSO): $\delta = 7.86$ (d, 2H; Ar-H, $J = 8$ Hz), 7.82 (d, 2H; Ar-H, $J = 8$ Hz), 7.17 (d, 2H; Ar-H, $J = 8$ Hz), 7.08 (d, 2H; Ar-H, $J = 8$ Hz), 4.56 (m, 2H; CH_2O), 4.11 (q, 2H; CH_2O, $J = 6$ Hz), 3.82 (m, 2H; CH_2N), 3.18 (s, 9H; CH_3N), 1.35 ppm (t, 3H; CH_3, $J = 4$ Hz); ^{13}C NMR (ppm, 400 MHz, DMSO): $\delta =$ 161.3 (Ar-C), 159.9 (Ar-C), 147.1 (Ar-C), 146.4 (Ar-C), 124.7 (Ar-C), 124.5 (Ar-C), 115.8 (Ar-C), 115.4 (Ar-C), 64.5 (CH_2O), 64.1 (CH_2N), 62.5 (CH_2O), 53.6 (CH_3N), 15.0 ppm (CH_3).

3.2 Photo-isomerization of azoTAB: Equilibrium *trans*:*cis* composition

trans-AzoTAB is the thermodynamically most stable isomer due to its planar geometry and aromaticity, therefore a solution of azoTAB kept in the dark for several days contains 100% of *trans*-azoTAB. Exposure of the solution to light will result in a mixture of *cis* and *trans* isomers, with an equilibrium composition that depends on the irradiation wavelength (Chevallier et al., 2011; Diguet et al., 2010; Hayashita et al., 1994). Typically, *trans-cis* isomerization of azoTAB is induced by irradiation at 365 nm (a value close to the wavelength of the π-π* absorption peak), and results in almost quantitative conversion of *trans*-azoTAB into the *cis* isomer. In comparison, the reversible *cis-trans* isomerization is induced by irradiation at 450 nm (n-π* absorption peak), and produces a mixture of *trans* and *cis*-azoTAB at the photostationary state (with a small excess of the *trans* isomer) because both isomers absorb light at 450 nm.

The composition of *trans* and *cis* isomers under UV and blue light can be easily determined by UV/vis spectrophotometry, and should be assessed before producing coacervates. These studies, detailed below, also allow to verify that the light sources subsequently used to control photoswitchable coacervates are effective to induce azoTAB photo-isomerization.

Different light sources can be used to isomerize azoTAB, including LEDs and benchtop UV lamps for global irradiation, or light sources directly installed on a microscope for local irradiation. A power per unit area of a few $mW\,cm^{-2}$ is sufficient to induce azoTAB isomerization. The power of the identified light source can be assessed using a power-meter when available, in which case the sensor of the power-meter should be located at the position of the sample.

3.2.1 Equipment

1. Analytical balance
2. Ultrasound bath
3. Heating bath or Eppendorf thermomixer
4. pH-meter
5. Micro-pipettes and pipette tips
6. Eppendorf tubes
7. 3 mL quartz cuvette
8. Small magnetic stirring bar
9. UV/vis spectrophotometer equipped with magnetic stirring
10. UV (365 nm) and blue (450 nm) LEDs. In the original study, LEDs from Thorlabs were used (Martin et al., 2019) (365 nm: reference M365D2, 1150 mW; 450 nm: reference M450D3, 1850 mW)

3.2.2 Protocols and analysis

3.2.2.1 Preparation of a 100 mM stock solution of *trans*-azoTAB

1. Dissolve 40.8 mg of synthesized azoTAB powder (M_{azoTAB} = 408.3 g/mol) in 950 μL of Milli-Q water in an Eppendorf tube.

Sonication, vortexing and heating (~65 °C) can be used to favor solubilization. The resulting orange solution should be clear.

2. Check the pH with a calibrated pH-meter and adjust if necessary to pH 8 with 0.1 M HCl or NaOH solutions, then fill to 1 mL with Milli-Q water to make a 100 mM azoTAB stock solution.
3. Store in the dark at room temperature for at least 3 days before use to ensure that all azoTAB molecules have relaxed to the *trans* isomer. Alternatively, the stock solution can be heated to 65 °C for at least 1 h to accelerate the *cis-trans* thermal relaxation using a heating bath or an Eppendorf thermomixer, then let to cool to room temperature in the dark.
4. Store at room temperature in the dark (e.g., by wrapping the Eppendorf tube in aluminum foil). The *trans*-azoTAB stock solution is stable for extended periods of time (months). The solution can be sonicated, vortexed and heated to re-dissolve any precipitate that may form over time (e.g., if the temperature in the room is too cold).

3.2.2.2 UV/vis measurements in the dark, and after UV and blue light irradiation

1. After baseline acquisition using pure water, record the UV/vis spectrum of a 30 μM *trans*-azoTAB solution prepared by dilution of the 100 mM stock solution with Milli-Q water in a quartz cuvette (wavelength range: 200–600 nm). It is important to perform this experiment in the dark (e.g., using foil, or in a dark room to prepare the solution) to make sure that all azoTAB molecules remain in their *trans* configuration.
2. Irradiate the cuvette with a UV light source (365 nm) under magnetic stirring for at least 5 min (the less intense the source, the longer the irradiation should be to ensure that equilibrium is reached). The irradiation can be performed from the top or the side of the cuvette using a LED (from the top, the irradiation can be performed directly in the UV/vis spectrophotometer if equipped with magnetic stirring).
3. Turn off the light source and immediately acquire the UV/vis spectrum of the UV-equilibrated solution (wavelength range: 200–600 nm).
4. Repeat the same procedure with blue light irradiation: irradiate the cuvette with a blue light source (450 nm) under magnetic stirring for at least 5 min, then turn off the light source and acquire the UV/vis spectrum of the blue-adapted solution (wavelength range: 200–600 nm).

3.2.2.3 Analysis: Determination of the *trans:cis* composition at the photostationary state

The UV/vis spectra acquired on the same azoTAB solution irradiated with UV and blue light are used to determine the fraction of *trans* and *cis* isomers at equilibrium. We assume that the dark-adapted solution contains 100% *trans*-azoTAB. This can be confirmed by checking that the UV/vis spectrum of the dark-adapted sample does not change over time.

Absorbances of the dark-, UV- and blue-adapted samples are considered to vary in proportion to the *trans:cis* composition, according to the following equations (Chevallier et al., 2011; Martin et al., 2017):

$$A^{dark}(\lambda) = \varepsilon_{trans}(\lambda).c.l = A_{trans}(\lambda) \tag{1}$$

$$\begin{aligned} A^{UV}(\lambda) &= \left[\varepsilon_{trans}(\lambda).x^{UV}_{trans} + \varepsilon_{cis}(\lambda).x^{UV}_{cis}\right].c.l \\ &= A_{trans}(\lambda).x^{UV}_{trans} + A_{cis}(\lambda).x^{UV}_{cis} \end{aligned} \tag{2}$$

$$\begin{aligned} A^{blue}(\lambda) &= \left[\varepsilon_{trans}(\lambda).x^{blue}_{trans} + \varepsilon_{cis}(\lambda).x^{blue}_{cis}\right].c.l \\ &= A_{trans}(\lambda).x^{blue}_{trans} + A_{cis}(\lambda).x^{blue}_{cis} \end{aligned} \tag{3}$$

where $A^{dark}(\lambda)$, $A^{UV}(\lambda)$ and $A^{blue}(\lambda)$ are the wavelength-dependent absorbances of the dark-, UV- and blue-adapted solutions, respectively; l is the pathlength; x^{UV}_{trans} and x^{UV}_{cis} are the fractions of *trans* and *cis* isomers in the UV-adapted sample; x^{blue}_{trans} and x^{blue}_{cis} are the fractions of *trans* and *cis* isomers in the blue-adapted sample; $\varepsilon_{trans}(\lambda)$ and $A_{trans}(\lambda)$ are the wavelength-dependent extinction coefficient and absorbance of *trans*-azoTAB, respectively; $\varepsilon_{cis}(\lambda)$ and $A_{cis}(\lambda)$ are the wavelength-dependent extinction coefficient and absorbance of *cis*-azoTAB, respectively; and c is the total azoTAB concentration.

The values of $\varepsilon_{trans}(\lambda)$ are obtained from the dark-adapted sample using Eq. (1). To obtain the values of $\varepsilon_{cis}(\lambda)$, and in turn the *trans:cis* composition under UV and blue light, an iterative process is used. Initial x^{UV}_{trans} and x^{blue}_{trans} values are calculated by assuming $\varepsilon_{cis}(\lambda^{UV}_{min}).\ x^{i}_{cis} << \varepsilon_{trans}(\lambda^{UV}_{min}).\ x^{i}_{trans}$ in Eqs. (2) ($i = UV$) and (3) ($i = blue$), where λ^{UV}_{min} is the wavelength at the minimum of absorption of the UV-adapted sample in the range $300 < \lambda < 450$ nm (here, $\lambda^{UV}_{min} = 375$ nm). At λ^{UV}_{min}, we can thus estimate x^{UV}_{trans} and x^{blue}_{trans} as:

$$x^{UV}_{trans} = \frac{A^{UV}\left(\lambda^{UV}_{min}\right)}{A^{dark}\left(\lambda^{UV}_{min}\right)} \tag{4}$$

$$x^{blue}_{trans} = \frac{A^{blue}\left(\lambda^{UV}_{min}\right)}{A^{dark}\left(\lambda^{UV}_{min}\right)} \tag{5}$$

The absorbance spectrum of the *trans* isomer weighed by the calculated x_{trans}^{UV} and x_{trans}^{blue} values is subtracted from the UV- and blue-adapted spectra, and normalized by x_{cis}^{UV} and x_{cis}^{blue}, respectively, to give a putative absorbance spectrum for the *cis* isomer, using Eqs. (2) and (3). An iterative process is then used to minimize the difference between the spectra calculated for the *cis* isomer from the UV- and the blue-adapted solutions in the range $300 < \lambda < 450$ nm. At the end of the process, the two calculated spectra for the *cis* isomer converge to a single one shown in Fig. 2B. Typical values of x_{trans} and x_{cis} obtained after this process are listed below.

	UV (365 nm)	Blue (450 nm)
x_{trans}	4%	58%
x_{cis}	96%	42%

4. Photoswitchable DNA coacervation

4.1 Preparation of *trans*-azoTAB/*ds*DNA coacervates

trans-AzoTAB/*ds*DNA coacervate micro-droplets are prepared at equimolar charge ratio, typically at 5 mM final concentration of *trans*-azoTAB and *ds*DNA (nucleobase concentration). Short DNA strands are used (typically <200 bp) because longer strands (~2000 bp) form solid-like precipitates in the presence of *trans*-azoTAB.

In our previous study, we further observed that droplets formed in the absence of salt were small, with an average hydrodynamic diameter of ~200 nm as determined by dynamic light scattering (Martin et al., 2019). Increasing the ionic strength to 100 mM using sodium chloride (NaCl) favored the formation of micrometer-sized droplets, consistent with the enhancement of coacervation reported in the literature upon addition of small amounts of salt (Perry, Li, Priftis, Leon, & Tirrell, 2014). In comparison, coacervation was prevented at much higher ionic strength (>500 mM NaCl), as expected from screening of attractive electrostatic interactions between *trans*-azoTAB and *ds*DNA. We therefore fixed the NaCl concentration to 100 mM to favor the formation of the largest coacervate droplets. We detail below the preparation of a suspension of such coacervate micro-droplets.

4.1.1 Equipment

1. Analytical balance
2. pH-meter
3. Micro-pipettes and pipette tips
4. Eppendorf tubes

4.1.2 Chemicals

1. Deoxyribonucleic acid, low molecular weight from salmon sperm, *ds*DNA (Sigma-Aldrich, reference 31149)
2. Synthesized azoTAB (see Section 3.1)
3. Sodium chloride (NaCl)
4. NaOH and HCl solutions (0.1 and 1 $mol\,L^{-1}$)

4.1.3 Protocol

4.1.3.1 Preparation of a 100 mM *ds*DNA stock solution

1. Dissolve 30.4 mg of *ds*DNA (0.1 mmol nucleobases, considering an average $M_{nucleobase}$ of 303.7 $g\,mol^{-1}$) in 700 μL of milli-Q water containing 40 μL of NaOH 1 $mol\,L^{-1}$ and gently mix by pipetting until the DNA is fully dissolved.
2. Check the pH and adjust to pH 8 by adding small volumes of a 1 $mol\,L^{-1}$ NaOH solution, then fill to 1 mL with Milli-Q water to make a 100 mM stock *ds*DNA solution (nucleobases concentration).
3. Split the solution into aliquots and store in the freezer at −20 °C until use.

4.1.3.2 Preparation of *trans*-azoTAB/*ds*DNA coacervates

1. Mix in the following order aliquots of stock solutions of *trans*-azoTAB (100 mM, see Section 3.2.2), NaCl (2 M) and *ds*DNA (100 mM nucleobases) in Milli-Q water in an Eppendorf tube to reach final concentrations of 5 mM *trans*-azoTAB, 5 mM *ds*DNA (nucleobase concentration) and 100 mM NaCl. (For instance: dilute 1 μL of 100 mM *trans*-azoTAB in 17 μL of water in an Eppendorf tube, then add 1 μL of 2 M NaCl followed by addition of 1 μL 100 mM *ds*DNA). Each addition should be followed by pipetting to ensure proper mixing. Upon addition of the last component (*ds*DNA), the solution should become turbid, which is indicative of the formation of coacervate droplets.
2. Leave in the Eppendorf tube for ca. 5 min. This allows the droplets to grow in size by coalescence.

4.2 Characterization of photoswitchable coacervates under global irradiation

Turbidity measurements and dynamic light scattering (DLS) can be used to characterize the equilibrium and kinetics behavior of photoswitchable coacervation under global irradiation of the droplets suspension (Fig. 1B). The light-modulated sequestration of fluorescently labeled solutes and dyes can be quantified by UV/vis spectrophotometry. These studies are detailed below.

4.2.1 Equipment

1. Eppendorf tubes, micropipettes and pipette tips
2. Low volume plastic cuvettes (minimum volume ~70 μL) of 1 cm pathlength
3. Zetasizer Nano-ZS instrument equipped with a 633 nm laser (Malvern)
4. UV/vis spectrophotometer
5. UV (365 nm) and blue (450 nm) LEDs

4.2.2 Protocols

4.2.2.1 General protocol for equilibrium DLS and turbidity measurements

The general procedure for equilibrium measurements consists in producing a suspension of *trans*-azoTAB/*ds*DNA coacervate droplets in the dark, measuring the desired property (turbidity or hydrodynamic radius), then irradiating the sample with UV (or blue) light to reach equilibrium and monitor again the desired property. These equilibrium studies are detailed below. NB: although not detailed here, turbidity and DLS equilibrium studies can also be used to screen the influence of the *trans*-azoTAB/*ds*DNA ratio and the ionic strength on the phase separation, in order to determine the optimal ratio, as well as the critical salt concentration at which coacervate disassemble.

1. Prepare a fresh *trans*-azoTAB/*ds*DNA droplets suspension (5 mM *trans*-azoTAB, 5 mM *ds*DNA, 100 mM NaCl, see Section 4.1.3) and incubate in a low volume plastic cuvette for ca. 15 min.
2. Measure the average hydrodynamic radius using a Zetasizer Nano-ZS instrument equipped with a 633 nm laser, or measure the absorbance at 700 nm using a UV/vis spectrophotometer. NB: for DLS measurements, the focal point of the laser should be adjusted to be as close as possible to the inner cuvette wall to limit multiple scattering in the turbid droplet suspension, and the backscattering detector should be used; for turbidity measurements, the absorbance is measured at a wavelength at

which neither *trans*- nor *cis*-azoTAB absorb (700 nm). The absorbance of the droplets suspension should be >1.

3. Irradiate the suspension for ca. 15 min from the side of the cuvette using a UV LED operating at 365 nm and by positioning the sample along the emission axis at a distance of 5 cm from the light source. Gentle mixing by taping or pipetting should be done regularly to ensure homogeneous irradiation. The solution should become clear, indicative of the dissolution of coacervate micro-droplets.
4. Measure the average hydrodynamic radius and the absorbance. The absorbance of the clear solution should be ~0.
5. Repeat steps 3 and 4 using a blue LED operating at 450 nm. The solution should become turbid under blue light, consistent with the reformation of coacervate micro-droplets. The absorbance of the droplets suspension should be >1.
6. Steps 3 to 5 can be repeated several times to demonstrate that the process is repeatable and reversible and does not show any apparent fatigue.

4.2.2.2 General protocol for kinetics turbidity measurements

Time-dependent turbidity measurements can also be performed under in situ UV or blue light irradiation to assess the kinetics of the photo-switchable phase separation behavior.

1. Prepare a fresh *trans*-azoTAB/*ds*DNA droplets suspension (5 mM *trans*-azoTAB, 5 mM *ds*DNA, 100 mM NaCl, see Section 4.1.3) and incubate in a low volume plastic cuvette for ca. 15 min.
2. Place the cuvette in a UV/vis spectrophotometer and start recording the evolution of the absorbance at 700 nm over time under magnetic stirring.
3. Irradiate the sample from the top using a UV LED (365 nm) and keep monitoring the decrease in absorbance at 700 nm until it reaches a plateau. NB: to ensure that the presence of the LED does not perturb the signal (e.g., if the lid of the UV/vis spectrophotometer needs to remain open when operating the LED), a blank using Milli-Q water should be first performed in the same conditions.
4. Change the irradiation wavelength to 450 nm (blue LED) and monitor the increase in absorbance at 700 nm over time until it reaches a plateau.
5. Steps 3 and 4 can be repeated at different light intensity. The characteristic time of droplet formation or dissolution under UV or blue light, respectively, is inversely proportional to the light intensity, as expected from the photo-isomerization of azoTAB.

6. Instead of using blue light in step 4 to induce the formation of coacervate droplets, the UV-adapted sample can also be stored in the dark at a fixed temperature. *Cis-trans* thermal relaxation indeed occurs in the dark, leading to the spontaneous formation of coacervate droplets over time (which can be observed by recording the absorbance at 700 nm over time in the dark). Since thermal relaxation is enhanced at higher temperatures, the kinetics of droplets formation can be controlled by varying the temperature of the sample, providing a simple approach for the temporal programming of coacervate assembly (Fig. 1A) (Martin et al., 2019).

4.2.2.3 Quantification of light-actuated capture and release of fluorescent solutes

The light-dependent amount of fluorescent solutes or dyes sequestered in the droplets are quantified by UV/vis spectrophotometry. These studies require the use of fluorophores/dyes that have a maximum absorption wavelength higher than 550 nm to avoid overlapping with azoTAB absorption range.

1. Prepare 100 μL of a fresh *trans*-azoTAB/*ds*DNA droplets suspension (5 mM *trans*-azoTAB, 5 mM *ds*DNA, 100 mM NaCl, see Section 4.1.3) and incubate in an Eppendorf tube for ca. 10 min.
2. Add a small volume of a stock solution of fluorescent solute or dye to the suspension of coacervate droplets to reach a final fluorophore/dye concentration of ~20 μM, and incubate for ~5 min. NB: the fluorophore/dye concentration should be sufficiently high to give a measurable signal by UV/vis spectrophotometry (this will depend on the molar extinction coefficient). To avoid diluting the droplets, use a relatively concentrated stock solution of solute.
3. Take a 30 μL aliquot of the coacervate droplets suspension and centrifuge at ~20,000 × *g* for 10 min. Collect the supernatant and record the UV/vis spectrum after dilution in water to get a maximum absorption peak comprised between 0 and 1. From that maximum absorption, the concentration of free fluorophore/dye in the supernatant can be determined.
4. Irradiate the remaining 70 μL of coacervate droplets suspension with UV light, then repeat step 3 to get the concentration of free fluorophore/dye in the UV-adapted sample.

5. Irradiate the remaining 40 µL of coacervate droplets suspension with blue light, then repeat step 3 to get the concentration of free fluorophore/dye in the blue-adapted sample.
6. Kinetics studies of dye release and capture can also be performed by preparing a larger volume of coacervate droplets suspension, and taking an aliquot at different time during UV or blue light exposure to perform steps 3 or 4 at various times.

4.3 Characterization of photoswitchable coacervates under local irradiation

Photoswitchable coacervate droplets can be observed and characterized under local irradiation by optical and confocal fluorescence microscopy. The droplets tend to adhere and spread on untreated glass coverslips, therefore a surface treatment has to be performed to avoid wetting and maintain the spherical shape of the droplets. We detail below how microscopy observations of coacervate droplets and their photoswitchable behavior are performed.

4.3.1 Equipment

1. Glass slides (~76 × 26 × 1 mm) and coverslips (24 × 24 mm, 0.13–0.17 mm thickness).
2. Coverslips staining jars (DWK Life Sciences Wheaton™ Columbia Jars).
3. UV-curing optical adhesives (NOA63 and NOA81 from Thorlabs).
4. Compressed air.
5. Inverted epifluorescence optical microscope with a ~60 × oil immersion objective equipped for UV and blue light excitation with a Hg arc lamp or LEDs (and appropriate filter cubes to select the desired wavelengths), and for monitoring red fluorescence.
6. Confocal fluorescence microscope with a ~60 × oil immersion objective equipped for UV (405 nm) and blue light (488 nm) excitation, and for monitoring red fluorescence.

4.3.2 Chemicals

1. 3-[methoxy(polyethyleneoxy)propyl]tri-methoxysilane; 90%, 6–9 PE units (abcr GmbH, reference AB111226)
2. Ethanol
3. Toluene

4.3.3 *Protocol*

4.3.3.1 Preparation of PEGylated capillary chambers

1. Wash coverslips with ethanol and dry using compressed air.
2. In a fume hood, place the ethanol-rinsed coverslips into a staining jar containing 5% *v*/v of 3-[methoxy(polyethyleneoxy)propyl]trimethoxysilane diluted in toluene.
3. Incubate for 24–48 h at room temperature.
4. Rinse the coverslips with toluene, ethanol then water (NB: avoid touching the surface of the coverslips), and dry with compressed air.
5. To assemble the capillary chambers: first, using the UV-curing glue NOA81, glue two glass coverslips at each edge of a glass slide; then, using the UV-curing glue NOA63, glue a PEG-functionalized coverslip on top of the two other coverslips to make a capillary chamber (as shown in Fig. 1C).

4.3.3.2 Local photocontrol of coacervate micro-droplets

1. Load 5 μL of a droplets suspension (see Section 4.1.3) into a pre-assembled capillary channel and let the coacervate droplets settle on the PEG-functionalized coverslip for ~5 min.
2. Image the droplets through the PEG-functionalized coverslip by optical microscopy using an inverted microscope and a ~60 × oil immersion objective. The droplets should remain spherical after settling on the coverslip. NB: Fusion and relaxation to a single droplet may be observed when two droplets come into contact, consistent with the liquid-like nature of the droplets. A movie of this process can be recorded, and the kinetics of fusion and relaxation exploited to determine the inverse capillary viscosity (η/γ) of the droplets (Mitrea et al., 2018).
3. To disassemble the droplets in situ, use the irradiation source directly installed on the microscope with appropriate filter cubes to select a wavelength range centered on ~365 nm. Simultaneous bright-field illumination (using a red filter to avoid altering the *trans*/*cis*-azoTAB equilibrium composition) should be used to observe and record in real time the dissolution of droplets. Alternatively, an external light source can be used to disassemble the droplets under bright-field observation.
4. To reversibly form coacervate droplets locally, use the irradiation source directly installed on the microscope and appropriate filter cubes to select a wavelength range centered on ~450 nm. Here again, simultaneous bright-field illumination (using a red filter to avoid altering the *trans*/*cis*-azoTAB equilibrium composition) should be used to observe and

record in real time the formation of droplets. Alternatively, an external light source can be used to reassemble the droplets under bright-field observation. NB: the droplets formed under blue light irradiation are smaller than the dark-adapted droplets due to the lower fraction of *trans*-azoTAB produced under blue light. When the blue light is turned off, coalescence between droplets can favor the formation of larger droplets that settle on the glass coverslip.

4.3.3.3 Light-induced release and uptake of fluorescently labeled solutes

Epifluorescence or confocal fluorescence microscopy can be used to assess the sequestration of fluorescent solutes and their release and capture upon light-actuated dissolution and reformation of coacervate micro-droplets. For these studies, the use of red-emitting fluorophores ($\lambda_{ex,em} > 550$ nm) is highly preferred to ensure that the excitation and emission spectra of the fluorophore do not overlap with the absorption of azoTAB.

1. Prepare a fresh *trans*-azoTAB/*ds*DNA droplets suspension (5 mM *trans*-azoTAB, 5 mM *ds*DNA, 100 mM NaCl, see Section 4.1.3) and incubate in an Eppendorf tube for ca. 10 min.
2. Add a small volume of a stock solution of fluorescent solute to the suspension of coacervate droplets and incubate for ~5 min to reach a final fluorophore concentration of ca. 5 μM. NB: to avoid diluting the droplets, use a relatively concentrated stock solution of solute.
3. Image the droplets by epifluorescence or confocal fluorescence microscopy using the appropriate excitation wavelength and wavelength emission range. Confocal fluorescence images can be used to determine the partition coefficient of the fluorescent solute in the coacervate phase as the ratio of the average fluorescence intensity per pixel in the coacervate droplets to the average fluorescence intensity per pixel in the background. NB: this analysis assumes that the fluorescence quantum yield of the fluorophore is the same in the coacervate phase and in the continuous aqueous phase.
4. Irradiate the sample with UV (365 nm) using the light source directly installed on the microscope, and acquire images. The fluorescence should become homogeneous throughout the images due to the dissolution of the coacervate droplets and associated release of the fluorescent molecules. NB: although it is not the optimal wavelength for *trans-cis* azoTAB photo-isomerization, the 405 nm source can be used for UV irradiation on the confocal microscope.

5. Irradiate the sample with blue light (450 nm) using the light source directly installed on the microscope (and appropriate filter cubes), and acquire images. Coacervate droplets should reform, and recapture the fluorescent molecules. NB: quantification of this light-induced release and capture process can be performed under global irradiation by UV/vis spectrophotometry (see Section 4.2.2).
6. Coacervate dissolution can also be performed at the single droplet level using the FRAP module on the confocal microscope. For this purpose, load a suspension of *trans*-azoTAB/*ds*DNA coacervate droplets in a capillary chamber and wait for the droplets to settle on the PEGylated coverslip for ~5 min. Using the FRAP module, define a region of interest around a single droplet and irradiate with 405 nm. The irradiation should be sufficiently long and intense to selectively disassemble the target droplet, but not too strong nor intense to avoid perturbing the neighboring droplets. This approach is used to release sequestered solutes locally around the disassembled droplet (Martin et al., 2019).

5. Conclusions

In this chapter, we have described a new strategy to produce photoswitchable coacervate micro-droplets capable of being reversibly dissolved with light. The droplets are formed by associative liquid-liquid phase separation between short *ds*DNA and a cationic azobenzene photoswitch. The latter undergoes *trans*-*cis* photoisomerization under UV light, converting the coacervate-forming *trans*-azoTAB into the coacervate-inactive *cis*-azoTAB species. The reversible process is activated with blue light irradiation. Significantly, the dynamics of light-induced droplet dissolution and reformation correlates with the release and recapture of guest solutes. This photoswitchable coacervate platform should therefore inspire the development of new approaches for the spatiotemporal control of biomolecule compartmentalization, and open perspectives for the regulation of signaling pathways in synthetic protocells and artificial membraneless organelles.

Acknowledgments

We acknowledge funding from IdEx Bordeaux (ANR-10-IDEX-03-02), an Investissement d'Avenir program of the French government, managed by the Agence Nationale de la Recherche.

References

Aumiller, W. M., Jr., & Keating, C. D. (2016). Phosphorylation-mediated RNA/peptide complex coacervation as a model for intracellular liquid organelles. *Nature Chemistry*, *8*, 129–137.

Berry, J., Brangwynne, C. P., & Haataja, M. (2018). Physical principles of intracellular organization via active and passive phase transitions. *Reports on Progress in Physics*, *81*, 0406601.

Brangwynne, C. P., Tompa, P., & Pappu, R. V. (2015). Polymer physics of intracellular phase separation. *Nature Physics*, *11*, 899–904.

Chevallier, E., Mamane, A., Stone, H. A., Tribet, C., Lequeux, F., & Monteux, C. (2011). Pumping-out photo-surfactants from an air-water interface using light. *Soft Matter*, *7*, 7866–7874.

Crowe, C. D., & Keating, C. D. (2018). Liquid-liquid phase separation in artificial cells. *Interface Focus*, *8*, 20180032.

Deshpande, S., Brandenburg, F., Lau, A., Last, M. G. F., Spoelstra, W. K., Reese, L., et al. (2019). Spatiotemporal control of coacervate formation within liposomes. *Nature Communications*, *10*, 1800.

Diguet, A., Mani, N. K., Geoffroy, M., Sollogoub, M., & Baigl, D. (2010). Photosensitive surfactants with various hydrophobic tail lengths for the photocontrol of genomic DNA conformation with improved efficiency. *Chemistry—A European Journal*, *16*, 11890–11896.

Douliez, J.-P., Martin, N., Gaillard, C., Beneyton, T., Baret, J.-C., Mann, S., et al. (2017). Catanionic coacervate droplets as a surfactant-based membrane-free protocell model. *Angewandte Chemie International Edition*, *56*, 13689–13693.

Estévez-Torres, A., Crozatier, C., Diguet, A., Hara, T., Saito, H., Yoshikawa, K., et al. (2009). Sequence-independent and reversible photocontrol of transcription/expression systems using a photosensitive nucleic acid binder. *Proceedings of the National Academy of Sciences of the United States of America*, *106*, 12219–12223.

Hayashita, T., Kurosawa, T., Miyata, T., Tanaka, K., & Igawa, M. (1994). Effect of structural variation within cationic azo-surfactant upon photoresponsive function in aqueous solution. *Colloid and Polymer Science*, *272*, 1611–1619.

Hondele, M., Sachdev, R., Heinrich, S., Wang, J., Vallotton, P., Fontoura, B. M. A., et al. (2019). DEAD-box ATPases are global regulators of phase-separated organelles. *Nature*, *573*, 144–148.

Hyman, A. A., Weber, C. A., & Jülicher, F. (2014). Liquid-liquid phase separation in biology. *Annual Review of Cell and Developmental Biology*, *30*, 39–58.

Le Ny, A.-L. M., & Lee, C. T. (2006). Photoreversible DNA condensation using light-responsive surfactants. *Journal of the American Chemical Society*, *128*, 6400–6408.

Love, C., Steinkühler, J., Gonzales, D. T., Yandrapalli, N., Robinson, T., Dimova, R., et al. (2020). Reversible pH-responsive coacervate formation in lipid vesicles activates dormant enzymatic reactions. *Angewandte Chemie International Edition*, *59*, 5950–5957.

Martin, N. (2019). Dynamic synthetic cells based on liquid-liquid phase separation. *Chembiochem*, *20*, 2553–2568.

Martin, N., Douliez, J.-P., Qiao, Y., Booth, R., Li, M., & Mann, S. (2018). Antagonistic chemical coupling in self-reconfigurable host-guest protocells. *Nature Communications*, *9*, 3652.

Martin, N., Li, M., & Mann, S. (2016). Selective uptake and refolding of globular proteins in coacervate microdroplets. *Langmuir*, *32*, 5881–5889.

Martin, N., Sharma, K. P., Harniman, R. L., Richardson, R. M., Hutchings, R. J., Alibhai, D., et al. (2017). Light-induced dynamic shaping and self-division of multipodal polyelectrolyte-surfactant microarchitectures via azobenzene photomechanics. *Scientific Reports*, 7, 41327.

Martin, N., Tian, L., Spencer, D., Coutable-Pennarun, A., Anderson, J. L. R., & Mann, S. (2019). Photoswitchable phase separation and oligonucleotide trafficking in DNA coacervate microdroplets. *Angewandte Chemie International Edition, 58*, 14594–14598.

Mitrea, D. M., Chandra, B., Ferrolino, M. C., Gibbs, E. B., Tolbert, M., White, M. R., et al. (2018). Methods for physical characterization of phase separated bodies and membraneless organelles. *Journal of Molecular Biology, 430*, 4773–4805.

Nakashima, K. K., Baaji, J. F., & Spruijt, E. (2018). Reversible generation of coacervate droplets in an enzymatic network. *Soft Matter, 14*, 361–367.

Nakashima, K. K., Vibhute, M. A., & Spruijt, E. (2019). Biomolecular chemistry in liquid phase separated compartments. *Frontiers in Molecular Biosciences, 6*, 21.

Nott, T. J., Petsalaki, E., Faber, P., Jervis, D., Fussner, E., Plochowietz, A., et al. (2015). Phase transition of a disordered nuage protein generates environmentally responsive membraneless organelles. *Molecular Cell, 57*, 936–947.

Perry, S. L. (2019). Phase separation: Bridging polymer physics and biology. *Current Opinion in Colloid & Interface Science, 39*, 86–97.

Perry, S. L., Li, Y., Priftis, D., Leon, L., & Tirrell, M. (2014). The effect of salt on the complex coacervation of vinyl polyelectrolytes. *Polymers, 6*, 1756–1772.

Rai, A. K., Chen, J., Selbach, M., & Pelkmans, L. (2018). Kinase-controlled phase transition of membraneless organelles in mitosis. *Nature, 559*, 211–216.

Shin, Y., Berry, J., Pannucci, N., Haataja, M. P., Toettcher, J. E., & Brangwynne, C. P. (2017). Spatiotemporal control of intracellular phase transitions using light-activated optodroplets. *Cell, 168*, 159–171.

Shin, Y., & Brangwynne, C. P. (2017). Liquid phase condensation in cell physiology and disease. *Science, 357*, eaaf4382.

Sollogoub, M., Guieu, S., Geoffroy, M., Yamada, A., Estévez-Torres, A., Yoshikawa, K., et al. (2008). Photocontrol of single-chain DNA conformation in cell-mimicking microcompartments. *Chembiochem, 9*, 1201–1206.

Wheeler, R. J., & Hyman, A. A. (2018). Controlling compartmentalization by non-membrane-bound organelles. *Philosophical Transactions of the Royal Society B, 373*, 20170193.

CHAPTER THIRTEEN

Enzymatic control over coacervation

Karina K. Nakashima[†], Alain A.M. André[†], and Evan Spruijt*
Institute for Molecules and Materials, Radboud University, Nijmegen, The Netherlands
*Corresponding author: e-mail address: e.spruijt@science.ru.nl

Contents

Abstract

The discovery of membraneless organelles (MLOs) formed by liquid-liquid phase separation raised many questions about the spatial organization of biomolecular processes in cells, but also offered a new tool to mimic cellular media. Since disordered and charged protein domains are often necessary for phase separation, coacervates can be used as models both to understand MLO regulation and to develop dynamic cellular-like compartments. A versatile way to turn passive coacervate droplets into

[†] Equal contributions.

Methods in Enzymology, Volume 646
ISSN 0076-6879
https://doi.org/10.1016/bs.mie.2020.06.007

active and dynamic compartments is by introducing enzymatic reactions that affect parameters relevant for complex coacervation, such as the charge and length of the components. However, these reactions strictly take place in a heterogeneous medium, and the complexity thereof is hardly addressed, making it difficult to achieve true control. In this chapter we help close this gap by describing two coacervate systems in which enzymatic reactions endow coacervate droplets with a dynamic character. We further highlight the technical challenges posed by the two-phase systems and strategies to overcome them.

Abbreviations

α	monomer charge of polyelectrolytes A or B in Fig. 1(iii)
ϕ	volume fraction of polyelectrolytes A or B in Fig. 1(iii)
σ	charge density of polyelectrolytes A or B in Fig. 1(iii)
τ	turbidity
Abs	absorbance
ADP	adenosine diphosphate
ATP	adenosine triphosphate
$\mathbf{B_{eq\ dilute\ phase}}$	equilibrium concentration of coacervate component B in the dilute phase
$\mathbf{B_{eq\ dense\ phase}}$	equilibrium concentration of coacervate component B in the dense phase
Ddx3	DEAD box helicase protein 3
Ddx4	DEAD box helicase protein 4
DMF	dimethylformamide
DNA	deoxyribonucleic acid
EDTA	ethylenediaminetetraacetic acid
F	free energy of demixing for a two-component (A and B) solution in Fig. 1(iii)
FL	fluorescence
HK	hexokinase enzyme
HPLC	high performance liquid chromatography
I	light intensity
IDP	intrinsically disordered protein
k_B	Boltzmann constant
K	generic rate constant
$\mathbf{k_{cat}}$	apparent rate constant of enzymatic reaction
$\mathbf{K_M}$	substrate concentration for 1/2 maximal enzymatic reaction rate
$\mathbf{K, K_p}$	partitioning equilibrium constant
LLPS	liquid-liquid phase separation
LPP	lambda protein phosphatase
MEG-3	protein maternally expressed gene 3
MLO	membraneless organelle
N	number of monomers per polyelectrolyte chain in Fig. 1(iii)
NCK	cytoplasmic protein NCK1
NMR	nuclear magnetic resonance
N-WASP	neural Wiskott-Aldrich syndrome protein
PDMS	polydimethylsiloxane
PEG	poly(ethylene glycol)
PEP	phosphoenolpyruvate

PLE	poly-L-glutamic acid
PLL	poly-L-lysine
PNPase	polynucleotide phosphorylase enzyme
poly-U	polyuridylic acid
PSL	poly-L-serine-L-lysine
PVA	polyvinyl alcohol
PyK	pyruvate kinase enzyme
rDNA	ribosomal DNA
RGG	arginine-glycine-glycine domain
RNA	riboxy nucleic acid
rRNA	ribosomal RNA
$T_{\%}$	percentual transmittance
TAMRA	tetramethylrhodamine
TEV	tobacco etch virus
UDP	uridine diphosphate
UMP	uridine monophosphate
V_{in}	volume of the coacervate phase after centrifugation
V_{out}	volume of the dilute phase after centrifugation
V'	volume of the coacervate phase after dissolution with NaCl
V_{NaCl}	volume of salt solution required to dissolve the coacervate pellet

1. Introduction

Coacervates are dense liquid droplets of macromolecules, that were described in the early 20th century by Bungenberg de Jong and Kruyt (1929). In the case of complex coacervation, associative interactions between multiple soluble molecules, of which at least one is a macromolecule (*e.g.*, protein, DNA, RNA and other polymers), results in demixing into a polymer dense and a polymer diluted phase. The best studied complex coacervates are based on multivalent electrostatic interactions (but the phenomenon is not limited to those), and therefore responsive to environmental changes, such as pH, salt concentration and temperature (Brangwynne, Tompa, & Pappu, 2015; Spruijt et al., 2010). Coacervates have been widely used as analogs for liquid-liquid phase separated bodies identified in cells recently. Many membraneless organelles (MLOs), such as nucleoli, P-bodies and stress granules, have been proposed to form through liquid-liquid phase separation (LLPS) (Brangwynne et al., 2009; Shin & Brangwynne, 2017), controlled by enzymes. However, realizing a similar degree of dynamic control over coacervation as is common in cells is not a trivial task. In this chapter we will present some tools to bring the expertise from physical chemistry to the study of coacervates as ideal *in vitro* models for MLOs and as dynamic compartments for cell mimicry.

In principle, any two oppositely charged macromolecules can undergo LLPS, but the area of the liquid phase in a phase diagram is affected by the strength of the interactions (Fig. 1). When interactions are too weak, the two-phase region is narrow and hard to find experimentally; when interactions are too strong, aggregation competes with LLPS at high concentration conditions. For complex coacervates, the strength depends, in addition to environmental factors, on two structural ones: the charge density and length of the poly-ions. Increasing charge density by adding small charged groups to the structure increases the electrostatic interaction energy

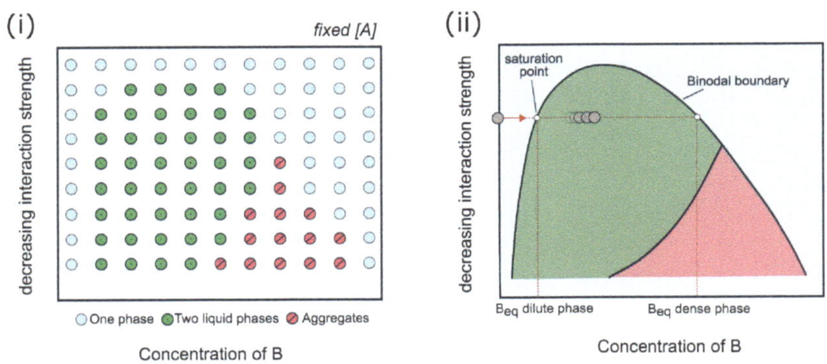

(iii) $$\frac{F}{k_B T} = \frac{\phi}{N} \ln \frac{\phi}{2} + (1-\phi)\ln(1-\phi) - \alpha(\sigma\phi)^{3/2}$$

entropic term — enthalpic term

Fig. 1 Phase diagram for a two-component (A and B) complex coacervate. (i) Phase diagrams can be presented with a discontinuous plot, where each point represents a sample that was prepared and characterized by microscopy. (ii) An approximate continuous phase diagram can be built by interpolating the phase boundaries from the diagram in (i), or by determining critical points with a titration experiment (see Section 4). When the concentration of a polyelectrolyte A is fixed, increasing the concentration of B beyond the saturation point results in phase separation. The horizontal line in (ii) connects the two coexisting phases; the lowest concentration (of B) being the dilute phase. For some combinations of A and B aggregation can happen and depending on the structure, the pink area can expand further into the green two-phase region. (iii) The boundary between the phases can be estimated by minimizing the free energy of demixing *F*. For a more detailed account of the theoretical approaches to complex coacervation, please refer to Brangwynne et al. (2015), Spruijt et al. (2010) and Perry and Sing (2015).

(represented in a mean-field Voorn-Overbeek model by the term $\alpha\sigma^{3/2}$) (Brangwynne et al., 2015; Overbeek & Voorn, 1957). Increasing the length will decrease the relative importance of the mixing entropy and also favor the demixed state (Fig. 1(iii)). These structural dependencies introduce the possibility to control LLPS through reversible chemical modifications, similar to the way cells control condensates. For the purpose of biomimicry such modifications are much more relevant than environmental changes, such as variations in pH, that go beyond physiological conditions (Koga, Williams, Perriman, & Mann, 2011). The challenge to achieve control translates then into finding good reaction candidates to move along the horizontal line in Fig. 1(ii). Once achieved, such dynamically controlled coacervates can be used as a platform to investigate MLO regulation and function and as a tool for spatiotemporal organization in the context of a synthetic cell.

For use as *in vitro* models of MLOs, coacervates must capture the essential features of cellular condensates. MLOs are dense liquid droplets containing various intrinsically disordered proteins (IDP) usually together with nucleic acids (Banani, Lee, Hyman, & Rosen, 2017; Harmon, Holehouse, Rosen, & Pappu, 2017). LLPS is driven by associative interactions between low-complexity domains within the disordered regions of proteins (Banani et al., 2017; Elbaum-Garfinkle et al., 2015; Mitrea, Chandra, et al., 2018; Mitrea, Cika, et al., 2018; Nott, Craggs, & Baldwin, 2016), or between oligomerization domains or nucleic acid binding domains and RNA (Mitrea, Chandra, et al., 2018; Mitrea, Cika, et al., 2018; Mitrea et al., 2016). The underlying interactions are generally responsive to environmental conditions including pH, crowding, ionic strength and temperature. However, cells could never rely solely on (passive) environmental changes to control MLO formation, as these factors should be fairly constant in the cellular environment. Hence, control of multiple MLOs independently would not be possible.

Instead, MLO formation is commonly controlled by subtle chemical modifications on side groups of the IDPs or by binding to regulatory proteins (Nott et al., 2015; Qamar et al., 2018; Rai, Chen, Selbach, & Pelkmans, 2018). Both control mechanisms may be linked to the suggested role of MLOs in buffering of RNA (Maharana et al., 2018). An example is the spatial patterning of P-granules. Phosphorylation of MEG-3 by anteriorly enriched kinases promotes granule disassembly and leads to a posterior-rich gradient (Wang et al., 2014). MLOs can also function as bio-reactors, where the phase separation is used to control partitioning of substrates and products (Case, Zhang, Ditlev, & Rosen, 2019; Li et al., 2012; Shin & Brangwynne, 2017). Reactions inside the MLO compartment could be enhanced as a result of higher concentrations (Case et al., 2019; Sokolova et al., 2013), but could also be limited when the substrate is not able to accumulate in the coacervate phase or when it interacts with the coacervate network (Banani et al., 2017; Shin & Brangwynne, 2017; Su et al., 2016). It is clear that enzymes play an important role in controlling intracellular phase separation, and they are therefore indispensable when studying coacervates as MLO models *in vitro*.

In the context of synthetic cells, dynamically controlled coacervates could serve as model systems for the complex, crowded intracellular milieu, helping solve traditional challenges in the field. Some benefits of coacervates over other synthetic cell compartments such as liposomes and polymersomes are: (i) coacervates reach a level of crowding closer to the cellular environment (Sokolova et al., 2013); (ii) transport of small molecules is not limited by a membrane, circumventing the need to incorporate, for instance, membrane proteins as pores (Mason, Buddingh, Williams, & van Hest, 2017); (iii) growth and division of coacervate droplets could be possible based on chemical reactions, if the underlying rates can be accurately matched to the coacervate phase characteristics (Te Brinke et al., 2018; Zwicker, Seyboldt, Weber, Hyman, & Jülicher, 2016). Some of these functions can rely on passive mechanisms, but to mimic cellular dynamics or life-like properties one has to look into active mechanisms, such as enzymatic control. But how do enzymes control coacervation in cells?

Enzymatic reactions can provide great control of coacervation because of their exceptional catalytic efficiency and well-defined conditions. It comes as no surprise that enzymes are heavily involved in the regulation of MLOs in cells. First of all, several examples have been reported where sidechains of proteins are modified by an enzyme, changing the charge density that is required for coacervation, which corresponds to modulating the first structural parameter. In the case of arginine methylation (*e.g.*, Ddx4 and FUS) or lysine acetylation (Ddx3), the modification leads to weakening of the electrostatic interactions, hence condensates dissolve (Nott et al., 2015; Qamar et al., 2018; Saito et al., 2019). In a related manner, enzymes can also be involved in regulating a specific recognition interaction. For instance, Li et al. reported that phosphorylation of tyrosine is required for phase separation within the N-WASP signaling pathway (Li et al., 2012): Nephrin contains three phosphorylation sites that can interact with

NCK; each NCK protein contains three binding domains that interact with N-WASP. Fully phosphorylated nephrin thus has effectively nine binding sites for interacting with N-WASP.

Secondly, it is likely that a reaction catalyzed by a transferase, ligase, or polymerase will favor phase separation, as the product in these cases has an increased length. This corresponds to the second structural parameter of length. With increasing length of a disordered protein or nucleic acid, the number of charged interaction sites increases and the relative importance of the mixing entropy decreases. An example can be found in the nucleolus, an extensively studied membraneless organelle both *in vitro* and *in vivo* (Berry, Weber, Vaidya, Haataja, & Brangwynne, 2015; Feric et al., 2016; Wei et al., 2017). It is the site of ribogenesis and consists of three type of condensates. The core is centered around the rDNA repeat on the chromosome. Transcription of rDNA results in the formation of long pre-rRNA molecules out of nucleotides. The pre-rRNA molecules are further processed in the second condensate region, called the dense fibrillar component. Finally, in the outer granular component, pre-ribosomal particles are assembled.

These examples of sophisticated enzymatic control over intracellular condensation suggest that control over less complicated coacervate systems should be possible. However, using enzymes to control coacervation adds extra complexity to the physical chemical characterization of systems at hand, as enzyme partitioning and differential reaction rates must be taken into account (Nakashima, Vibhute, & Spruijt, 2019). In the following paragraphs we first discuss the fundamentals of enzymatic reactions in two phases, followed by two types of enzymatic networks to control coacervation through the structural factors introduced above.

2. Enzymatic reactions in two phases

Finding a suitable enzymatic reaction to increase the length or charge density of a particular molecule can be a challenging task on itself, especially for proteins, but even if such a reaction has been found, it still does not warrant full control over the system. Once the enzyme has catalyzed formation of enough of the phase separating form of a (macro)molecule (this is sometimes referred to as "droplet material"), the system enters a two-phase regime; initially the amount of dense phase (coacervates) may be negligible, but if concentrations, and thus the degree of supersaturation, are high enough, the volume can be comparable to that of the dilute phase and it

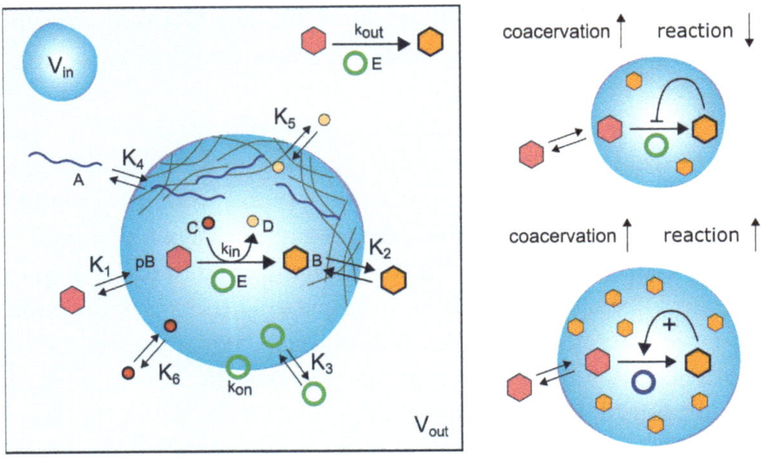

Fig. 2 Relevant factors for the reaction kinetics of a biomolecular reaction pB + C → B + D taking place in a two-phase system. An enzyme **E** converts a precursor **pB** into droplet material **B**, that complexes with **A** to form coacervates. Rate constants (k_{in}, k_{out}, k_{on}) might differ depending on the environment of the reaction and partitioning coefficients (K_i) affect the concentration and co-localization of all molecules involved. If the enzymatic reaction suffers from product inhibition, coacervation might limit its progression because of product accumulation; if product activation is possible, coacervation might lead to autocatalysis.

can be physically separated (Nott et al., 2015; Spruijt et al., 2010). The components of the enzymatic reaction, including the enzyme, substrate and possible cofactors, distribute over the two phases, and although there is free exchange between droplets and solution, the droplets represent a new chemical micro-environment. It is crucial to keep in mind the several added variables (Fig. 2): (i) additional rate constant in the droplet (k_{in}), (ii) new reactant and enzyme concentrations inside and outside of the droplets, (iii) partitioning coefficient of all components. These variables all affect the rate, location and outcome of the reaction. Moreover, it is relevant to control the rate at which the system enters the area under the binodal (Fig. 1) in order to avoid spinodal decomposition. If one of the purposes of enzyme-driven LLPS is to achieve spatiotemporal control over condensation, the parameters shown in Fig. 2 must be measured and known.

Specifics of enzymatic kinetics also come into play. Recent studies combining enzymes and LLPS rarely mention K_M and k_{cat} determination (exceptions are Davis et al., 2015; Yewdall et al., 2019), which could provide insight in how condensates affect biochemical reactions. Extreme differences in partitioning coefficients may result in a situation in which

Michaelis-Menten conditions no longer apply, making it difficult to modulate the kinetics to achieve the desired behavior (such as droplet growth, highly dependent on reaction rates (Zwicker et al., 2016)). Given how sensitive enzyme activity can be to the precise buffer conditions and other environmental factors, it may seem surprising that these reactions still proceed at all beyond the phase saturation point (Fig. 1(ii)), especially if the enzymes are taken up inside the droplets. For example, it has been taken for granted that pyruvate kinase (PyK) converts ADP to ATP in presence of poly-L-lysine to form coacervates (Nakashima, Baaij, & Spruijt, 2018), but it is not known how PyK overcomes the expected inhibition by ATP, which is now present at high concentrations inside the droplets.

We propose that enzyme-driven LLPS can benefit from a more detailed chemical characterization, and that approaching such systems as a two-phase reaction system is crucial for full control. This poses many analytical challenges as not all techniques are suitable for inhomogeneous mixtures. First, there is no straightforward method to quantify the extent of coacervation; one alternative is to measure the volume of the dense phase, but that cannot be done in a continuous manner. The most often used observable is turbidity (fraction of light scattered at a wavelength where the sample is transparent and non-absorbing, normally 500–700 nm) (Kaibara, Okazaki, Bohidar, & Dubin, 2000), which is an indirect measurement of the amount of droplets, and not a proof of the formation of liquid condensates. Still, for small timescales (minutes) it may correlate well to the progress of a reaction that results in the formation of droplet material. For longer observation periods, the gravitational settling of the droplets, coalescence or their adhesion to the walls of the well plate where the measurement is normally performed, start to interfere by decreasing turbidity even if the amount of droplet material remains the same. It is important to note this will also affect absorbance measurements performed in emulsions at any wavelength.

Turbidity measurements in a plate reader must be paired with microscopy to demonstrate the liquid nature of the condensates, as thin solid aggregates in suspension may be confused with an emulsion of liquid droplets based on turbidity alone. Microscopy is the most reliable technique to quantify components inside and outside the coacervate droplets, with the advantage of analyzing the two phases separately, while turbidity is a bulk measurement. The drawback is that covalent fluorescent labels are needed and their choice is crucial as to not drastically affect the structure (charge density or length) (Quinn et al., 2015) and the enzymatic activity. An additional requirement of microscopy is that, in order to monitor coacervates as single droplets, with a stable position over time, the observation surface must be treated to minimize wetting. With regular and non-interacting surfaces (*e.g.*, PVA, PEG), the midplane of the droplets directly on the glass can be analyzed for extended periods, as a way to prove the condensation process and to estimate the volume of coacervate phase.

Typical reaction-monitoring techniques—chromatography, mass spectrometry, NMR—can be used for the emulsion as a whole, or for the phases separately. The former will always require a homogenizing step to prevent fluctuations in the signal across the sample or aggregation during injection. This makes it challenging to study fast kinetics of reactions in coacervates, but it may be the only available alternative for small molecule-based coacervates, when fluorescent labeling is the most disturbing. It is also possible and sometimes essential to separate the coacervate phase and the dilute phase by centrifugation. However, for small-scale experiments, with a limited amount of protein, it will generate a supernatant and a tiny coacervate pellet, which must be diluted or dissolved again before analysis. In theory, this approach yields more information about the parameters displayed in Fig. 2, but it introduces another challenge of knowing the initial volume of the pellet with precision.

3. Control over coacervation by chain length and charge density

In order to describe a strategy as general as possible, we will use two emblematic examples of enzymatically controlled coacervation, one of each category previously described. For elongation-driven coacervation, we chose the system based on poly-U-spermine complexation reported by Aumiller, Pir Cakmak, Davis, and Keating (2016) and Marianelli, Miller, and Keating (2018), and the reaction catalyzed by polynucleotide phosphorylase shown in the work of Spoelstra, van der Sluis, Dogterom, and Reese (2020). For charge density-driven coacervation, we will apply our methods to the poly-L-lysine (PLL)-ATP coacervates driven by pyruvate kinase reported earlier by our group (Nakashima et al., 2018). For an overview of alternative systems to which our methods can also apply, we start by discussing some recent work that can be classified as charge or length-based.

3.1 Length-based coacervation

Controlling the chain length of macromolecules is a powerful tool to control coacervation. Two routes have been reported in literature. First, coacervate dissolution can be achieved through shortening the chain length through a protease. Semenov et al. have shown a system where a poly-L-lysine-L-serine (PSL) was cleaved by trypsin (Semenov et al., 2015). Trypsin is a proteolytic enzyme that catalyzes the hydrolysis at the C-prime end of lysine and arginine residues in proteins. PSL is a positively charged polypeptide that forms coacervate droplets with, for instance, poly-L-glutamic acid. Trypsin degraded the polymer into smaller fragments of different lengths, but all were too short to phase separate with poly-L-glutamic acid at the pH and salt concentration used. As a result, the coacervate droplets dissolved and turbidity disappeared. Another example in which a protease was employed, involved the cleavage of tandem and triplet RGG-domains by Tobacco etch virus (TEV) protease (Schuster et al., 2018). Schuster et al. used the RGG-domain of the P-granule protein LAF-1, which is able to form coacervates by itself under very mild conditions (Elbaum-Garfinkle et al., 2015; Schuster et al., 2018). By coupling two or three of these RGG-domains with a TEV-recognition motif in the linker, they could obtain stable coacervates *in vitro*, without addition of any negatively charged polymer, such as RNA. The TEV-protease cleaved the tandem/triplet repeat, yielding short RGG species and resulting in dissolution of the coacervates. This example shows that also cleavage of larger protein domains can be used to control coacervation.

Second, increasing the length of a biomolecule already prone to phase separation can widen the two-phase region in its phase diagram and enhance the degree of coacervation. Banerjee et al. formed coacervates based on a short arginine-rich peptide, $(RRXXX)_3$ where X could be any uncharged, polar residue, with either poly-U or RNA (Banerjee, Milin, Moosa, Onuchic, & Deniz, 2017). They used a T7 polymerase to synthesize the RNA from nucleotides *in situ* using a DNA template, which resulted in enhanced coacervation. However, when the RNA concentration exceeded a certain threshold, coacervates started to dissolve as a result of the shift in the charge ratio of RNA to cationic peptide. Similarly, poly-U can form coacervates in combination with spermine as shown by Aumiller et al. (2016) and Marianelli et al. (2018). These coacervates have inspired Spoelstra et al. and Deshpande et al. to use a polynucleotide phosphorylase (PNPase) for active coacervation (Deshpande et al., 2019; Spoelstra et al., 2020). PNPase consumes nucleotide diphosphates (NDPs) in a reaction where a nucleotide monophosphate (NMP) is added to the 3′-end of an oligomeric seed RNA, hereafter referred to as RNA primer (Fig. 3A) (Beljanski, 1996). At low concentrations the RNA primer (U_{20} was used by Spoelstra et al.)

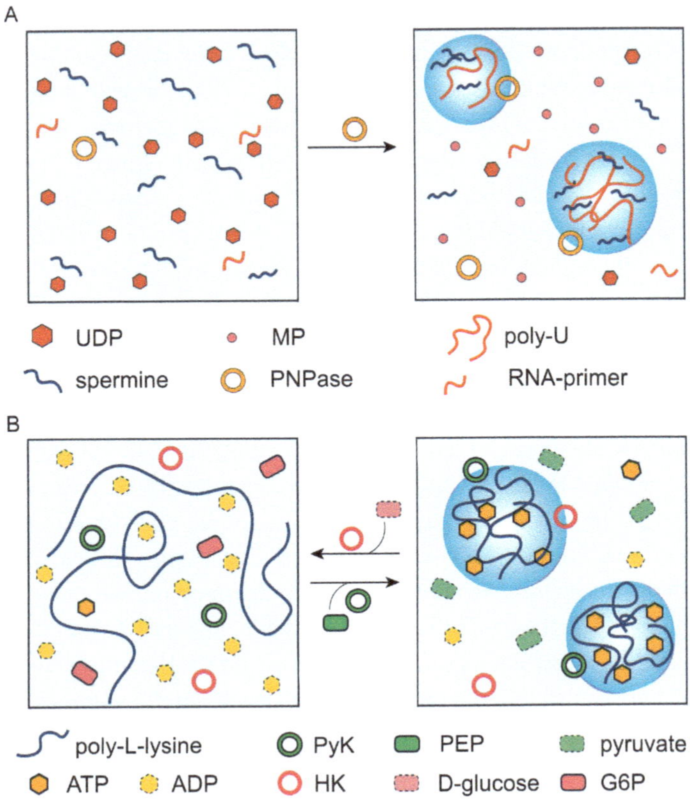

Fig. 3 Schematic representation of two model systems to control coacervation. (A) Length-based system. PNPase consumes UDP and incorporates UMP to a 20–24 nucleotides-long RNA primer, releasing a monophosphate (MP). The product is a long and polydisperse poly-U RNA-polymer which forms coacervates with spermine. (B) Charge-based system. PyK converts ADP and PEP to ATP and pyruvate, resulting in PLL-ATP coacervates. The reverse reaction, conversion of ATP and D-glucose to ADP and glucose-6-phosphate (G6P), is catalyzed by HK and dissolves the coacervates.

was unable to form coacervate droplets. As PNPase catalyzed the polymerization of NDPs, long RNA strands were formed and droplets emerged in solution as a result of nucleation and growth of spermine-RNA coacervates. Interestingly, PNPase was also able to catalyze the reverse reaction at high phosphate concentration, although the dissolution of coacervates took about three times longer than the assembly reaction (Spoelstra et al., 2020).

In Sections 4.2.1 and 4.3.1 we will focus on the latter example of length-controlled coacervation. We will discuss the parameters that should be taken into account when designing such a system, and we give a detailed protocol to realize dynamic control over coacervation in this system.

A few small adaptations were made compared to Spoelstra et al.: instead of using U_{20} as RNA primer, here we show this system can work for any type of oligo-RNA by using a $(ACUG)_6$ repeat as RNA primer.

3.2 Charge-density based coacervation

Changing length is not always a feasible route to control coacervation dynamics. Models based on subtle changes in charge-density might be better candidates to mimic the cellular control. We discuss two systems here that allow for both formation and dissolution of coacervates, controlled by enzymatic reactions.

One of the first reported examples of enzymatic control over coacervation came from Aumiller and Keating. They used a positively charged peptide repeat $[RRASL]_{2–3}$ to form complex coacervates with a long poly-U (Aumiller & Keating, 2015). Inspired by naturally occurring post-translational modifications, they used protein kinase A (PKA) to phosphorylate the serine residues in the peptide. The addition of effectively two negative charges per phosphate group decreased the net positive charge on the peptide, thereby weakening the interaction with poly-U and causing dissolution of the coacervates. They were able to reverse this process by dephosphorylating the peptide using lambda protein phosphatase (LPP). By controlling the addition of chemicals, such as EDTA which inhibits LPP, and ATP which is required by PKA for phosphorylation, this system shows great potential for out-of-equilibrium control.

A variation on this reversible phosphorylation system has been proposed by Nakashima et al. (Fig. 3B). In this scheme, coacervation is controlled by the concentrations of ADP and ATP, and their conversion is regulated by two enzymes (Nakashima et al., 2018). Pyruvate kinase (PyK) converts ADP to ATP by transfer of a phosphate group from phosphoenolpyruvate (PEP), thereby producing pyruvate. The ATP that is produced in this reaction formed coacervates with poly-L-lysine (PLL) that was present in solution. The reverse reaction, conversion of ATP back into ADP is linked to the phosphorylation of glucose to glucose-6-phosphate, catalyzed by hexokinase (HK). Without changing the initial ATP/ADP concentration, coacervation can be controlled by addition of the two co-substrates (PEP and glucose). This system was shown to be fully reversible and able to undergo repeated cycles of condensation and dissolution. In Sections 4.2.2, 4.3.2, 4.4 and 4.5 we will further explain how this system has been characterized.

4. Protocols

Here we present a detailed workflow to carry out the enzymatically controlled coacervate formation, protocols to observe and monitor the coacervate droplets, and protocols to quantify rates and partitioning (Fig. 4). We chose the two systems described above as examples, with a general strategy consisting of: determining the phase diagrams of the precursor and of the product (see Section 4.2); probing the enzymatic reaction in presence of the poly-ions with turbidity measurements (see Section 4.3); evaluating the control and kinetics of the reaction with microscopy measurements (see Section 4.4); and finally, characterizing the two phases separately (see Section 4.5). We start by describing the chemicals used and how they can affect the results; we then describe the protocols we used and highlight critical steps; and finally, in Section 5, we outline the steps to analyze the results.

4.1 Materials

Unless otherwise indicated all reagents were used as received from the manufacturer. All aqueous solutions were prepared in MilliQ water obtained from a MilliPore purification system, with an electrical resistivity of 18.2 MΩ cm at room temperature. Buffers were passed through a syringe filter (Nylon 0.22 μm) to prevent aggregation due to undissolved particles. Alexa labeling kits were ordered from Thermo Fisher, *N,N′*-bis(3-aminopropyl)-1,4-butanediamine (spermine) tetrachloride from TCI, and oligonucleotides Cy5-$[ACUG]_6$ and $[ACUG]_6$ from IDT. All other compounds mentioned below have been purchased from Sigma Aldrich.

Adenosine diphosphate disodium salt (ADP), adenosine triphosphate trisodium salt (ATP) and uracil diphosphate disodium salt (UDP) were separately dissolved in MilliQ water to prepare stock solutions of 0.50 M. Phosphoenolpyruvate potassium salt (PEP-K) was dissolved directly in 1.0 M HEPES buffer to make a 0.50 M PEP stock. The use of potassium salt avoided the need for other K^+ sources to activate pyruvate kinase. We found poly-L-lysine (PLL, HBr adduct) with the molecular weight in the range of 15–30 kDa to be ideal for our conditions, whereas chains of 52 kDa average (Alamanda Polymers) were more prone to aggregation with ATP, and chains of 4 kDa did not form condensates. A stock of 0.10 M PLL was prepared based on the molecular weight of the monomeric units, Lys-HBr (209 g mol^{-1}).

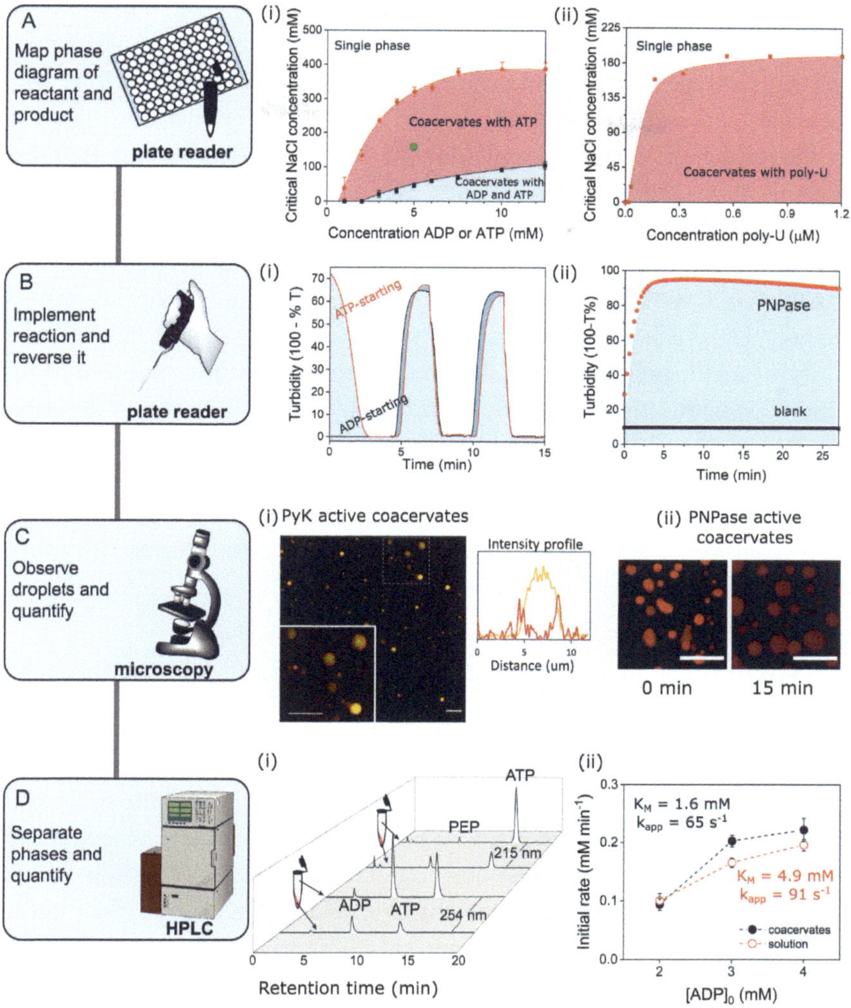

Fig. 4 General workflow to determine enzymatic control over coacervation and an overview of the results obtained. (A) Phase diagrams obtained from titration experiments for (i) PLL-ADP/ATP and (ii) spermine-poly-U mixtures. For further experiments (B–D) we chose the working salt concentration indicated by the green circle in A(i) for PLL-ADP/ATP, or no additional NaCl for spermine-poly-U. (B) Turbidity plots for reaction mixtures. B(i) shows the reversibility of phase separation upon addition of glucose (pink arrows) and PEP (green arrows) to the combined PyK/HK reaction mixtures. B(ii) depicts the increase in turbidity when the PNPase reaction is carried out. (C) Microscopic evidence of liquid condensates of (i) PLL-ATP and (ii) spermine-poly-U. In C(i), PLL was labeled with TAMRA and is shown in yellow, clearly coinciding with the droplets; PyK labeled with Alexa 647 is shown in red and can be seen in the interface of the droplets. The plot profile of a droplet in the insert highlights the *(Continued)*

The concentrations we report are therefore of monomer, even when we refer to it as PLL. The stocks of 0.50 M concentration should be kept at −20 °C and used in the course of a month; the dilutions to 0.1 M kept at −20 °C were normally used within a week.

Polyuridylic acid potassium salt (poly-U, M_W 600–800 kDa, corresponding to about 2000–3200 bases), UDP, and RNA were dissolved in nuclease free water. Nuclease free water was prepared by autoclaving MilliQ water for 20 min at 121 °C. These stock solutions (see below for concentrations) were stored at −20 °C and used in the course of 3 months, with exception for the poly-U, which normally was used within a week.

Pyruvate kinase (PyK from rabbit muscle type VII, stocks prepared from a buffered aqueous glycerol solution) and hexokinase (HK from *Saccharomyces cerevisiae*, stocks prepared from lyophilized powder) stocks were kept at 4 °C, and dilutions in MilliQ water were prepared freshly in minimal volume, at the start of the experimental day. Polynucleotide phosphorylase (PNPase from *Synechocystis* sp.) stocks were kept at −80 °C, and during the day of the experiment diluted PNPase was stored at −20 °C.

4.2 Determining the coacervation window

Before implementing enzymatic control, we started by testing whether the conversion from reactant to product results in coacervation. We determined the phase diagram of the reactant and of the product under the same conditions, in order to find the region where the reactant does not form coacervates, but the product does—our "coacervation window" in Fig. 1. The experiment involves preparing the coacervates and determining the interaction strength at which they dissolve—we chose to do it by measuring turbidity during a salt titration, finding the critical salt concentration for each coacervate composition.

Fig. 4—Cont'd partitioning difference. For C(ii), poly-U labeled with Cy5 is seen in red also partitioned inside the droplets. (D) Chromatogram and reaction profiles obtained from quantitative analysis of the PLL-ATP mixtures. D(i) contains the chromatograms of the coacervate samples used to determine the partitioning of PEP and ADP. PEP and ADP have similar retention times, but the first is monitored at 215 nm, while ADP can also be detected at 254 nm. D(ii) is the result of the kinetic analysis of the PyK reaction mixture over time, after screening different initial concentrations of ADP and obtaining the initial velocity. *The diagrams in A(i) and B(i) were adapted from a previous paper in our group Nakashima, K. K., Baaij, J. F., & Spruijt, E. (2018). Reversible generation of coacervate droplets in an enzymatic network.* Soft Matter, 14*(3), 361–367. doi:10.1039/c7sm01897e.*

4.2.1 Coacervation window of oligo RNA and poly-U with spermine

Stock solutions	PNPase buffer (4×)	Passive poly-U-spermine coacervates
10% wt spermine tetrahydrochloride (494 mM)	400 mM Tris-HCl	1 × PNPase buffer
1.0 wt% poly-U (<16 μM)	pH 9.0	1.0 wt% spermine
100 mM UDP	4.0 mM EDTA	0.0010 wt% up to 0.20 wt% poly-U
100 μM Cy5-[ACUG]$_6$	20 mM $MgCl_2$	
1.2 M NaCl		

All in nuclease free water.

a. A dilution series of poly-U was prepared in a 96-well plate (flat, transparent bottom). Each well contained 60 μL of the reaction mixture. The poly-U concentration was varied from 0.0010 to 0.20 wt%, while the total volume was kept constant by adding MilliQ water. In wells containing more than 0.010 wt% poly-U, we observed a turbid mixture by eye and by plate reader (experiments described here were carried out at 30 °C on a Tecan Spark M10 multimode plate reader equipped with 0.5 and 1.0 mL injectors). To determine turbidity levels of individual wells, the absorbance was measured at 520 nm (bandwidth 3.5 nm, and a settle time of 50 ms).

b. A small volume (about 5 μL) was collected from each well and imaged with bright-field microscopy at a typical magnification of 40 × to confirm that no aggregates were present. The pH was checked with a paper strip.

c. The plate reader was subsequently prepared for the turbidity titration. The low-volume injector was first rinsed with MilliQ water and primed with the titrant solution, in this case 1.2 M NaCl.

d. The absorbance was followed at 520 nm (same settings as under **a**) in a repeated cycle of 2 μL injections (100 μL/s jet speed) of NaCl and linear shaking (1 mm amplitude, 1440 rpm frequency).

e. The injections were performed until the absorbance reached the value of a blank well (about 10 times).

f. This procedure was repeated by varying the concentrations of spermine to ultimately determine the optimal coacervate conditions.

Note: *when you do not have access to an automated injector, this procedure could also be performed manually with a micro pipette.*

4.2.2 Coacervation window of ADP and ATP with PLL

Stock solutions	Passive ADP/ATP-PLL coacervates
100 mM ADP	100 mM HEPES pH 7.4
100 mM ATP	5.0 mM PLL
100 mM PLL (15–30 kDa)	5.0 mM $MgCl_2$
0.50 M HEPES pH 7.4	1.0–10 mM ADP/ATP
100 mM $MgCl_2$	
0.50 and 1.0 M NaCl	

a. ADP-PLL mixtures (passive coacervates) were prepared with a fixed concentration of PLL, as described above, but with a varying concentration of ADP from 1.0 to 10 mM in steps of 1.0 mM. The samples were directly prepared in a 96-well plate (flat, transparent bottom). The volume was kept constant at 100 μL by adding MilliQ water. For most wells, some turbidity was visible upon gentle mixing with the pipette.

Note: *a high concentration of PLL was chosen for this screening to make sure that the nucleotides were limiting for coacervation;* Mg^{2+} *concentration was chosen to not limit PyK kinetics. The range of concentrations was chosen based on PyK common reaction conditions.*

b. A small volume, 5 μL, was collected from each well and imaged with bright-field microscopy at a typical magnification of 40 × to confirm that no aggregates were present. The pH was checked with a paper strip.

c. The plate reader (see details in Section 4.2.1**c**) was subsequently prepared for the turbidity titration. The low-volume injector of the plate reader was rinsed with MilliQ water and primed with the 0.50 M NaCl solution. Measurements were performed as described in Section 4.2.1**d** and **e**. For complete disappearance of turbidity, a maximum of 50 μL of titrant was needed (bringing the final volume up to 150 μL). The experiment was performed in triplicate.

d. For the phase diagram of ATP and PLL, the ADP stock was replaced with ATP. 1.0 M NaCl was used as titrant to minimize the total added volume. Importantly, all other concentrations and conditions were kept the same as in **a**. Steps **b** and **c** were repeated.

4.3 Conducting the enzymatic reactions to form and dissolve coacervates

After analyzing the data from the titrations and determining the phase diagram, we chose a "safe" NaCl concentration inside the coacervation window. For the UDP-spermine system we chose to add no salt; for the ADP-PLL, we used 130 mM NaCl. We then measured the turbidity of the reaction mixtures for enzyme-controlled coacervation over time, after the addition of the substrate or the enzyme.

4.3.1 PNPase reaction in presence of spermine

Stock solutions	PNPase buffer (4×)	PNPase reaction mixture
10% wt spermine (494 mM) 1.0 wt% poly-U (<16 μM) 100 mM UDP 100 μM Cy5-[ACUG]$_6$	400 mM Tris-HCl pH 9.0 4.0 mM EDTA 20 mM $MgCl_2$	1× PNPase buffer 5.0 μM oligo RNA (Cy5-[ACUG]$_6$ or [ACUG]$_6$) 15 mM UDP 1.0 wt% spermine 1 μM PNPase *(added just before measurement)*

a. A 30 μL PNPase reaction mixture was prepared without the enzyme and kept on ice. The PNPase enzyme was added just before the start of the turbidity measurement on the plate reader, and mixed gently by pipetting. Turbidity was measured every 10 s at 520 nm (bandwidth 3.5 nm) over a time course of 30 min.

b. The same preparation (**a**) was repeated for microscopy, adjusting the volume to 10 μL. We used an inverted confocal microscope (Leica TCS Sp8X), equipped with a HC PL APO 100×/1.40 (oil) CS2 objective and white light laser (20% power), 649 nm was used to excite Cy5-labeled RNA. Emission was recorded using a HyD detector in the range of 653–780 nm.

c. To study the influence of the PNPase enzyme, we varied the enzyme concentration between 0.50 and 2.0 μM, and the UDP concentration between 5.0 and 30 mM.

4.3.2 PyK and HK reactions in presence of PLL

Stock solutions	PyK reaction mixture	HK reaction mixture
100 mM ADP	100 mM HEPES pH 7.4	100 mM HEPES pH 7.4
100 mM ATP	5.0 mM PLL	5.0 mM PLL
100 mM PLL (15–30 kDa)	5.0 mM $MgCl_2$	5.0 mM $MgCl_2$
0.50 M HEPES pH 7.4	5.0 mM ADP	5.0 mM ATP
100 mM $MgCl_2$	130 mM NaCl	130 mM NaCl
100 mM PEP	10 units mL^{-1} PyK (approx. 80 nM)	5.0 units mL^{-1} HK (approx. 250 nM)
100 mM D-glucose	1% /v/v of Alexa-labeled PyK (see Section 4.4.1)	5.0 mM D-glucose (*added just before measurement*)
1.0 M NaCl	1% v/v of TAMRA-labeled PLL (see Section 4.4.2)	
	5.0 mM PEP-K	

a. The reaction mixture of PyK was prepared directly in a 96-well plate. PEP was added just before the start of the turbidity measurement on the plate reader, and homogenized by gentle pipetting.

b. Turbidity was measured every 10 s at 520 or 550 nm depending on filter available for a time course of 30 min, with gentle shaking in between readings. The temperature was fixed to 30 °C.

c. The reaction mixture of HK was prepared in a 96-well plate. Glucose was added just before the start of the turbidity measurement on the plate reader and homogenized by gentle pipetting. Turbidity was measured under the same conditions as step **b**.

Note: *due to settling of coacervate droplets, when starting with ATP and PLL it is important to be fast to bring the system to a measurement, or the initial turbidity of different samples will vary too much.*

d. To perform cycles of coacervation-dissolution, PEP and glucose were added alternatively to the well, and the absorbance was measured at 550 nm. For manual addition, include a "plate out," "plate in" step in the measurement cycle. The substrate was added once turbidity has reached a maximum (glucose) or a minimum (PEP).

e. The addition of substrate experiment was repeated in a microscope chamber adjusting the total volume to 20–30 μL. We used a confocal microscope (Leica TCS Sp8), equipped with a HC PL APO 20×/0.75 (dry) CS2 objective and 552 and 638 solid state lasers (for rhodamine and Alexa 647 labeled components, respectively). Emission was recorded using a PMT detector at 570–700 and 650–700 nm.

4.4 Studying coacervate formation at a single droplet-level

From turbidity as a bulk measurement of coacervation, we move on to observing coacervate droplets individually by fluorescence microscopy. This involves fluorescent labeling of at least one of the polyelectrolytes and also of the enzymes, passivating the observation surface and preparing reaction chambers suitable for microscopy. From this microscopy-based study of coacervate formation, information about the droplets size and shape, and the partitioning coefficient of labeled components can be obtained.

4.4.1 Fluorescent labeling of enzymes

Labeling solutions
$10\,mg\,mL^{-1}$ Alexa-647 NHS ester in DMF (*ca.* 10 mM)
1.0 M $NaHCO_3$ pH 8.3
$2\,mg\,mL^{-1}$ PyK (50% aqueous glycerol solution, approx. 40 μM)

a. Enzymes were labeled using a Thermo Fisher Alexa-647-NHS-ester labeling kit and the accompanying instructions.

b. To a 500 μL of enzyme solution (around $2\,mg\,mL^{-1}$ in a 50% glycerol solution), 50 μL of 1.0 M $NaHCO_3$ (pH 8.3) was added and mixed gently. Subsequently, 14 μL of the dye stock was added and incubated in a thermoshaker for 2 h at room temperature (20 °C).

c. Excess dye (unbound) was removed with a disposable centrifugal membrane filter unit (*e.g.*, VivaSpin concentrator). For PyK and HK, a molecular weight cut-off of 10 kDa was used (less than half the molecular weight of the enzyme). The membrane filter unit should be "blocked" first with a Tween-20 solution (0.1 wt%), to minimize irreversible adsorption of the enzyme or protein to the membrane, and rinsed five times with MilliQ water, and then with 10 mM phosphate buffer (pH 7).

d. After washing, the labeled protein was added to the centrifugal filtration tube and centrifuged at low speed (500 rpm) until it was concentrated to 250 μL. An equal volume of glycerol was added before storing at 4 °C.

Note: *the recommended storage conditions may vary depending on the type and stability of the enzyme. For example, the PNPase enzyme solution is stored at −80 °C as recommended by the manufacturer.*

4.4.2 Fluorescent labeling of poly-L-lysine

Labeling solutions
100 mM carboxytetramethylrhodamine (TAMRA) in DMF
100 mM 1-ethyl-3-(3-dimethylaminopropyl)carbodiimide (EDC) in DMF
100 mM N-hydroxysuccinimide (NHS) in DMF
200 mM PLL in MilliQ

a. PLL was labeled in a similar protocol as Section 4.4.1, using TAMRA as a labeling agent, together with EDC and NHS as activators. 50 μL of each stock described previously (TAMRA, EDC, NHS) were mixed and added to 430 μL of 0.20 M PLL solution in water. The reaction was left overnight in a thermoshaker at room temperature.

b. Excess dye was removed by concentrating the reaction mixture in a Centricon filter unit with a molecular weight cut-off of 5 kDa. Steps **c** and **d** in Section 4.4.1 were repeated for this sample, replacing the phosphate buffer with 10 mM HBr and eliminating the addition of glycerol.

4.4.3 Passivation of glass surfaces

General	PVA coating	PEGylation
Glass bottom petri dish (Cell-vis) Borosilicate cover glass (24 × 50 mm, thickness No. 1.5 from VWR) Ethanol 70% O_2-plasma or ozone cleaner	5.0 wt% Poly-vinyl alcohol (PVA, 87–89% hydrolyzed, 13–23 kDa from Sigma Aldrich) in MilliQ 5 min @RT	30 mg mL^{-1} mPEG-silane (M_n 5000 Da, from JenKem Technology) in toluene 1 h @65 °C

a. The stock solutions were prepared according to whether PVA or mPEG-silane were chosen. We observe that different coacervate compositions require different coatings for droplet stabilization.

b. The cover glass was cleaned with distilled water, 70% ethanol and MilliQ water, and then dried using pressurized air or nitrogen.

c. The glass surface was cleaned/activated using a plasma or ozone cleaner, according to the manufacturer's instructions for cleaning glassware. Plasma treatment will result in removal of any leftover contaminants on the glass surface, and expose surface hydroxyl groups that are required for modification.

d. The cover glass was incubated with either the mPEG-silane or the PVA solution, covering the surface with excess solution. We used 1 h at 65 °C (oven) for mPEG-silane, and 5–30 min at room temperature for PVA.

e. Subsequently, the glass surface was washed with copious amounts of MilliQ water and dried with compressed air or nitrogen. If the solution has dried during the reaction, the wash may require sonication. The glass surfaces were kept at 65 °C inside the oven until 1 h before use, or in a covered petri dish, and used in the course of 2 weeks.

4.4.4 Preparation of microscopy chambers

The required scale of the reaction was determined, depending on the total volume that must be added during the experiment. For example, an experiment in which coacervates are formed and dissolved alternatingly requires a larger volume than an experiment to determine the partitioning of a single enzyme. Here, we present two versions of microscopy chambers (Fig. 5).

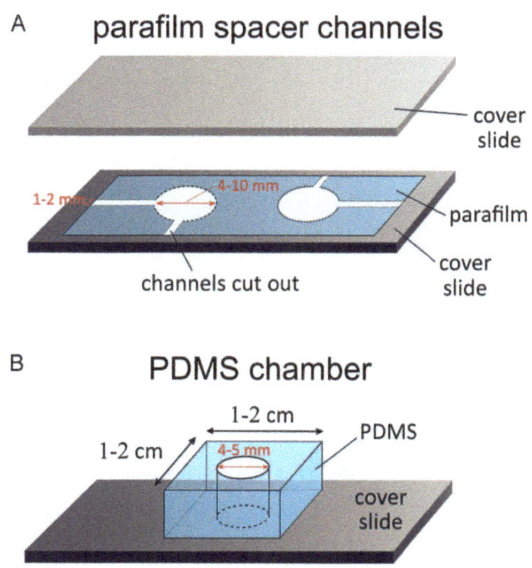

Fig. 5 Imaging chambers for fluorescent microscopy. (A) Parafilm-based chambers. (B) PDMS-based chambers.

We used the parafilm-sandwich layout for experiments in **a** and **b**, and PDMS chambers for experiments in **c** and **d**.

Parafilm chambers	PDMS chambers
PVA- or PEG-passivate coverslip (see Section 4.4.3)	Unmodified coverslip
Unmodified coverslip parafilm	Sylgard base and curing agent for silicone elastomer 184 kit (Dow Corning)

a. For up to 10 μL samples we made chambers from a parafilm-sandwich layout, as evaporation is minimized (Fig. 5). A rectangular piece of parafilm was taken (*ca.* 5 cm according to the coverslip length) and a small area was cut out with a puncher and a scalpel. We recommend the design shown in Fig. 5. The parafilm was then placed on top of the *passivated* face of the coverslip (observation surface), and covered with another coverslip.

b. To complete sealing, it was incubated for 10–15 min at 90 °C. Note: that longer incubation will improve the seal between the parafilm and the glass surfaces, but also lead to increased shrinking of the parafilm, which could cause trouble when adding your sample.

c. For up to 50 μL samples, we made the chambers from cured polydimethylsiloxane (PDMS). A mixture of Sylgard 184 base and curing agent in a 10:1 mass ratio was made, to obtain a moderately stiff silicone rubber. This mixture was placed in a petri dish Ø 35 mm inside a desiccator to remove air bubbles. Vacuum was applied for about 30 min. Holes of the desired volume were created by placing metal cylinders (Ø 4–5 mm, h 3–5 mm) into the silicone mixture (Fig. 5). The silicone mixture was baked at 65 °C for 90 min to obtain crosslinked PDMS that can be cut into smaller pieces.

d. The solidified PDMS was separated from the petri dish, and the cylinders were removed. The PDMS was then cleaned with detergent, MilliQ water, 70–100% ethanol, and finally dried in the oven at 65 °C.

e. The PDMS was cut in pieces smaller than the cover glass surface and stored on a piece of filter paper after covering the top and bottom sides with Scotch tape.

f. During the plasma or ozone cleaning of the glass surface (*before passivation*), the PDMS chambers were included, with the plain face (previously at the bottom of the petri dish) upwards. The glass and the face of the PDMS that were exposed to plasma were bonded by pushing them gently together. For sealing, the glass with PDMS was incubated for 45 min at 90 °C.

g. The passivating agent (PVA or mPEG-silane) was added to the wells, and proceed as in Section 4.4.3**d** and **e**.

4.4.5 Determination of partitioning coefficients with fluorescence microscopy

Stock solutions	Passive ATP-PLL coacervates
100 mM ATP	100 mM HEPES pH 7.4
100 mM PLL (15–30 kDa)	5.0 mM PLL
0.50 M HEPES pH 7.4	5.0 mM $MgCl_2$
100 mM $MgCl_2$	5.0 mM ATP
1.0 M NaCl	130 mM NaCl
	1% v/v of Alexa-labeled PyK (see Section 4.4.1)
	1% v/v of TAMRA-labeled PLL (see Section 4.4.2)

a. ATP-PLL coacervates were prepared in volumes ranging from 0.10 to 1.0 mL.

b. Two variations of the passive coacervates were prepared: one with lower ATP concentration (*e.g.*, 3.0 mM) and one with higher PLL concentration (*e.g.*, 20 mM). This is to verify that K_p is independent of the concentration.

c. The samples were injected into the microscopy chambers as described above. The well was covered with a circular coverslip to minimize evaporation.

d. An image was recorded of a well filled with coacervate mixture but without fluorophore with the same laser settings to use as a blank. A blank was recorded for every filter cube or emission wavelength used.

4.5 Quantifying components inside and outside of the coacervate droplets

In this step we determine the concentrations of unlabeled components in the mixture using HPLC with UV/Vis detection. In general, this is required for small molecules or precursors, such as ADP, PEP and ATP in our examples. The determination of the partitioning coefficients in this case requires: (i) centrifugation of coacervate samples and separation of the phases; (ii) estimation of the volume of each phase; (iii) dissolution and dilution to appropriate concentrations; (iv) HPLC analysis of the diluted samples. Additional quantitative data is obtained by following the PyK reaction over time, which involves quenching a sample from the reaction mixture at regular time intervals and preparing it for HPLC as outlined below.

4.5.1 Separation of the coacervate and dilute phase and volume estimation

Stock solutions	Passive ATP-PLL coacervates
100 mM ADP	100 mM HEPES pH 7.4
100 mM PEP	5.0 mM PLL
100 mM ATP	5.0 mM $MgCl_2$
100 mM PLL (15–30 kDa)	5.0 mM ATP
100 mM $MgCl_2$	130 mM NaCl
0.50 M HEPES pH 7.4	5.0 mM PEP or ADP
1.0 M NaCl	

a. 0.10–1.0 mL of ATP-PLL coacervates were prepared as described in Section 4.1. In order to measure the partitioning of PEP and ADP, those components were included, separately, in the ATP-PLL coacervate mixture the same concentration as in the reaction mixture (5.0 mM). Mix it by vortexing.

b. The sample was centrifuged at a low speed for an extended period. We used 3000 rpm and 30 min after observing that for shorter spinning times, the system had not reached a constant concentration (the determined concentration in the dilute phase were still slowly decreasing after 15 min due to very small coacervate droplets that had not settled yet). The low speed prevents the accumulation of a dense phase film (pellet) at the side wall of the eppendorf tube.

c. The dilute phase (supernatant) was collected carefully with a pipette by stopping just above the interface between the dense coacervate phase (bottom phase) and the dilute phase (top phase), to avoid contamination of the dilute phase with the coacervate phase. The collected dilute phase was transferred to a separate eppendorf tube. The amount was measured by setting the volume of the automatic pipette to an estimated value and attempting to aspirate all fluid with the pipette; the set volume was adjusted until no liquid was left, and no air was aspirated. This was set as the volume of the dilute phase, $\mathbf{V_{out}}$. For the ATP-PLL system, $\mathbf{V_{out}}$ was very close to the total volume of the mixture.

d. The remaining dilute phase from the centrifuged tube was slowly collected using a thin pipette tip (0.20–10 μL). The volume of the pellet left behind (V_{in}) was measured by first dissolving it by adding a known volume of 1.0 M NaCl ($\mathbf{V_{NaCl}}$). This decreased the viscosity of the coacervate phase and facilitated easy handling by pipetting. Step **c** was repeated for the dissolved coacervate solution, obtaining V′. Calculate $\mathbf{V_{in}}$ as $\mathbf{V' - V_{NaCl}}$. For ATP-PLL coacervates prepared as stated, V_{in} is typically 1% of the total volume.

e. The typical concentrations determined in the ATP-PLL mixtures (5.0 mM in each component) are in the 0–10 mM range for the dilute phase, and 30–50 mM range for the dense phase. For that, we recommend for a first attempt, respectively, 100 × and 500 × dilution prior to HPLC injection.

4.5.2 HPLC analysis of the separate phases

Column	**Eluents**
Shim-pack WAX-1, 3 μm particles, 4.0 × 50 mm (anion exchange) 45 °C, 1 mL/min	20 mM phosphate buffer pH 7.0 **(A)** 480 mM phosphate buffer pH 7.0 **(B)**
Stock solutions	**Gradient program**
1.0 M KH_2PO_4 (500 mL) 1.0 M K_2HPO_4 (500 mL)	0–100% B in 20 min 100% B for 5 min 100–0% B in 5 min 0% B for 5 min

a. Each of the phases were diluted around 100 × before injection to reach concentrations around 100 μM (this may require several optimizations as concentrations are unknown *a priori*).

b. Detection at 215 and 254 nm (for PLL and PEP, and ADP and ATP, respectively) were used. The separation of the nucleotides required normal-phase HPLC with an anion exchange column in a gradient elution, as described in the chart above.

c. Using a calibration curve (peak area *vs* concentration), we determined the concentrations of the desired components considering dilutions (from $\mathbf{V_{out}}$ to the vial; from $\mathbf{V_{in}}$ to $\mathbf{V'}$ and then to the vial). If the final analyzed volume (with concentrations $A_{in,measured}$ and $A_{out,measured}$) was the same for both phases, the partitioning coefficient (K_p) of A was calculated as: $K_p = [A]_{in,measured} \times V_{out} / [A]_{out,measured} \times V_{in}$.

4.5.3 HPLC analysis of the reaction mixture over time

Stock solutions	PyK reaction mixture	PyK reaction mixture without PLL
100 mM ADP	100 mM HEPES pH 7.4	100 mM HEPES pH 7.4
100 mM PLL (15–30 kDa)	5.0 mM PLL	5.0 mM ADP
0.50 M HEPES pH 7.4	5.0 mM ADP	130 mM NaCl
100 mM $MgCl_2$	130 mM NaCl	10 units mL^{-1} PyK (approx. 80 nM)
100 mM PEP	10 units mL^{-1} PyK (approx. 80 nM)	5.0 mM PEP
1.0 M NaCl	5.0 mM PEP	5.0 mM $MgCl_2$ (added at $t=0$)
10% v/v acetic acid	5.0 mM $MgCl_2$ (added at $t=0$)	0.2% acetic acid (when quenching only)
	0.2% acetic acid (when quenching only)	

a. A 1.0 mL PyK reaction mixture was prepared as described in Section 4.1, including the PEP but leaving out the $MgCl_2$ (to make sure that the enzyme remains inactive until the start of the reaction). Adding the PEP before aliquoting minimized the effect of small variations among the samples on the final reaction rate.

b. The reaction mixture was split in 10 samples of 95 μL each and added to separate eppendorf tubes. 5 μL of the 100 mM $MgCl_2$ stock was added to each of the sample tubes, which were placed in a thermoshaker at 25 °C.

c. One tube at a time was quenched by adding 2 μL of 10% (v/v) acetic acid (the final pH should be around 3 and the turbid mixture should turn clear). The results from the turbidity measurement in protocol Section 4.3.2 were used to estimate the appropriate time intervals, keeping in mind the initial time points are crucial for K_M and k_{cat} determination.

d. The samples were diluted around 100 × and we applied steps **b** and **c** in Section 4.5.2 to obtain the kinetic profile of ATP concentration.
e. The same procedure was used with a sample without PLL, as a "solution phase control."
f. Substrate concentrations were varied to obtain a typical Michaelis-Menten plot for coacervates and solution phase. For the PyK reaction, ADP could be varied from 2 to 5 mM while still working in the coacervation window.

5. Analysis

The protocols we proposed above provide all the data required to obtain the principal properties of a dynamic two-phase system: phase diagram, partitioning coefficients and reaction profiles. We now discuss how these properties can be determined from the raw turbidity data, microscopy images and chromatograms.

5.1 Phase diagrams from turbidity data

A first step to define the conditions under which a candidate enzyme could induce coacervation or dissolution is to plot the phase diagram(s). Although turbidity is not a direct and decisive measure of coacervation, it allows to determine the approximate dilute branch of the binodal in a fast and simple way. The disadvantage of the titration method is that the starting components (*e.g.*, polyelectrolytes) get diluted, so it is important to maximize titrant concentration and minimize additions. To confirm that turbidity is caused by the nucleation of liquid droplets, it is important to use microscopy, as aggregates in solution will also result in increased turbidity—though over time, turbidity will have much more noise for suspensions than for emulsions.

Turbidity is the level of transparency of a mixture, or the fraction of light that is not transmitted, but scattered. Turbidity (τ) is commonly defined for a sample without absorption as:

$$\tau = -\ln\left(\frac{I}{I_0}\right) \tag{1}$$

It can be calculated from the measured extinction ($\text{Abs} = -\log_{10}(I/I_0)$), provided that no absorption of light occurs at the detection wavelength. It should therefore be measured at a wavelength far from absorption bands, which is why we chose 520 or 600 nm. From the absorbance value, turbidity can be calculated using the following equation:

$$\tau = 2.3 \times \text{Abs} \tag{2}$$

Information about the turbidity of samples is often reported as a derivative of transmittance ($T_{\%}$), or the percentage of light that is transmitted:

$$T_{\%} = 100\% \times \left(\frac{I}{I_0}\right) = 100\% \times 10^{-\text{Abs}} \tag{3}$$

The transmittance ranges from 0% to 100%, and a plot of $(100 - T_{\%})$ indicates the amount of scattered light, and is sometimes also called turbidity. To determine the critical salt concentration, a plot of $(100 - T_{\%})$ as a function of the added salt concentration (NaCl) is made. Note that we did not take into account the total ionic strength of the solution, which would include counterions from ADP, ATP and PLL, but just the salt added. Dilution during titration was taken into account. At the critical concentration (C_s), the turbidity reaches zero (after baseline correction). If the turbidity did not reach zero, C_s can also be found from a linear fit of the points in the steepest part of the turbidity profile, and taking the x-intercept. We used this approach for our determination of the phase diagrams of both ADP-PLL and ATP-PLL. Usually the steepest decrease was found about 3–5 points before turbidity stabilized at its minimal value close to zero.

To obtain the phase diagram in Fig. 4A, C_s is plotted as a function of polyelectrolyte concentration—poly-U or ADP or ATP. It is important to notice the difference between the length- and the charge-controlled systems. In the case of spermine-poly-U, the substrate mixture contains no polymeric species (UDP, an RNA primer and spermine), and no stable coacervates can be formed under the PNPase reaction conditions. The product of the PNPase reaction, poly-U, does form coacervates with spermine, and C_s can be determined. However, since the exact length of the final product is unknown, the phase diagram can only be determined for a poly-U of a known molecular weight, and it should be used as an estimate of the PNPase product. These provisions apply to many elongation reactions, which can result in products with a distribution of lengths instead of a single composition (Banerjee et al., 2017). Transcription of a genetic construct is a notable exception to this. In the ATP/ADP-PLL case, a clear window between the two-phase region of each phase diagram is found.

Some modifications of our method are possible. First, we fixed the concentration one of the two interacting species (spermine and PLL, respectively), so that the phase diagram is technically obtained for different charge ratios (f_+, f_-). One could also fix poly-U, ADP or ATP concentration in order to find the optimal spermine or PLL concentrations (respectively), assuring that the system is always limited by the dynamic component. Second, we performed a salt titration (in which the titrant has a high ionic strength) to determine the critical salt concentration, but an alternative water titration (in which the titrant is MilliQ water or a low-salt buffer) of a sample prepared above the critical salt concentration normally gives the same value, highlighting the reversibility of LLPS. Finally, instead of C_s as the interaction parameter, temperature and pH could also be used.

Several factors will affect the coacervation window: (i) components in the enzyme stock may affect the interaction strength, (ii) substrates, products, and cofactors can also contribute to the overall ionic strength and narrow the window, and (iii) temperature strongly affects the diagram. It is important to realize that all components of a candidate reaction might have an influence on the process; ions in the buffers and secondary reaction products (such as pyruvate in the PyK reaction) have shown strong effects in solubilizing coacervates. As shown in Fig. 4A(ii), the phase diagram of poly-U with spermine is obtained at the exact conditions of the PNPase reaction, which requires alkaline conditions (pH 9). In our case, spermine is a stable positively charged molecule in a wide pH range, but deprotonation reduces the net charge and likely narrows the two-phase region.

Based on the phase diagrams, a working salt concentration and polyelectrolyte concentration were selected. The marked point in Fig. 4A indicates where coacervation is possible for a nearly 100% efficient reaction, and with some margin for the effect of secondary products (pyruvate, orthophosphate) and undetermined products (wide range of poly-Umers).

5.2 Reaction monitoring with turbidity

In Fig. 4B we show that plate reader experiments can be used to easily verify the reproducibility of the forward and reverse reaction of the PyK and HK controlled reactions. The green and orange arrows indicate substrate addition. For some of the enzymes the reaction is found to proceed very quickly, such as the PNPase reaction in Fig. 4B(ii). However, these measurements only show the turbidity over time, or the turbidity in the presence of salt. To visualize that the mixture really contains coacervates, we have to go to microscopy experiments.

5.3 Partitioning coefficients from fluorescence microscopy

Fluorescence microscopy on passivated glass surfaces produce a detailed picture of the two-phase system. As can be seen in Supplementary videos 1 and 2 in the online version at https://doi.org/10.1016/bs.mie.2020.06.007, coacervates nucleate after an incubation period of 5 min and move in and out of the focal plane. For quantitative measurements we use the focal plane near the glass, where coacervates settle and are less mobile, Fig. 4C(i and ii). Due to the different sizes of the coacervates (ranging from 1 to 10 μm radius at the final reaction time), wide-field fluorescence microscopy can be advantageous to remove the variation in droplet midplanes.

Light intensity profiles of both excitation wavelengths were plotted using ImageJ software for the ATP-PLL coacervates. A clear difference in partitioning can be observed between PLL and PyK (Fig. 4C(i)). While TAMRA-labeled PLL completely co-localizes with the observed coacervates from transmission images, PyK labeled with Alexa-647 accumulates at the coacervate interface. The partitioning coefficient (K_p) is determined from the ratio of emission intensity between the inner coacervate region and its surroundings. We use the integrated intensity (I) in a fixed squared area, and correct it for the emission of a blank (same excitation settings, no fluorophore).

$$K_P = \frac{I_{in} - I_{blank}}{I_{out} - I_{blank}} \tag{4}$$

Note that the blank emission can drastically affect K_p. For example, for a droplet with integrated Alexa-647 fluorescence intensity of 100, and an intensity in the surrounding solution of 10, a blank intensity of either 1 or 9 will result in a K_p of 10 or 91. The difference in midplanes also requires that coacervates of different sizes and frame positions, are used for a reliable determination of K_p. Finally, K_p must be measured under equilibrium conditions, and it is therefore crucial to perform the measurement with different droplet incubation times to check for variation.

Using this method, we find a partitioning coefficient of *ca.* 100 for PLL (PLL^{TAMRA}) in ATP-based coacervates. In this case we assume the labeling does not affect partitioning, but that is not necessarily true. Especially in the case of small molecules, such as ATP, attachment of a fluorescent label can strongly affect its partitioning (Deshpande et al., 2019). Also in the case of enzymes, the degree of labeling of the enzyme, and the hydrophobicity of the label (which can be accessed by its membrane interaction parameter), can completely change properties, including activity, which we did not test for here. One way to find out if labeling affects partitioning is to repeat the labeling with a different fluorophore, for example, with a different charge and/or hydrophobicity, and compare the partitioning coefficients for both labels.

5.4 Concentrations from classical analytical tools

Typical chemical tools for quantification can be used for coacervates if the proper sample preparation steps are taken. NMR and UV/Vis spectroscopy can be performed in the emulsion, but we observe better reproducibility if the phases are separated or the emulsion is dissolved. For partitioning coefficients, macrophase separation is needed; for bulk reaction monitoring, we suggest dissolving the coacervates prior to chromatographic analysis.

Fig. 4D(i) shows the chromatograms of the treated dense and dilute phases of ATP-PLL coacervates. ADP and ATP can be clearly separated with our method, and also PEP and PLL can be detected at 215 nm. After applying the dilution factors, we obtain the partitioning coefficients in Table 1.

An important source of error that remains is the calculation of the coacervate volume. The coacervate volumes are usually small, and relative uncertainties in their quantification can be large. In addition, in our approach we assume that the volume changes upon mixing a high ionic strength solution with a dense coacervate phase are negligible. In our experience, the potential error arising from this assumption is smaller than the typical uncertainties introduced in any of the alternative methods to determine ultrasmall volumes of liquids with unknown densities (Aumiller et al., 2016).

Table 1 Partitioning coefficients obtained using different methodologies.

Component	K_p	Determination method
PLL	98 ± 16	FL-microscopy
PyK enzyme	Interfacial	FL-microscopy
ATP	52 ± 18	HPLC
ADP	18 ± 4	HPLC
PEP	0.8 ± 0.5	HPLC

The accuracy of HPLC also produces a reliable kinetic profile of ATP formation, that can be used to obtain K_M and k_{cat}. Because it requires quenching and dissolution prior to analysis, it is difficult to obtain several time points in the linear range of the kinetic curve, so we recommend fitting the exponential curve to derive the initial velocities. The Michaelis-Menten plot shown in Fig. 4D(ii) reveals that coacervation hardly affected the overall reaction, which can be due to the small number of coacervates. K_M and k_{cat} values can be obtained from the linearization of the plot (and considering an enzyme concentration of 80 nM) and although not remarkably different, they demonstrate that it is possible to perform a typical enzyme-kinetics study with coacervates droplets using our method. Microscopy can be misleading in terms of the ratio between coacervates and solution, but our volume measurements confirm that ATP-PLL coacervates take up no more than 1% of the total mixture volume (10–13 μL for 1000 μL), similar to the volume fraction of PNPase coacervates (1–2 μL for 100 μL). These results also match the micrographs showing that after 10 min, the nucleation of new coacervates seems to stall and only coalescence events are observed. Additional information could be obtained by carrying out the kinetic analysis in the two phases separately, which would be less challenging for coacervates of higher volume fraction, or for fluorogenic enzymatic reactions.

6. Summary and conclusion

To summarize, we have given elaborate protocols and guidelines to study enzymatically controlled coacervation. Dynamic control by enzymes is desirable as coacervates are being used both as promising models to study membraneless organelles, and as synthetic cell models. The transition from passive coacervate droplets toward active and dynamic coacervates brings some additional requirements that have to be taken into account. In this chapter we used a charged-based coacervate system based on ATP-poly-lysine and a length-based coacervate system based on elongation of an oligo-RNA primer by UDP, which phase separates with spermine, as two examples, but the protocols should not be limited to these two systems. The workflow presented here is general and may help in assessing the feasibility of other candidate enzymes as active modulators of coacervation.

Acknowledgment

The authors acknowledge financial support from the Netherlands Organization for Scientific Research (NWO).

Competing interests

The authors declare no competing interests.

References

Aumiller, W. M., Jr., Pir Cakmak, F., Davis, B. W., & Keating, C. D. (2016). RNA-based coacervates as a model for membraneless organelles: Formation, properties, and interfacial liposome assembly. *Langmuir, 32*(39), 10042–10053. https://doi.org/10.1021/acs.langmuir.6b02499.

Aumiller, W. M., & Keating, C. D. (2015). Phosphorylation-mediated RNA/peptide complex coacervation as a model for intracellular liquid organelles. *Nature Chemistry, 8*(2), 129–137. https://doi.org/10.1038/nchem.2414.

Banani, S. F., Lee, H. O., Hyman, A. A., & Rosen, M. K. (2017). Biomolecular condensates: Organizers of cellular biochemistry. *Nature Reviews Molecular Cell Biology, 18*(5), 285–298. https://doi.org/10.1038/nrm.2017.7.

Banerjee, P. R., Milin, A. N., Moosa, M. M., Onuchic, P. L., & Deniz, A. A. (2017). Reentrant phase transition drives dynamic substructure formation in ribonucleoprotein droplets. *Angewandte Chemie International Edition in English, 56*(38), 11354–11359. https://doi.org/10.1002/anie.201703191.

Beljanski, M. (1996). De novo synthesis of DNA-like molecules by polynucleotide phosphorylase in vitro. *Journal of Molecular Evolution, 42*(5), 493–499. https://doi.org/10.1007/bf02352279.

Berry, J., Weber, S. C., Vaidya, N., Haataja, M., & Brangwynne, C. P. (2015). RNA transcription modulates phase transition-driven nuclear body assembly. *Proceedings of the National Academy of Sciences of the United States of America, 112*(38), E5237–E5245. https://doi.org/10.1073/pnas.1509317112.

Brangwynne, C. P., Eckmann, C. R., Courson, D. S., Rybarska, A., Hoege, C., Gharakhani, J., et al. (2009). Germline P granules are liquid droplets that localize by controlled dissolution/condensation. *Science, 324*(5935), 1729. https://doi.org/10.1126/science.1172046.

Brangwynne, C. P., Tompa, P., & Pappu, R. V. (2015). Polymer physics of intracellular phase transitions. *Nature Physics, 11*(11), 899–904. https://doi.org/10.1038/nphys3532.

Bungenberg de Jong, H. G., & Kruyt, H. R. (1929). Coacervation (partial miscibility in colloid systems). *Proceedings of the Royal Academy of Sciences at Amsterdam, 32*, 849–856.

Case, L. B., Zhang, X., Ditlev, J. A., & Rosen, M. K. (2019). Stoichiometry controls activity of phase-separated clusters of actin signaling proteins. *Science, 363*(6431), 1093–1097. https://doi.org/10.1126/science.aau6313.

Davis, B. W., Aumiller, W. M., Jr., Hashemian, N., An, S., Armaou, A., & Keating, C. D. (2015). Colocalization and sequential enzyme activity in aqueous biphasic systems: Experiments and modeling. *Biophysical Journal, 109*(10), 2182–2194. https://doi.org/10.1016/j.bpj.2015.09.020.

Deshpande, S., Brandenburg, F., Lau, A., Last, M. G. F., Spoelstra, W. K., Reese, L., et al. (2019). Spatiotemporal control of coacervate formation within liposomes. *Nature Communications, 10*(1), 1800. https://doi.org/10.1038/s41467-019-09855-x.

Elbaum-Garfinkle, S., Kim, Y., Szczepaniak, K., Chen, C. C., Eckmann, C. R., Myong, S., et al. (2015). The disordered P granule protein LAF-1 drives phase separation into droplets with tunable viscosity and dynamics. *Proceedings of the National Academy of Sciences of the United States of America, 112*(23), 7189–7194. https://doi.org/10.1073/pnas.1504822112.

Feric, M., Vaidya, N., Harmon, T. S., Mitrea, D. M., Zhu, L., Richardson, T. M., et al. (2016). Coexisting liquid phases underlie nucleolar subcompartments. *Cell, 165*(7), 1686–1697. https://doi.org/10.1016/j.cell.2016.04.047.

Harmon, T. S., Holehouse, A. S., Rosen, M. K., & Pappu, R. V. (2017). Intrinsically disordered linkers determine the interplay between phase separation and gelation in multivalent proteins. *Elife, 6*, e30294. https://doi.org/10.7554/eLife.30294.

Kaibara, K., Okazaki, T., Bohidar, H. B., & Dubin, P. L. (2000). pH-induced coacervation in complexes of bovine serum albumin and cationic polyelectrolytes. *Biomacromolecules*, *1*(1), 100–107. https://doi.org/10.1021/bm990006k.

Koga, S., Williams, D. S., Perriman, A. W., & Mann, S. (2011). Peptide-nucleotide microdroplets as a step towards a membrane-free protocell model. *Nature Chemistry*, *3*(9), 720–724. https://doi.org/10.1038/nchem.1110.

Li, P., Banjade, S., Cheng, H. C., Kim, S., Chen, B., Guo, L., et al. (2012). Phase transitions in the assembly of multivalent signalling proteins. *Nature*, *483*(7389), 336–340. https://doi.org/10.1038/nature10879.

Maharana, S., Wang, J., Papadopoulos, D. K., Richter, D., Pozniakovsky, A., Poser, I., et al. (2018). RNA buffers the phase separation behavior of prion-like RNA binding proteins. *Science*, *360*(6391), 918–921. https://doi.org/10.1126/science.aar7366.

Marianelli, A. M., Miller, B. M., & Keating, C. D. (2018). Impact of macromolecular crowding on RNA/spermine complex coacervation and oligonucleotide compartmentalization. *Soft Matter*, *14*(3), 368–378. https://doi.org/10.1039/c7sm02146a.

Mason, A. F., Buddingh, B. C., Williams, D. S., & van Hest, J. C. M. (2017). Hierarchical self-assembly of a copolymer-stabilized coacervate protocell. *Journal of the American Chemical Society*, *139*(48), 17309–17312. https://doi.org/10.1021/jacs.7b10846.

Mitrea, D. M., Chandra, B., Ferrolino, M. C., Gibbs, E. B., Tolbert, M., White, M. R., et al. (2018). Methods for physical characterization of phase-separated bodies and membrane-less organelles. *Journal of Molecular Biology*, *430*(23), 4773–4805. https://doi.org/10.1016/j.jmb.2018.07.006.

Mitrea, D. M., Cika, J. A., Guy, C. S., Ban, D., Banerjee, P. R., Stanley, C. B., et al. (2016). Nucleophosmin integrates within the nucleolus via multi-modal interactions with proteins displaying R-rich linear motifs and rRNA. *eLife*, *5*, e13571. https://doi.org/10.7554/eLife.13571.

Mitrea, D. M., Cika, J. A., Stanley, C. B., Nourse, A., Onuchic, P. L., Banerjee, P. R., et al. (2018). Self-interaction of NPM1 modulates multiple mechanisms of liquid-liquid phase separation. *Nature Communications*, *9*(1), 842. https://doi.org/10.1038/s41467-018-03255-3.

Nakashima, K. K., Baaij, J. F., & Spruijt, E. (2018). Reversible generation of coacervate droplets in an enzymatic network. *Soft Matter*, *14*(3), 361–367. https://doi.org/10.1039/c7sm01897e.

Nakashima, K. K., Vibhute, M. A., & Spruijt, E. (2019). Biomolecular chemistry in liquid phase separated compartments. *Frontiers in Molecular Biosciences*, *6*, 21. https://doi.org/10.3389/fmolb.2019.00021.

Nott, T. J., Craggs, T. D., & Baldwin, A. J. (2016). Membraneless organelles can melt nucleic acid duplexes and act as biomolecular filters. *Nature Chemistry*, *8*(6), 569–575. https://doi.org/10.1038/nchem.2519.

Nott, T. J., Petsalaki, E., Farber, P., Jervis, D., Fussner, E., Plochowietz, A., et al. (2015). Phase transition of a disordered nuage protein generates environmentally responsive membraneless organelles. *Molecular Cell*, *57*(5), 936–947. https://doi.org/10.1016/j.molcel.2015.01.013.

Overbeek, J. T. G., & Voorn, M. J. (1957). Phase separation in polyelectrolyte solutions. Theory of complex coacervation. *Journal of Cellular and Comparative Physiology*, *49*(S1), 7–26. https://doi.org/10.1002/jcp.1030490404.

Perry, S. L., & Sing, C. I. (2015). PRISM-based theory of complex coacervation: Excluded volume versus chain correlation. *Macromolecules*, *48*, 5040–5053. https://pubs.acs.org/doi/abs/10.1021/acs.macromol.5b01027#.

Qamar, S., Wang, G., Randle, S. J., Ruggeri, F. S., Varela, J. A., Lin, J. Q., et al. (2018). FUS phase separation is modulated by a molecular chaperone and methylation of arginine cation-pi interactions. *Cell*, *173*(3), 720–734 e715. https://doi.org/10.1016/j.cell.2018.03.056.

Quinn, M. K., Gnan, N., James, S., Ninarello, A., Sciortino, F., Zaccarelli, E., et al. (2015). How fluorescent labelling alters the solution behaviour of proteins. *Physical Chemistry Chemical Physics*, *17*(46), 31177–31187. https://doi.org/10.1039/c5cp04463d.

Rai, A. K., Chen, J. X., Selbach, M., & Pelkmans, L. (2018). Kinase-controlled phase transition of membraneless organelles in mitosis. *Nature*, *559*(7713), 211–216. https://doi.org/10.1038/s41586-018-0279-8.

Saito, M., Hess, D., Eglinger, J., Fritsch, A. W., Kreysing, M., Weinert, B. T., et al. (2019). Acetylation of intrinsically disordered regions regulates phase separation. *Nature Chemical Biology*, *15*(1), 51–61. https://doi.org/10.1038/s41589-018-0180-7.

Schuster, B. S., Reed, E. H., Parthasarathy, R., Jahnke, C. N., Caldwell, R. M., Bermudez, J. G., et al. (2018). Controllable protein phase separation and modular recruitment to form responsive membraneless organelles. *Nature Communications*, *9*(1), 2985. https://doi.org/10.1038/s41467-018-05403-1.

Semenov, S. N., Wong, A. S., van der Made, R. M., Postma, S. G., Groen, J., van Roekel, H. W., et al. (2015). Rational design of functional and tunable oscillating enzymatic networks. *Nature Chemistry*, 7(2), 160–165. https://doi.org/10.1038/nchem.2142.

Shin, Y., & Brangwynne, C. P. (2017). Liquid phase condensation in cell physiology and disease. *Science*, *357*(6357), eaaf4382. https://doi.org/10.1126/science.aaf4382.

Sokolova, E., Spruijt, E., Hansen, M. M., Dubuc, E., Groen, J., Chokkalingam, V., et al. (2013). Enhanced transcription rates in membrane-free protocells formed by coacervation of cell lysate. *Proceedings of the National Academy of Sciences of the United States of America*, *110*(29), 11692–11697. https://doi.org/10.1073/pnas.1222321110.

Spoelstra, W. K., van der Sluis, E. O., Dogterom, M., & Reese, L. (2020). Non-spherical coacervate shapes in an enzyme driven active system. *Langmuir*, *36*, 1956–1964. https://doi.org/10.1021/acs.langmuir.9b02719.

Spruijt, E., Westphal, A. H., Borst, J. W., Cohen Stuart, M. A., & van der Gucht, J. (2010). Binodal compositions of polyelectrolyte complexes. *Macromolecules*, *43*(15), 6476–6484. https://doi.org/10.1021/ma101031t.

Su, X., Ditlev, J. A., Hui, E., Xing, W., Banjade, S., Okrut, J., et al. (2016). Phase separation of signaling molecules promotes T cell receptor signal transduction. *Science*, *352*(6285), 595–599. https://doi.org/10.1126/science.aad9964.

Te Brinke, E., Groen, J., Herrmann, A., Heus, H. A., Rivas, G., Spruijt, E., et al. (2018). Dissipative adaptation in driven self-assembly leading to self-dividing fibrils. *Nature Nanotechnology*, *13*(9), 849–855. https://doi.org/10.1038/s41565-018-0192-1.

Wang, J. T., Smith, J., Chen, B. C., Schmidt, H., Rasoloson, D., Paix, A., et al. (2014). Regulation of RNA granule dynamics by phosphorylation of serine-rich, intrinsically disordered proteins in C. elegans. *Elife*, *3*, e04591. https://doi.org/10.7554/eLife.04591.

Wei, M. T., Elbaum-Garfinkle, S., Holehouse, A. S., Chen, C. C., Feric, M., Arnold, C. B., et al. (2017). Phase behaviour of disordered proteins underlying low density and high permeability of liquid organelles. *Nature Chemistry*, *9*(11), 1118–1125. https://doi.org/10.1038/nchem.2803.

Yewdall, N. A., Buddingh, B. C., Altenburg, W. J., Timmermans, S., Vervoort, D. F. M., Abdelmohsen, L., et al. (2019). Physicochemical characterization of polymer-stabilized coacervate protocells. *Chembiochem*, *20*(20), 2643–2652. https://doi.org/10.1002/cbic.201900195.

Zwicker, D., Seyboldt, R., Weber, C. A., Hyman, A. A., & Jülicher, F. (2016). Growth and division of active droplets provides a model for protocells. *Nature Physics*, *13*(4), 408–413. https://doi.org/10.1038/nphys3984.

FSC
www.fsc.org
MIX
Papier | Fördert gute Waldnutzung
FSC® C083411

Kolibri 360 GmbH
Ferdinand-Jühlke-Straße 7
99095 Erfurt, Deutschland
produktsicherheit@kolibri360.de